THE JEWEL HOUSE

THE JEWEL HOUSE

Elizabethan London and the Scientific Revolution

Deborah E. Harkness

Yale University Press
New Haven & London

Published with assistance from the Louis Stern Memorial Fund.

Set in ElectraLH and Trajan type by The Composing Room of Michigan, Inc.
Printed in the United States of America.

Library of Congress Cataloging-in-Publication Data

Harkness, Deborah E., 1965–
The Jewel house : Elizabethan London and the scientific revolution /
Deborah E. Harkness.
p. cm.
Includes bibliographical references and index.
ISBN 978-0-300-11196-5 (hardcover)   ISBN 978-0-300-14316-4 (paperback)
1. Science—England—London—History—16th century.   2. Natural history—
England—London—History—16th century.
3. Science, Renaissance.   4. London (England)—Social conditions—16th century.
5. London (England)—Social life and customs—16th century.   6. London (England)
—Intellectual life—16th century.   I. Title.
Q127.G4H37 2007
509.421'09031—dc22
2007002683
A catalogue record for this book is available from the British Library.

For Karen Halttunen
— the sine qua non —

It is the manner of men first to wonder that any such thing should be possible, and after it is found out to wonder again how the world should miss it so long.

— Francis Bacon, after Titus Livius, *Valerius Terminus* (1603)

CONTENTS

ACKNOWLEDGMENTS

During the years it took to research and write this book, I was privileged to receive the support of the Huntington Library, the National Science Foundation Grant 80813, the University of California at Davis, the University of Southern California, the John S. Guggenheim Foundation, and the National Humanities Center. Additional thanks go to the National Endowment for the Humanities (which supported my time at the Huntington Library), the University of California at Davis Chancellor's Fellows Program, and the Mellon Foundation (which supports the John Sawyer Fellowship I received at the National Humanities Center).

While writing I enjoyed the intellectual benefits associated with being part of not one but several intellectual communities. My students Celeste Chamberland, Michele Clouse, Brooke Newman, and Christina Ramos helped me to keep my perspective. They are all fine historians, and it has been a privilege and pleasure to work with them over the years. At the University of California at Davis my colleagues Joan Cadden, Fran Dolan, Margie Ferguson, and Cathy Kudlick were enormously supportive of this project. At the University of Southern California, I am surrounded by gifted urban and cultural historians who have been generous with their enthusiasm and feedback, and I especially thank Phil Ethington, Joan Piggott, and Vanessa Schwartz for their interest in a city and time far distant from those that they study. At USC I am fortunate to be among a diverse cadre of British historians, and I am deeply grateful for the intellectual companionship of Judith Bennett, Lisa Bitel, Cynthia Herrup, Philippa Levine, Peter Mancall, and Carole Shammas — each of whom has influenced this project. My special thanks go to Cynthia Herrup, who told me when I began this project that studying science in the age of Elizabeth was probably sufficient for a single book, and who shared her insightful comments with me when the manuscript was completed.

The National Humanities Center made it possible for me to take fifteen file tubs of notes and transform them into a book. My fellow fellows of the National Humanities Center class of 2004–5 were an inspiration, especially my colleagues in the Life Stories Seminar, Julia Clancy-Smith, Ed Curtis, Tom Kaiser, Lisa Lindsay, Gregg Mitman, Cara Robertson, and Tim Tyson. Kent Mullikan, Lois Whittington, and Bernice Patterson put a smile on my face as soon as I walked in the door. The library staff of Liza Roberts, Betsy Dain, Jean Houston, and their crack team of book pagers at Duke and the University of North Carolina helped me to have an inconceivable number of books and articles at my fingertips. Marie Brubaker provided invaluable assistance with photocopying, and Karen Carroll was an expert copy editor for the preliminary manuscript. Phillip Barron and Joel Elliott fixed my computer more times than they should have had to in a just world. Jana Johnson and Kristen Rosselli offered up the hospitality for which the South is justly famous. And, for more than a year at the North Carolina Research Triangle, Tom Cogswell, Lynda Coon, Maura Nolan, Cara Robertson, Ding-Xiang Warner, and Georgia Warnke provided me with that most essential ingredient of productive scholarship: friendship.

The following institutions and organizations provided me with an opportunity to present my work, and I thank my hosts and audiences for their probing comments when earlier versions of this book were presented: the American Historical Association Annual Meeting; Cornell University; Drew University and the New Jersey Shakespeare Festival; Duke University Medieval and Renaissance Studies; Durham University Thomas Harriot Seminar; Five-College Renaissance Studies Faculty Colloquium; History of Science Society Annual Meeting; the Huguenot Society and His Royal Highness the Prince of Wales; Huntington Library Early Modern British Seminar; Huntington Library Renaissance Seminar; Johns Hopkins University History of Medicine Program; Mount Holyoke College; the National Humanities Center; New College, Oxford University; North Carolina Graduate Colloquium in Medieval and Renaissance Studies; Pomona College; Princeton University; Renaissance Society of America Annual Meeting; Stanford University; University of California at Berkeley; University of California at Davis Chancellor's Club; University of California at Davis Cross-Cultural Women's History; UCLA Center for Seventeenth- and Eighteenth-Century Studies; UCLA Science Studies Colloquium; University of Chicago Eric Cochrane Symposium; University of Mississippi Renaissance — Early Modern Studies Symposium; University of North Carolina Renaissance Workshop; University of San Diego Science Studies Colloquium; Santa Clara University; University of Southern California Department of History; and the USC-Huntington Early Modern Studies Institute.

No work of history can be completed without the patient assistance of scores of librarians. I would like to particularly thank the librarians and archivists of the following institutions for supporting my research: the Bodleian Library's Duke Humfrey's Reading Room Staff; the Corporation of London Records Office; the Family History Centre, London; the Greater London Records Office; London Guildhall Library Manuscript Services; the London Metropolitan Archives; Magdalen College, Oxford; Pembroke College, Cambridge; and St. Bartholomew's Hospital Archives, London. Two libraries deserve special mention: the Huntington Library and the British Library. The rare book and manuscript staffs at both institutions were especially helpful, despite my never-ending requests for more books and manuscripts.

Over the years I have been blessed with friends and colleagues who were willing to have me talk at great length about Elizabethan London and the characters who walked those streets. I extend my special thanks to Margaret Jacob, who believed in this project when I did not, and pushed me to challenge "big" ideas and take greater intellectual risks. Adrian Johns, Alexander Marr, and Joseph Ward have long been this book's most enthusiastic supporters, as well as constant sources of inspiration to the author, and I owe them an incalculable debt. Ari Berk, Stephen Clucas, Florike Egmond, Paula Findlen, Mary Fissell, Vanessa Harding, Robert Hatch, Lynette Hunter, Katy Park, Bill Sherman, Nigel Smith, and Bob Westman all saved me from embarrassing errors and clarified my thinking on a wide range of issues. A Folger Library seminar led by Steven Shapin on early modern autobiography provided me with a fruitful environment in which to rethink some of what I thought I knew about reading and writing Elizabethan lives. Margery Wolf and Mac Marshall undertook an anthropological intervention when I most needed it, explaining and exemplifying what "deep hanging out" is all about. The friendship of Hilla Ahvenainen and Margaret Smith has sustained me in more ways than one, and I honestly don't know what I would do without them; I hope I never have to find out. My parents remain convinced that all this picking at early modern texts is worth it, which is amazing to me. Their support and confidence in me is as unflagging now as it has ever been—and I love them for it.

Several people deserve thanks for transforming my manuscript into a completed book. My agent, Sam Stoloff of the Frances Goldin Literary Agency, thought this project had promise when it was just a title, and encouraged me to think in fresh ways about it. Writing this book would have been a far lonelier business without you, Sam! At Yale University Press I have been fortunate to work with two supportive editors, Larisa Heimert and Christopher Rogers. Two anonymous readers for the Press offered probing criticisms and greatly improved the quality of the final version. Three anonymous readers for Yale read the initial book proposal, and shaped the final version in ways that they will immediately recognize. My manuscript editor, Dan Heaton, was everything that an author hopes for and more: he was conscientious, consistent, careful, and kind. He was also good-humored, which made the process easier to bear. Karen Halttunen, Cynthia Herrup, Margaret Jacob, and Alexander Marr all read the complete manuscript and shared their questions and suggestions with me. As always, any faults that remain in the book after such generous assistance has been given are the responsibility of the author.

Finally, this book is dedicated to Karen Halttunen, without whom it would never have been undertaken or reached completion. From the earliest days of research when I optimistically predicted writing a short book quickly, to the final days of writing and revising, Karen has been an enthusiastic supporter and critic. She walked every street of the City with enthusiasm as we looked for traces of a vanished London, enjoyed a number of "vacations" within arm's reach of the Guildhall Library, where I insisted on spending every waking hour, listened to endless rambling stories about obscure Elizabethans at the dinner table, and put up with prolonged absences as I made yet another "final" trip to the London archives. She brought her own formidable historical sensibility into every conversation we had about this project. Both this book — and my life — would be considerably less rich had she not contributed so much to them for so many years.

Conventions

Throughout the book, London appears in the feminine following the early modern convention. When capitalized, City refers to London's corporate identity, which, even today, is distinct from metropolitan London.

All early modern texts quoted have been silently modernized in their spelling, grammar, and punctuation. No words have been added without the use of square brackets. Words deleted within the quotation are indicated by ellipses. An exception to this convention has been made for book titles, whose spelling and capitalization is consistent with the *English Short Title Catalogue* for ease of reference. Within titles, I have only switched "i" to "j" and "u" to "v" where appropriate.

Early modern English names are notoriously variable in their spelling. I have used the following criteria for spelling names:

If she or he was included in the *English Short Title Catalogue*, I use the spelling in the *ESTC*.

If a figure is in the Oxford *Dictionary of National Biography* and has no books catalogued in the ESTC, I use the *DNB* spelling.

If she or he is an author known equally by both a vernacular and Latin name, I use the vernacular.

If she or he is an author known primarily by a Latin name, I use the Latin form for ease of reference.

If she or he is well known but none of the rules above applies, I use the spelling that occurs most commonly in the scholarly literature.

If she or he is not mentioned in the scholarly literature but left a will, I use the spelling that was used in probate.

If none of the above rules applies, I use the spelling that was most common in the sources that I consulted.

Elizabethans began their new year at the Feast of the Annunciation on 25 March, and did not change from the Julian to the Gregorian calendar when it was proposed in 1582. To avoid chronological confusion, I have converted Elizabethan dates for January, February, and portions of March to bring them in line with modern conventions by altering the year, but have not converted them from Julian to Gregorian by adjusting the day of the month. Thus the Elizabethan date 15 January 1575 appears here as 15 January 1576.

I capitalize guild affiliations, like "Barber-Surgeon," but not occupational designations. Thus if someone is a "Mercer," the capitalization means that he belonged to that guild, not that he dealt in cloth; if someone is a "surgeon," that was his occupation, but he was not necessarily a member of the Barber-Surgeons' Company.

A Note About "Science"

Some readers may take issue with my use of the apparently anachronistic collective term *science* to describe the varied Elizabethan interest in nature as it was expressed in London. After all, the *Oxford English Dictionary* does not report *science* — as a term designating the mathematical and physical study of nature — as having come into use until the middle of the nineteenth century.[1] I myself was trained to call the early modern interest in nature *natural philosophy*, since it was not laboratory-based, experimental, palpably modern science. But natural philosophy (I had also been taught) was an elite set of interests, founded in Aristotelian and anti-Aristotelian currents, informed by the new scholarship of humanism, and practiced by gentlemen and scholars with the free time and material resources required to contemplate nature. This definition did not seem to fit the background of the Londoners whose stories appear in this book or to convey their aspirations to put nature to productive and profitable ends.

Science was emerging in sixteenth-century English vernacular usage as an umbrella term to cover scores of such smaller, more easily described interests in specific aspects of the natural world as viticulture, alchemy, mining, and mathematics. From as early in Elizabeth's reign as 1559, when William Cuningham characterized mechanical invention as science, to the end, when Ralph Rabbards signed off on his translation of an alchemical text, "yours in the furtherance of science," the term was used to denote both a study of the natural world and a manipulation of the natural world for productive and profitable ends. Sometimes it was linked explicitly with nature, as when Cuningham identified cosmography as the "most excellent of all [the] other natural science[s]" and Thomas Charnock's *Breviary of Natural Philosophy* listed the names and uses for "all vessels and all instruments . . . in this science" of alchemy. John Securis helpfully defined the term in his *Detection and querimonie of the daily enormi-*

ties and abuses co[m]mited in physick: "science is a habit . . . [a] ready, prompt and bent disposition to do any thing confirmed and had by long study, exercise, and use."[2]

A wide range of Elizabethans used the collective term *science* to describe their interest in properties of the natural world or their efforts to manipulate and control those properties. In works published or republished during the age of Elizabeth, Leonard Digges used the mathematical sciences to identify the contemporary interest in astronomy, astrology, instrumentation, arithmetic, and geometry, as did John Dee, John Blagrave, and William Bourne. The queen voiced her support for "all good sciences and wise and learned inventions," and letters patent were issued under her name for technological experts in the "science" of furnace manufacturing as well as the "science" of glassmaking. Humfrey Baker promised that he had "more ready knowledge in this science" of mathematics than most others in England. Donald Lupton, writing about London, called it a center of all "sciences, arts, and trades," indicating that by the early seventeenth century there were some who were already teasing out the differences between these forms of inquiry and practice. The surgeon George Baker commended Italian and French writers for putting their knowledge of surgical science in the vernacular. His fellow surgeon William Clowes placed medicine and surgery among the sciences and instructed his readers to observe the boundaries that divided science from science. And Francis Bacon looked forward to the time when Englishmen would develop "sciences unknown."[3]

My research uncovered that English vernacular writers on natural history, medicine, mathematics, instrumentation, mechanics, and chemistry used the word *science* throughout the Elizabethan period both as a collective term and to denote individual sciences. The word appears most often among mathematical and medical authors; but it also creeps into the crown's language for the letters patent issued for technological inventions, into the works of popular authors like Donald Lupton who were not writing on science but on the city of London, and even into the queen's official pronouncements on inventions and monopolies. By contrast, very few of the Elizabethans I studied described their work as natural philosophy. Nor did they use any of the other terms that have been suggested to me as more historically sensitive alternatives to the word *science*, including *productive knowledge, imitating nature*, or *utility for the common good*. And so I had to make a choice whether to use the collective term at least some of my historical subjects used to describe their efforts or to adopt a term that none of them used but which most historians of science insist is correct and more historically nuanced. I decided to use *science*, just as my subjects did, because this struck me as the least anachronistic alternative.

During the Elizabethan period, residents of London were developing an urban sensibility toward the natural world that historians today might fruitfully call *vernacular science*. As this study shows, vernacular science was based in urban ways of knowing and evaluating nature that came from the densely overlapping social worlds of the City. As such it was distinct from the approach to natural knowledge taught at the university, or the genteel culture shared by gentlemen and gentlewomen at court who might be interested in the natural world. Like vernacular architecture, vernacular science bore some resemblance to the high-culture natural philosophy and intellectual pretensions of both the university and the court, but it had different priorities (observation was privileged over tradition, for example), different forms of expression (sharing experiments was preferred to sharing theoretical knowledge), and different values (dispute and questioning was accepted as a routine aspect of the business of science).

Throughout this book you will find evidence of a contemporary sixteenth-century use of this collective term. References to these instances are indexed under "*science*, contemporary usage of the term." Extrapolating from these examples, I also use *science* throughout as a shorthand collective term to denote Elizabethan Londoners' interest in understanding and exploiting nature even when the individuals in question did not do so. Quentin Skinner has argued that concepts can exist before the word that has come to designate them. This case is a bit more complicated, since the word *science* came into usage, along with some concept of the natural sciences, before we have been willing to admit that it did. This odd state of affairs is evident in the often mentioned, but seldom quoted, earliest definition for science in the *Oxford English Dictionary*. In 1867 W. G. Ward is given credit for coining this particular collective usage for *science*, when he wrote, "We shall . . . use the word 'science' in the sense which Englishmen so commonly give to it; as expressing physical and experimental science, to the exclusion of theological and metaphysical." It would seem that what was common among Englishmen in the nineteenth century was equally common in the sixteenth century.

Similar problems of anachronistic terminology developed when it came time to discuss individual students of nature and to describe their activities in a way that would be consistent and meaningful to a modern reader as well as accurate within an early modern Elizabethan context. Because I found no Elizabethan usage of the term *scientist* to describe a student of nature, I do not use the word unless I am discussing a nineteenth-, twentieth-, or twenty-first-century figure. Instead, I use specific early modern designations — such as *alchemist, surgeon, apothecary,* and *mathematician* — whenever possible. These terms were in use in the Elizabethan City, based on references I found in London records. Apothecaries, surgeons, and physicians were recognized occupational headings, and I also found one City record for an "alchemist," one for a "wizard," and a handful of references to "mathematicians." I use the modern terms *technology, technicians,* and *engineers* to refer collectively to avenues of inquiry and individuals who appeared in Elizabethan records only in connection with very specific activities: "clockmaker," "furnace maker," and "maker of gun-stocks," to give a few examples. There were no contemporary City records that referred to botanists or zoologists, although some Elizabethans did use the word *botanographer* to describe plant hunters and expert gardeners. For ease and consistency I followed one of my subjects, Thomas Moffett, in using the term *naturalist* to describe men and women interested in plants, animals, fossils, and ancient forms of life. When discussing these areas of natural knowledge, I use the term *natural history* even though none of my historical subjects did so.

The passages I have indexed which include contemporary uses of the word *science* are representative of the Elizabethans who used the term but are in no way exhaustive of all the examples that I found in the books and manuscripts of the period. Readers will appreciate that to make an exact catalogue of every Elizabethan use of the word *science* would make this a very different book. So, too, would any attempt to go back to medieval manuscripts and forward to the seventeenth and eighteenth centuries to see how the use of the word *science* changed in earlier and later vernacular English writings. Both of these endeavors would trace historical developments over time, and both would be worthy studies, but my concern here was to describe the study of nature at a particular place and at a particular time.

THE JEWEL HOUSE

LO

THAMESIS   South   We

1. S. Paul.   4. Arundel houss.   7. Durham houss.   11. Cheape Crosse.   11. Cole harbour.   16. S. Anthonies.   19. the Dutch Church.   11. Ludco H.
2. S. Brides.   5. Somset.   8. Iork houss.   11. Bow Church.   14. S. Laurence.   17. S. Laurence Poultneys.   10. S. Michaelis.   11. Rethmony.
3. Baynards Castle. 6. Devbey houss.   9. Kinges Pallace.   12. The Stilliarde.   15. Guild Hall.   18. The Exchange.   11. S. Peter.   14. S. Helli.

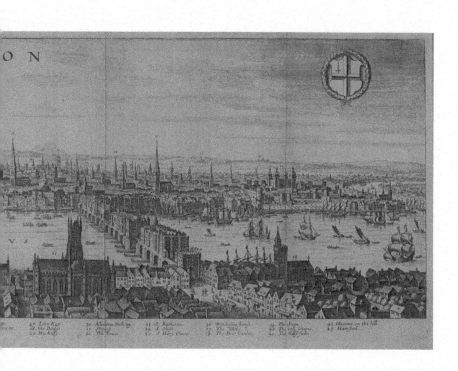

Figure 1 (overleaf). From Matthäus Merian's panorama of London (1638) it is possible to gain some sense of how imposing the Elizabethan skyline was. St. Paul's cathedral and the Tower of London bracket the eastern and western edges of the City. Reproduced with the permission of the Guildhall Library, City of London.

PRELUDE

London, 1600: The View from Somewhere

Standing on the south bank of the River Thames in 1600 and looking from Southwark to the ancient walled City of London, the viewer would have been struck by four features of her skyline: the monumental, crenellated stone fortress of the Tower of London to the east; the round, half-timbered "O" of the Globe Theater to the south; the truncated spire of St. Paul's cathedral in the west, rising up from the rectangular bulk of the enormous medieval church, still charred from a stroke of lightning that had blown off the top; and the sun glinting off the golden grasshopper that hovered over the smooth stone façade and arched colonnades of Gresham's new Royal Exchange to the north (see Figure 1). These four buildings marked London's distinct skyline and symbolized the political, cultural, religious, and economic power of the metropolis. Elizabethan London was truly the capital of early modern England, the vital, cosmopolitan center of the country's life. Anyone who has struggled to understand the Elizabethan City as a cultural, social, economic, political, or spatial entity will find comfort in the fact that contemporary residents, too, found it both exhilarating and bewildering. "She is grown so great, I am almost afraid to meddle with her," wrote Donald Lupton in 1632, continuing, "she is certainly a great world, there are so many worlds in her."

While London's skyline would have struck any visitor, the real vibrancy of the City — the energy symbolized by the Tower of London, the Globe, St. Paul's, and the Royal Exchange — rested in her people. On her dark, congested streets Londoners lived and worked, argued and worshiped, struggled and thrived. London grew from a small urban center of some 50,000 in 1550 to the second-largest city in Europe by 1600, with more than 200,000 residents. "This city of London is not only brimful of curiosities, but so popular also that one simply cannot walk along the streets for the crowd," wrote Swiss visitor Thomas Platter in 1599. Elizabethan Londoners were sophisticated and cosmopolitan, living cheek by jowl with immigrants from France, the Netherlands, Spain, Portugal, and Italy. The City's residents included Africans, Ottoman Turks, and Jewish *conversos*. Her foreign population was both a great asset and a great source of anxiety. The immigrants brought new trades like pinmaking and glassmaking, as well as new ideas for waterworks and other engineering projects, but they also increased the stress on already overburdened job and housing markets. Life on the City's streets, below the church spires and under the walkways of the Exchange, was both creative and competitive—the ideal environment for cultural and intellectual change.

Among the bustling crowds were hundreds of men and women who studied and exploited nature. Though they lacked a single building like the Globe Theater to draw the eye of a passing stranger, at street level they made up a recognizable and important feature of London life. These naturalists, medical practitioners, mathematicians, teachers, inventors, and alchemists not only actively studied the natural world; they were also interested in how that study could benefit human lives. During the age of Elizabeth, London nurtured the development of an empirical culture—the culture of the Scientific Revolution. While members of the royal court occupied themselves with threats foreign and domestic, and the universities of Oxford and Cambridge still debated the authority of ancient texts, the residents of London were busy constructing ingenious mechanical devices, testing new medicines, and studying the secrets of nature. There would have been no Scientific Revolution in England without the intellectual vitality present in Elizabethan London, for she provided later scientists with its foundations: the skilled labor, tools, techniques, and empirical insights that were necessary to shift the study of nature out of the library and into the laboratory.

To understand how London helped bring about such a change, it is helpful to return to her streets. St. Paul's cathedral, built in the Middle Ages to signify London's devotion to God, had stood for centuries as the most important landmark in the square-mile center of what was known simply as "the City" (Figure 2). Elizabeth I had launched numerous schemes, including a public lottery, in an effort to rebuild the damaged church. Despite its diminished height, St. Paul's remained the City's ecclesiastical center in 1600. Outside, preachers in open-air pulpits urged throngs of Londoners to repent and mend their ways, raising their voices to be heard over the booksellers and printers who had made the churchyard precincts their home for the past century. Where once it had been the religious hub of the City, by Elizabeth's reign the cathedral had become the intellectual epicenter of the realm, the source not only of religious debate but also of news sheets, broadsides, and thousands of printed books that spread the ideas of the Renaissance to eager readers. Buyers haunted the stalls outside St. Paul's purchasing used copies of Francis Bacon's *Essays* and first imprints of John Marston's latest play, or picking through bins of cheap old almanacs and star charts for hair-raising illustrations of local wonders like a two-headed calf, and to scoff at once-frightening prognostications made humorous by hindsight. And students of nature flocked there to buy vernacular books on medicine and surgery, imported foreign botanical works, and the mathematical instruments that were often sold along with handbooks that helped explain their use.

There was a high degree of literacy within London, helped along by a system of grammar schools that taught basic skills to City children. Reading, writing, and arithmetic were advantageous in London's competitive markets and in the global trade networks to which they were linked. Sir Thomas Gresham's major gift to the City, the Royal Exchange, was built as a lasting monument to London's position in these markets, and it quickly became the center of economic life in the City. Modeled after Antwerp's famous Bourse, the Royal Exchange housed shops full of expensive luxury items, offices where elite members of the Barber-Surgeons' Company plied their trade, covered walkways where spices, drugs, and cloth were bought and sold, and a grand open courtyard where gossips met and would-be lovers formed romantic liaisons. As the Exchange's popularity grew, market stalls opened up outside the gates, including shops that sold mathematical instruments and hastily erected platforms where itinerant medical practitioners peddled their potions and lotions. One Elizabethan proclaimed that it often seemed as if all of London could be found in the Exchange — lords and ladies, tradesmen and their wives, servants, apprentices, and thieves. Within the Royal Exchange and in the streets surrounding the building you could have heard every language from Arabic to Swedish being spoken by the merchants and foreign immigrants who formed an integral part of London life.

Not all of London's citizens were buying books, but an illiterate person did not lack opportunities to hear the news, make friends with foreigners, or learn of recent economic and political developments at home or abroad. A visit to the Royal Exchange, St. Paul's churchyard, or the theaters on the south bank of the Thames provided any Londoner, literate or illiterate, high or low, foreign or native born, with easy access to news and information. The area around the Globe in Southwark was particularly important as a metropolitan cultural center. New plays by William Shakespeare, Thomas Dekker, and their contemporaries were

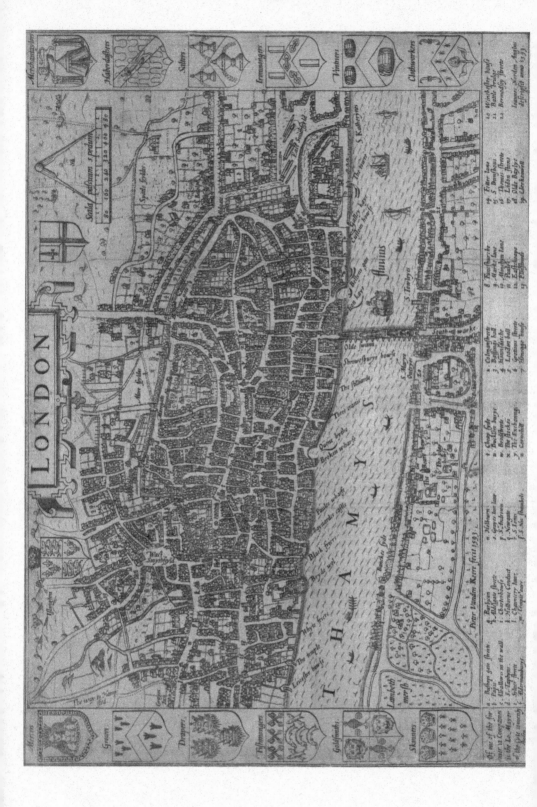

LONDON

enacted there and stimulated debate and comment among the audience. Ben Jonson, in and out of prison repeatedly throughout the 1590s, published his satirical send-up of London characters called *Every Man Out of His Humour* in 1600. While the 1599 production of the play at the Globe was warmly received by courtiers traveling down the Thames from Whitehall who enjoyed figuring out which of their friends were being lampooned, the play was seen by some City residents as lacking in both plot and taste. But the area around the Globe also housed a large hospital at St. Thomas's church which was known for its surgical staff, the workshops of several large-scale instrument makers who constructed parish clocks for the City and palace clocks for the royal court, and dozens of immigrant brewers who ran strange distillation and fermentation apparatus that astonished and intrigued the locals.

In 1600 the City's residents still clung tenaciously to London's Roman foundations, the uneven stone walls that encompassed roughly a square mile of territory on the north bank of the Thames. Yet the steady influx of people forced London to push relentlessly outward, and suburban sprawl began to encompass lands that had been gardens (such as the open fields to the west that would in the next centuries be developed into the residential and shopping district Covent Garden), industrial areas (such as the artillery foundries and glasshouses to the northeast of the City walls by the Tower of London), and the lands once held by the now banished Catholic Church. London's rapid expansion underscored her strategic and political importance to the realm, and Elizabeth I took care to treat her capital and its residents with an astute combination of firmness and respect. Yet the City remained difficult to govern due to her cobbled-together nature, the contrast between rich and poor that existed on every street, and her diverse population. While the political ideal of the City was one of peace and harmony, in reality London was a conglomeration of distinct neighborhoods anchored by more than one hundred churches and scores of trade organizations. Each individual and every corporate body wanted to advance their own causes and maintain their hard-won privileges. Overcrowding, the presence of foreign immigrants, poverty, public health crises, and civic unrest made this loose conglomeration contentious and politically explosive. Monarchs, including Elizabeth I, not surprisingly preferred to remain outside the City, maintaining no

---

Figure 2. Map of London in 1593. Around the margins are the emblems of the twelve major livery companies. A key at the bottom helped readers identify key landmarks, such as the Royal Exchange and St. Paul's. From John Norden's *Speculum Brittaniæ* (1593). Reproduced with the permission of the Henry E. Huntington Library.

official residence within London's walls. Palaces in nearby Westminster and Greenwich put the queen within easy reach of the City in case of serious internal troubles, and within easy reach of the Tower of London in case a foreign invasion was threatened. Such threats were frequent in the half-century before 1600, especially in the years leading up to and away from the great naval battle of the Spanish Armada in 1588, when rumors of impending danger spread like wildfire in the City's streets. Most foreign threats stemmed from the continuing religious antagonisms between Protestant England and her Catholic neighbors Scotland, France, and Spain. In 1600, despite England's victory against Spain, the rumor mill still suggested that there was a Jesuit hiding in every cupboard and a Spanish assassin lurking around every corner. And in fact espionage was a common feature of life in London, with French agents, Italian double-agents, and English spies frequenting the alleys and taverns to gather news and intelligence.

Within these landmark buildings, on the streets around them, behind shopfronts, and upstairs in residences throughout the City, men and women were studying and manipulating nature. This book is about these minor vernacular figures and their small successes, trial-and-error progress, and mundane aspirations. It is about the powerful partnership that existed in London between collaboration and competition, which often led to a heated but amiable discussion of ideas about nature in English rather than a publication of them in Latin. It provides an account of a relatively brief period in London's history and of the men and women who studied the natural world and tried to find better ways to harness its power and control its processes. They pursued this course by examining their own experiences as well as by repeatedly testing and verifying the experiences of their friends and rivals, thus taking steps toward experimentation. In Elizabethan London we can see how students of nature eagerly embraced the new print culture that was available to them but preserved the vibrant manuscript culture of the medieval period in their notebooks and recipe collections. By sketching out this vital world and exploring the ways in which the City of London functioned as a center for inquiry into and debates about nature, I am contributing to an ongoing historical project to situate the work of a small handful of acknowledged scientific geniuses within the densely social communities of practice that surrounded them.

Readers who want to learn more about Isaac Newton, Robert Boyle, Robert Hooke, Edmond Halley, and the other geniuses of the Scientific Revolution — and it is important to do so — may find this book unsatisfying. Here the most well-known figure you will encounter is Francis Bacon (1561–1626). Though not a genius in the sense of Newton or Boyle, he was a visionary. Born and bred in the age of Elizabeth, he lived just outside London's walls for much of his adult life in the western suburbs of the City. Popularly regarded as the father of modern science for his argument that science should be an organized activity pursued for the benefit of humanity, Bacon was deeply interested in the natural world and her mysterious workings and in *The New Atlantis* (1627) shared his vision for how science could be more functional and productive. Given his fascination with the natural world and his commitment to putting its control into human hands, he should have found the City an exciting and intriguing place. Instead, Bacon found the City's interest in nature deeply troubling. It was too plebeian, too democratic, and too vernacular for his taste. Bacon belonged to a higher level of the social order than did most of the City residents who made medicines, planted botanical gardens, or conducted experiments. How, he wondered, could he transform London's doubtlessly energetic — but to his mind inchoate and purposeless — inquiries into nature into a tool of state that could benefit the commonwealth?

Bacon responded to this question by constructing an ideal house of science — named Salomon's House after the archetype of wisdom in the Bible, King Solomon — that contained the scientific, medical, and technical activities he found most fruitful. The final pages of *The New Atlantis* were dedicated to a fantastic description of Salomon's House and the studies of nature that would take place there. These included efforts to explore the earth's mineral resources, to discover the most fruitful cultivation techniques, and to peruse the heavens with mathematical instruments. The study of the human body would take place in hospitals and anatomical theaters while others struggled to develop new medical cures. In his imaginary house of science, Bacon put all of these activities within a single, hierarchical institution overseen by a single, well-educated man.

In all other respects, however, Salomon's House already existed in the City of London — and Bacon and his contemporaries knew it. Every activity Bacon describes taking place in Salomon's House was already taking place in the City; every goal for seeking out profitable natural knowledge was already being pursued by one or more Londoner. It was not until the end of the seventeenth century, when the memories of the Elizabethan interest in nature had faded and the Royal Society had been established, that people began to look back on Bacon as a prophet of a newly empirical science. And their view has shaped our view. The intervening centuries have not been kind to Elizabethan London's interest in the natural world, and our knowledge of it has been so slight that we, too, saw Salomon's House as a blueprint for what science could *become* rather than a description of what science already *was*. This study puts Bacon's often-cited, seemingly prophetic blueprint for scientific work into its proper context, with the result that Bacon becomes less of a prophet and more a man in search of a position heading up this remarkable activity.

In the following pages I present multiple views of science as it was practiced during Bacon's life in the City of London. Instead of focusing on a single institution or a particular kind of inquiry into nature, I am interested in the great variety and communal character of London's science — the very qualities that most disturbed Bacon. In Bacon's London, ordinary, foreign- and native-born Londoners engaged in collaboration and competition over matters of natural knowledge with an urban, rather than a gentlemanly, sensibility. The urban sensibility that shaped the study of nature in Elizabethan London emerged from the densely overlapping obligations, ties, and community affiliations that were part of City life during the early modern period. London thrived when its residents embraced complicated partnerships between opposing urban forces — partnerships between private ambitions and public interests, English natives and foreign immigrants, market forces and collegiality, variety and coherence. While gentlemen like Bacon may have seen this urban sensibility as chaotic, it nevertheless shaped how Londoners approached the natural world. Not limited to artists, skilled craftsmen, or members of the City's formal guild and livery company structure, this urban sensibility was shared by most Londoners. City residents belonged to complex social networks that included family members, occupational acquaintances, neighbors, and other friends. In London, likeminded students of nature were drawn together through face-to-face interactions, residential proximity, and a shared desire to exchange books, specimens, techniques, and tools. Through thick description and the mapping of social and intellectual networks among individuals and between communities of practitioners, I sketch a picture of urban science that is striking in both its complexity and its functionality.

London, as a house of science and a prototype of a modern laboratory, worked. Despite the absence of a single institution to order and control it, her urban sensibility helped London practitioners successfully investigate nature, mediate conflicts over knowledge claims, collaborate on projects, expertly adjudicate disputes over methods, train new practitioners, seek financial support from civic and court figures, and negotiate their way through the challenges of studying nature in a crowded urban setting. With respect to science, the urban sensibility shared by Londoners had three important features. First, men and women living in the City expected that their work would be publicly known even if it were not published, because it would be studied and evaluated by other Londoners, especially those involved in similar occupations or trades. Trade associations like the Grocers' Company and the Barber-Surgeons' Company, for example, were deeply involved in overseeing the quality of medical goods and services produced in their members' shops as well as in policing those individuals outside the company who might impinge on their honor or privileges. Second, London's urban sensibility fostered a belief that residents had specific types of expertise that could and should be exploited to benefit particular individuals and the City as a whole. London was home to all kinds of experts — from ale makers to zookeepers — who could be called on by students of nature to provide specialized assistance in their inquiries. And one did not have to call very far to catch their attention, since London's compact size facilitated exchange and interaction. Third, London's urban sensibility confirmed that work done in collaboration with others was both necessary and desirable in a thriving city. Whether it involved working together with other parish residents to get an unhealthy ditch cleaned out, joining forces to prosecute a particularly egregious offender of civic ordinances, or sharing the responsibilities for policing apprentices with other members of your guild, collaboration was a vital component of getting business done in early modern London.

It was London's urban sensibility, along with new trade networks and her growing population, which made the City an ideal place for new ideas about the natural world to emerge. Shouting to be heard over the din from market stalls on nearly every street, working in cramped backyards over furnaces and smelting ovens, and operating out of storefronts in the Royal Exchange and other merchant neighborhoods, hundreds of men and women of all nationalities engaged in the work of science, medicine, and technology in London. Some were poor immigrants, like "Dutch Hans," a German metalworker who traded his knowledge of the properties of molten lead for beer in a crowded pub near the Globe. Others were fixtures of London's trade organizations, like the Barber-Surgeon George Baker, who extracted teeth, set bones, and performed surgical procedures in his shop at the gates of the Royal Exchange. The Antwerp native Lieven Alaertes and Londoner Thomasina Scarlet established lucrative medical practices in the City, specializing in treating obstetrical and gynecological complaints despite the best efforts of London's College of Physicians to force women out of the medical market. A Venetian merchant and alchemist, Giovan Battista Agnello, operated a dangerous blast furnace in one crowded neighborhood without any complaints from his neighbors, while one of the queen's physicians, the Portuguese converso Roderigo López, was famous, long before accusations that he had tried to poison the queen made him notorious, for seeing patients accompanied by an entourage of African servants.

On London's streets, in her shops, and within the more private enclaves of house and garden, these women and men shared information and expertise with colleagues and collaborators, argued about knowledge claims and procedures, waged war with competitors, and struggled to come to terms with a confusing and rapidly changing world. In the Middle Ages, ancient authorities like Aristotle and Galen had been the ultimate arbiters in disputes about nature and science. In the sixteenth century, however, such conflicts were more difficult to settle. Nicolaus Copernicus and Johannes Kepler had sketched out a new heaven, and explorers were rapidly charting an entirely new world full of flora and fauna, like the alligator and the tomato, that defied description and challenged the ancient encyclopedias that Elizabethan Londoners still consulted for information. To be in London during the second half of the sixteenth century was to be in a state of heady confusion when it came to natural knowledge and questions of science. Every ship that put in at a London dock might contain new materials that needed to be classified and understood, each new book rolling off the presses at St. Paul's could contain a radical idea about the natural world, and the experiments undertaken in London had, at any moment, the potential to bring long-held beliefs into question.

Yet the Elizabethans, for all their energetic and enthusiastic work studying the natural world, were responsible for only a few scientific breakthroughs. How, then, can their stories help us to understand the Scientific Revolution? Their significance lies not in the elucidation of new formulas or the construction of new cosmological systems, but in the ways that they organized their communities and settled disputes; the value they placed on the acquisition of various literacies (including mathematical, technical, and instrumental literacies); and the practices they developed that led to an increasingly sophisticated hands-on exploration of the natural world. These contributions, I argue, laid the social foundations for the Scientific Revolution in England and did the groundwork that was required so that a man like Boyle knew whom to ask, and what to ask for, when he sought out a man to assist him in his air pump experiments. Every seemingly isolated "great man" of science in the early modern period was surrounded by a "great mass" of workers, assistants, and technicians. In this book I help explain where this great mass came from, and how they developed their skills and knowledge.

By examining six emblematic cases in Elizabethan London, we will survey the social foundations of the Scientific Revolution as they were laid in the City, and come to know the people and the practices that inspired Bacon to conceive Salomon's House. Vivid portraits will emerge of individual practitioners and the challenges they faced when attempting to come to terms with the complexities of the natural world. Networks of intellectual exchange and communities of inquiry can be mapped onto the terrain of Elizabethan London in ways that illuminate the blind alleys and surprising twists and turns taken as science became the field of knowledge we recognize today. The drive to educate London citizens and make them mathematically literate, for example, created an atmosphere of intense competition as well as fruitful collaboration that was both functional and highly organized, despite its vernacular qualities. And a new emphasis on the hands-on study of nature created demand for technical knowledge about distillation apparatus, medical formulas, grafting techniques, and experimental practices.

The foundations of the Scientific Revolution in Elizabethan London depended on three interrelated social endeavors: forging communities, establishing literacies, and engaging in hands-on practices. In the first two chapters I focus on how communities interested in the natural world coped with the stresses and strains of studying and manipulating nature in an urban environment. Anxieties about new ideas and the influx of foreign practitioners into the City informed how groups negotiated knowledge claims and disputes, and shaped the kind of natural knowledge that was produced. In "Living on Lime Street: 'English' Natural History and the European Republic of Letters" I usher readers into the world of Dutch, French, Flemish, and English naturalists — including botanists, apothecaries, and entomologists — in the downtown neighborhood of Lime Street. Living in close proximity to one another, these friends and business associates enjoyed the face-to-face exchange of ideas as well as correspondence with other naturalists throughout Europe. The Lime Street naturalists and their work were so internationally famous that they caught the attention of an avid student of nature, who was especially interested in plants, the English Barber-Surgeon John Gerard. Gerard lived outside the City's walls, however, and was never able to become a central figure in their cosmopolitan world. So Gerard found another community for his scholarly efforts: readers. Unlike the Lime Street naturalists, Gerard embraced the world of print, and through his published books he began a full-scale assault on Lime Street's credibility and reputation.

Two Barber-Surgeons similarly emphasized the power of the printed word when they tried to expand their role in London's competitive medical marketplace. In "The Contest over Medical Authority: Valentine Russwurin and the Barber-Surgeons" I examine a conflict between foreign and English practitioners that began in the City's streets and ended up in the press. When the German surgeon Valentine Russwurin established a medical stall outside the gates of London's Royal Exchange, he prescribed Paracelsian treatments and therapies to his patients, such as the use of mercury to treat the New World disease syphilis. But important members of the City's Barber-Surgeons' Company were also eager to promote Paracelsian cures and use them in their medical practices. Resentful of Russwurin's success, and anxious about the status of Paracelsian medicine as his patients began to show serious side effects, a handful of Barber-Surgeons decided to exploit the City's love-hate relationship with foreigners and foreignness. The English Barber-Surgeons began a concerted publishing campaign that polarized London's medical marketplace. Through competition and collaboration, Londoners began to form recognizable communities of expert practitioners during the age of Elizabeth, and successive generations of these communities continued to provide resources and to shape the study of nature well into the seventeenth century.

From a focus on communities, we turn to a focus on literacies. To thrive in the Elizabethan City, Londoners needed to have (or be perceived as having) some special expertise and the most powerful forms of expertise often related to one or more forms of literacy. While we think of literacy as the ability to read and perhaps write, Elizabethan Londoners were eager to gain a broader range of literacies. They wanted to know mathematics — to use arithmetic to keep accounts and geometry to fashion instruments. The ability to manipulate and use such instruments was another kind of literacy, as I explain in Chapter 3. In "Educating Icarus and Displaying Daedalus: Mathematics and Instrumentation in Elizabethan London," we explore how instrument makers, in concert with mathematical educators and civic leaders, put the acquisition of mathematical and instrumental literacy at the forefront of London's educational priorities. Emphasizing the utility of mathematics, and the development of accurate problem-solving skills, these Londoners wanted the City to stand at the vanguard of Europe when it came to the practical application of mathematical knowledge. Though conflicts broke out over whether mathematics was a suitable subject of study for ordinary Londoners, mathematics had the backing of some of the City's most powerful citizens. Their support ensured that the community of mathematicians and technicians thrived, challenging the ancient idea that the life of the mind and the life of the hands must be kept distinct.

In Chapter 4 I take London's concern for multiple forms of literacy and examine the development of an Elizabethan "Big Science" that depended upon citizens with instrumental and technical literacy as well as the financial and political support of the crown and City investors. Elizabeth may not have been a direct source of patronage along the lines of the Holy Roman Emperor Rudolf II or the Medici princes, but she was a shrewd businesswoman who used her secretary of state, William Cecil, and his staff of informants to ensure that England would not be left behind technologically. In "'Big Science' in Elizabethan London" I examine how Elizabeth promoted science by offering inventors monopolies on technical developments, including corn mills, fuel-efficient ovens, water pumps, and methods of processing metals. Competitors resorted to espionage and theft in order to secure the crown's letters patent, and Cecil brokered disputes over issues of technical expertise and intellectual property, relying on a network of technically literate informants to keep him apprised of the details of each case.

An intense interest in exploring and exploiting the natural world, when coupled with increased mathematical literacy and flourishing technological projects, led to the development of hands-on, experimental practices in Elizabethan London. In the final two chapters I make detailed use of manuscript notebooks to explore how experimental practices were pursued, recorded, and questioned in the City. In "Clement Draper's Prison Notebooks: Reading, Writing, and Doing Science" I focus on the manuscripts kept by an Elizabethan merchant imprisoned for debt. Like the illustrious Francis Bacon, Draper was a visionary, but for him the ideal intellectual community was a commonwealth of experimental knowledge where practitioners shared ideas and experiments irrespective of wealth, class, or university training. Draper's notebooks illuminate how elastic the sense of community could be in a city like London, and how this elasticity made even the most humble contributions potentially valuable in the study of nature. Recording every snippet of text and experimental lore that came his way in the King's Bench prison, Draper's notebooks clarify how writing was itself a valuable form of work during the early days of experimental culture.

We end by accompanying Elizabethan London's most curious student of nature, Hugh Plat, as he amassed an enormous amount of experimental information from City residents. For Plat, London was a "jewel house" of experimental expertise, a treasure trove of natural knowledge, and he saw himself as the lapidary responsible for testing, polishing, and authoritatively recounting the best examples of this work for future generations. In "From the Jewel House to Salomon's House'" I compare Plat's vernacular approach to experimental practice and natural knowledge with that of Francis Bacon. Plat believed that all classes should participate in science, and he valued his grocer's opinions on the virtues of plants as highly as he did a university-educated botanist. Both deserved a place in London's jewel house of science, provided that Plat could verify and test their information. Bacon's conception of the ideal workspace, Salomon's House, was more hierarchical and bureaucratic. Whereas Hugh Plat was deeply embedded in the messy, decentralized world of Elizabethan London science, Bacon was an outsider to that world and held himself aloof from it, severely criticizing its values, practices, and personnel.

While these stories present unfamiliar people like Valentine Russwurin and put us in uncomfortable places like the King's Bench prison, they recover a moment in the history of the Scientific Revolution that has been not irretrievably lost but only misplaced. My intention was to write a book that could serve as a kind of Baedeker's guide to a strange and unfamiliar City in an effort to help readers explore this overlooked landscape and its inhabitants by immersing them in the sights, sounds, smells, and personalities of science as it was understood in London during the Elizabethan period. At the end of the book, for readers interested in what I think these stories have to offer historians of science and the urban experience, there is a coda in which I try to situate my work in the context of other scholarship in the history of science in general and early modern science in particular. Scholars who have thumbed through this introduction in vain for references to academic arguments about the Scientific Revolution and feel it essential to understand my methodology and point of view at the outset will want to read that coda now. But I hope that many readers will be willing instead to turn the page and slip among the naturalists on Lime Street, where our tour of the City starts with James Garret, a Flemish apothecary with some very bad news for a London publishing house. Garret will be the first of many guides who will point out the buried but still discernible social foundations of the Scientific Revolution in Elizabethan London.

1

LIVING ON LIME STREET

"English" Natural History and the
European Republic of Letters

In 1597 James Garret, a Flemish apothecary who lived and worked in London, paid a visit to the shop of Bonham and John Norton, one of the City's busiest and most prestigious publishers. The Nortons specialized in expensive, large-format books, and they were in the process of readying John Gerard's massive manuscript of *The herball or Generall historie of plantes* for the presses (Figure 1.1). It was the most ambitious English-language publication on the subject that had ever been attempted, and its release promised to make Gerard a household name as ladies bought up copies so that they could trace the illustrations in their embroidery and consult its medical lore, and as plant fanciers purchased the item to be conversant with all the newest discoveries of flora from around the globe. The shop would have been abuzz with activity: apprentices carrying paper, journeymen setting the movable type in trays and inking them before loading them into large, imposing presses. Garret would have had to duck to avoid jostling damp sheets of paper pegged up on lines to dry, then swerve to keep from knocking into customers perusing the Nortons' latest releases, such as Plutarch's *Lives of the noble Grecians and Romanes* (1595) and a new edition of Monardes's *Joyfull newes out of the new-found worlde* (1596). What Garret was about to reveal, however, would bring part of the Nortons' operations to a sudden, though temporary, halt.[1]

Garret presented the Nortons with the unwelcome news that John Gerard's costly new *Herball* was a mess of inaccuracies, plagiarized passages, and improperly placed illustrations. According to the Flemish apothecary, Gerard's most egregious fault was his crude use of Mathias de L'Obel's (1538–1616) taxo-

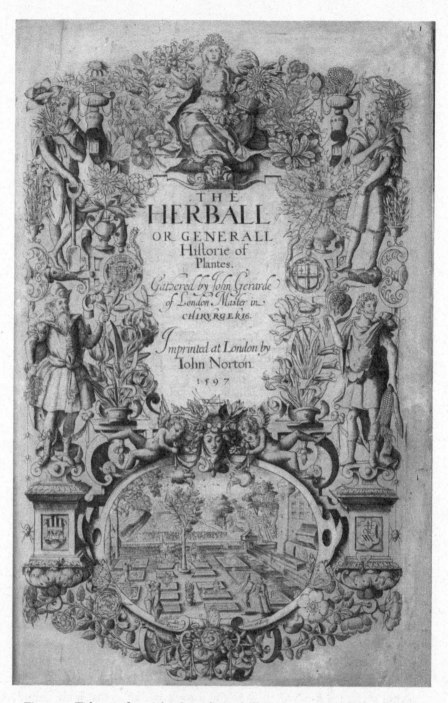

Figure 1.1. Title page from John Gerard's *Herball* (1597). Its ornate appearance was an indication to the potential buyer of the grandeur and costliness of the publication. Reproduced with the permission of the Henry E. Huntington Library.

nomic system for plants, which he had developed along with Pierre Pena in the *Stirpium adversaria nova* (1571). Garret complained that Gerard's new *Herball* was not in fact new at all. Instead, it had been cobbled together — not terribly effectively — by combining L'Obel's work with a translation of a well-known Flemish herbal by Rembert Dodoens that most who were interested in the subject already owned. Garret was L'Obel's neighbor in the parish of St. Dionysius Backchurch, and the two men shared a passion for plants as well as a common immigrant status. To sit by and watch an English Barber-Surgeon like Gerard take credit for the painstaking work of his friend was too much for Garret to bear.

The Elizabethan period was an age when the standards for plagiarism were notoriously low, and Garret's claim that *The herball*'s author had liberally borrowed from other published books would not have been enough for the Nortons to call a halt to the printing process. But the Nortons did want to make a profit on this expensive new book of Gerard's, so Garret's charge that the book was inaccurate was cause for concern. Who would be willing to spend so much money on a book that they could not rely upon to accurately identify common medicinal plants and the more exotic plant specimens pouring into London from the New World and beyond? Garret's observations that there were mistranslations of the Flemish herbal that served as Gerard's source, that the illustrations of plant specimens did not always appear alongside the correct descriptions, and that some illustrations were even inserted upside down had the potential to cut the Nortons' audience dramatically and make them a public laughingstock. Because *The herball* was a costly publishing endeavor involving hundreds of pages and expensive woodcuts, the Nortons became understandably alarmed at the fiscal implications of Garret's message.

After taking a long, hard look at the manuscript, the Nortons realized that they had to do something to amend it. *The herball* was not yet in a fit state to roll off the presses and into the hands of London's eager book buyers. Their solution was to hire Garret's neighbor, Mathias de L'Obel — the same man who had already (albeit unwittingly) made a substantial contribution to the work — to proofread the translations, fix the mismatched illustrations, and right its other textual wrongs. L'Obel began the laborious work of editing straight away and made numerous corrections in the manuscript before John Gerard, tipped off to what was going on in the Nortons' print shop, barged in and had him dismissed. Gerard was outraged at the interference of a foreigner in his great English work and cast aspersions on L'Obel's heavily accented use of English. L'Obel was internationally renowned for his expertise with plants and had the added prestige of possessing a doctorate from France's premier medical institution, the University of Montpellier, but this mattered little to the English Barber-Surgeon. L'Obel was just as furious with Gerard, whose studies he had furthered by providing access to rare plant specimens and, even more important, to his numerous friends who occupied high positions in royal gardens and university medical schools throughout Europe.[2]

The Nortons finally went ahead with the publication of Gerard's *Herball*, and it sold well despite its length, cost, and the fact that the quality of the work plummeted dramatically in Book III, where the author turned to the subject of grasses. Those who were charitably inclined could attribute this failing to Gerard's intellectual exhaustion, while those less so could point out that the change in tone marked the spot where L'Obel was dismissed from his thankless job of editing the manuscript. Today, however, the efforts that Garret made on behalf of his friend L'Obel are largely forgotten, consigned to a dustbin of publishing anecdotes. Few have ever heard of James Garret, and Mathias de L'Obel's name is known only to serious students of botany and curious gardeners who wonder how the *lobelia* got its name. *The herball*'s author, John Gerard, on the other hand, is remembered as Elizabethan England's premier naturalist. In his own time, however, his reputation was mixed, and the publication of *The herball* marked not the apotheosis of England's first great botanist but the development of a schism in London's natural history community.

Until the publication of *The herball*, London's naturalists revolved like satellites around the parish of St. Dionysius Backchurch in the center of the City where Garret and L'Obel were neighbors. There, on a twisting street only a tenth of a mile long, the two men played a key role in a natural history community that included English and foreign members. It was a wealthy area, and most of the community's members were successful merchants, physicians, apothecaries, and City politicians. Other London naturalists, such as Gerard, paid frequent visits to Lime Street to examine the natural curiosities housed in cabinets there, and to walk the paths of the many fine gardens that graced the parish. They could also post letters to their foreign correspondents, since the postmaster for the Dutch community lived on Lime Street and had an excellent network of friends who sped those letters on their way across Europe no matter how incompetent the political system, vicious the religious dispute, or violent the warfare. Members of the Lime Street natural history community were interested in collaboration and welcomed men like John Gerard into their circles with warm collegiality. Lime Street may have been one of the Republic of Letters' more remote European outposts, but its members conducted business within the parameters of a code of conduct that was rigidly enforced to promote the exchange of information and the courteous acknowledgement of friendly contributions to one's own studies. When John Gerard put *The herball* together, he violated the rules that governed this closely knit community of students of nature. From 1597 Gerard was an outcast and had to rely on competition rather than collaboration, the patronage of a few court figures rather than the support of a European network of naturalists, and by continuing to set his work out in print rather than engaging in the lengthy (and typically unpublished) collaborative projects that were representative of the Lime Street community.

Gerard may have suffered for a few years from the sting of being shunned by the Lime Street naturalists, but in the long run they paid a higher price: they have been largely forgotten. Gerard's *Herball* has become a monument of early botanical knowledge that has overshadowed the Lime Street community. Not all the blame for this can be put at Gerard's door, however. While the Lime Street naturalists worked at their collaborative projects and wrote letters to like-minded students of nature, Gerard published his book — imperfect though it was — and it now serves as a touchstone for historians of botany. Some of the Lime Street naturalists did publish, but their works were highly specialized and never received the popular acclaim that Gerard enjoyed. So while members of the Lime Street community are, as we will see, often mentioned individually in historical studies of the Republic of Letters, humanism, and natural history, they do not often appear as part of a coherent and vital intellectual group. And because many of them were immigrants, some members of the community are left out of narratives about the development of "English" science. To fully understand the significance of the Lime Street naturalists, however, we must see them as they saw themselves: as an important community interested in natural history with links to other naturalists in England and on the European continent. To do so, we need first to understand how the community was constituted and how it functioned. Then we will be in a better position to examine their intellectual interests and the debates they engaged in with other leading naturalists at home and abroad. Finally, we will explore how this vital, dynamic group of collectors, plant hunters, and students of nature were lost to historical view through a grand publishing venture and the ambitions of its author.

FRIENDS AND FAMILY: THE LIME STREET COMMUNITY

In 1606 James Cole (1563–1628), a silk merchant and Flemish immigrant, married his second wife, Louisa de L'Obel, the daughter of the famous Flemish naturalist and physician Mathias de L'Obel, in the city of London. It was a moment that the local community and the European Republic of Letters were eager to celebrate. The groom was congratulated by friends at home and abroad for choosing a wife who was not only "beautiful and loving" but also the daughter of a "more diligent searcher after plants than Dioscorides." One friend writing to Cole from the Continent treated the match as an ideal arrangement for the groom and his new father-in-law. He did not mention the bride at all, but commented only that the marriage must make Cole especially pleased because "you have obtained a father-in-law with whom you can continually converse on a part of your studies."[3] When the Cole family and the L'Obel family became entwined in marriage, the ceremony gave formal permanence to the warm relationship that had existed between them for years in the central London neighborhood of Lime Street.

Elizabethans were struck by the area's prosperity. John Stow, in his *Survey of London* (1598), described Lime Street as an expanse "of fair houses for merchants and others" that twisted and turned from St. Dionysius Backchurch and Fenchurch Street in the south to the parish church of St. Andrew Undershaft and Cornhill Street in the north.[4] Today the landscape is altered from Stow's time. St. Dionysius Backchurch is gone: the medieval church fell prey to the Great Fire of London in 1666, and the rebuilt church was demolished in the nineteenth century. Though St. Andrew Undershaft remains, its once lofty tower is now dwarfed by the neighboring Lloyd's building, and its walls are ringed by motorcycles and scooters parked by City workers. A modern visitor to London can still walk the narrow street, however, which maintains its Elizabethan layout, and can make a pilgrimage to Stow's monument in St. Andrew Undershaft, where the lord mayor of London goes once every three years to put a fresh quill pen in the hands of a stone effigy dedicated to Elizabethan London's best-known historian.

Lime Street had a relatively low population density by Elizabethan standards, and was blissfully free from the apartment-style dwellings known as tenements — old houses converted into multihousehold residences. Behind high walls and through great gates were gardens and a tennis court, and everywhere there was an audible hum of activity. Lime Street's numerous surgeons and apothecaries welcomed patients into their shops and sent apprentices and servants scurrying into nearby markets for supplies. Members of the Pewterers' Company walked the street to enter their guild's hall for ceremonial dinners and to attend meetings. Artists and builders took a shortcut down a small alley on the west side of the street and ended up in the Leaden Hall, whose attics were used to construct and store props and scenery for City pageants and festivals, high above the main floor where the more mundane business of weighing wool and grain took place.

Elizabethan Lime Street boasted the cosmopolitan assortment of residents that was fast becoming the norm in London: English, French, Flemish, and Italian. The southern end of Lime Street, down by St. Dionysius Backchurch, in-

tersected with Lombard Street, once the haunt of London's European merchants who met there twice a day to exchange news and information until 1568,
when they were expected to transact their business under the protected porticoes
of the Royal Exchange.[5] Many of these Europeans had become permanent residents of London, most to avoid religious persecution, some because of economic opportunity. These immigrants, or "Strangers" as they were called in
Stow's time, never became a sizable part of the City's population — at most, the
Strangers constituted slightly more than 5 percent of London's population — but
few residents were unaware of their presence.[6] The merchants, skilled craftsmen, and simple tradesmen who came from exotic locales like Venice and
Antwerp made an impression on London's English residents that was far deeper
than their relatively small numbers might suggest.

While any Elizabethan visitor could have grasped these visible and audible
features of Lime Street easily and quickly, other significant neighborhood activities would have been harder to discern. For behind garden walls, inside the
apothecary shops, and within the well-appointed houses of the merchants lived
an important community of naturalists. Lime Street was the English outpost of a
Europe-wide network of students of nature — including plant hunters, gardeners, rock and fossil collectors, and scholars interested in animals and insects —
who eagerly studied the marvelous and manifold properties of the animal, vegetable, and mineral kingdoms. Both Strangers and English citizens from the
neighborhood around Lime Street made contributions to natural history, enjoyed an active correspondence with other humanists in Europe-wide networks
of correspondence and exchange, and forged intellectual alliances that tran-
scended linguistic and national boundaries.

The Elizabethan City was riddled with neighborhoods like Lime Street,
where common interests shaped social and intellectual life and an urban sensibility emerged that blended cosmopolitanism with nascent nationalism, competition with collaboration, and theoretical learning with practical experience. Because of the ephemeral nature of most social and intellectual interaction, which
relies heavily on face-to-face conversation, it is hard for historians to establish
who belonged to a given community or neighborhood, not to mention what
those individuals gossiped and thought about as they met over dinner, or in an
apothecary's shop while buying a medical elixir to ease their aches and pains.[7]
What makes Lime Street unusual is not that it existed but that its existence is well
documented in correspondence, bequests in wills, and passing references in
printed natural history texts published in England and abroad. These materials
enable us to see the Lime Street neighborhood and the naturalists who lived
there as friends and intellectual colleagues. We are able to peep into the houses

and shopfronts along Lime Street, where animal and insect specimens were studied and classified, fossils were examined and displayed in curiosity cabinets, and rare plants were cultivated and propagated.

It was not easy to forge a natural history community in early modern Europe. You needed an elusive combination of ingredients that, after being mixed and compounded, only then resulted in the intellectual vitality apparent on Lime Street. First and foremost you needed to find other men and women who shared your passion for deciphering the curious natural world through the study of plants, antiquities, and zoological specimens.[8] Once they were located, mundane issues could stunt the development of an otherwise promising community. Naturalists needed ample financial resources so that they could acquire specimens for their own studies, the books that kept them informed of far-flung developments in natural history, and the leisure to engage in the work itself. Not every member of a community needed to be fabulously wealthy, but at least one individual had to be able to finance such a costly and labor-intensive endeavor. And a great natural history community required that most valuable of all urban commodities — space. Space was needed to cultivate gardens, since so much of natural history focused on the study of plant specimens, and unoccupied land came at a high premium in an early modern city like London.

In order to truly flourish, a community of naturalists like the one on Lime Street also needed to be able to forge connections with others who might have access to different plants, animals, and mineral specimens. These specimens were the intellectual capital of the Republic of Letters, an early modern Euro — a form of intellectual currency that could travel freely and be exchanged easily no matter where the specimens originated, or where they concluded their journey. Rare plants, cultivated and propagated in Lime Street gardens, became important items of exchange, accompanying letters to learned naturalists in Italy, Germany, France, and the Netherlands. Through these exchanges the Lime Street community established its identity and reputation within the wider Republic of Letters.[9] Such high-stakes correspondence demanded that at least one member of the community be able to read and write Latin — and other European languages were desirable commodities, as well. Linguistic facility was necessary for the community to remain familiar with influential printed natural history texts, many of which were available only in Latin. Intellectual relationships with other naturalists often depended on how "learned" a group was, and this estimation (like the plants grown in their gardens) became valuable capital in the Republic of Letters.

The Lime Street community had these important resources — passion, wealth, space, connections, and learning — and mobilized them all in the pursuit of natural history. At the heart of the community was a man who possessed each of these commodities in abundance: the silk merchant James Cole. Cole lived at the southern end of Lime Street, near St. Dionysius Backchurch, and commanded considerable wealth. His ordinary, English-sounding name hid a Flemish immigrant background with distinguished antecedents: his mother, Elizabeth Ortels, was the sister of the celebrated Antwerp cartographer known by his Latinate nom de plume, Abraham Ortelius. Cole's mother had worked until her marriage in her famous brother's cartography shop, mounting the maps on linen and hand coloring them according to the wishes of each buyer. Her son, christened Jacob Coels in Antwerp in 1563, became James Cole only after his family fled religious persecution in the Netherlands and joined the hundreds of Huguenot refugees living in London.[10] From his youth, the family seems to have seen in James the intellectual curiosity and abilities of his uncle, and the two became close through occasional visits and regular correspondence.

Their letters reveal that James Cole was a well-read Latinist with an avid interest in plants, fossils, old coins, and other curiosities. Cole does not appear to have been formally educated, either at the grammar school or university level, and while contemporaries saw him as an autodidact, his early love of learning may have been fostered by his well-educated mother. By the age of twelve he was studying Greek and receiving simple Latin letters from Ortelius, who also sent his young nephew parcels of books. In his mid-twenties Cole went to Antwerp to stay with his uncle and aunt, Anna. "He does not pass his time uselessly," Ortelius wrote proudly to James's father back in London; "he studies, he writes. He learns every day, which I observe with pleasure." Cole's studies were not limited to natural history, and he also delved into matters of history, theology, and philosophy. "I found him passionate and knowledgeable about ancient things," the antiquarian Marquardus Freherus reported to Ortelius.[11]

Cole was a highly respected scholar in his own community, and a published author who wrote a wide-ranging assortment of treatises popular enough to be reprinted and translated, including a work in praise of the study of plants, a description of the plague in London, and religious texts. A devoted student of history, Cole transformed his knowledge of ancient Roman festivals into a useful political tool when he helped to design the Dutch Church's festivities celebrating the coronation of England's new king, James I, in 1603–4. Working with his friends the physician Raphael Thorius and the architect Conraet Jansen, Cole penned elegant verses praising the new king. In his printed works, Cole also displayed his learning, as well as his love of versifying. His *Syntagma herbarum encomiasticum* (1606, 1614, 1628) was a brief discourse in verse on the joys of plant study and collection. The perils of the plague made a deep impression on Cole, and he vividly described the progress of the disease across London during the late summer and autumn of 1603 in *The State of London During the Great Plague* (*Den staet van London in hare Groote Peste* [1606]). A rare eyewitness account of the sights, sounds, and smells of an early modern plague epidemic, Cole's work explained how doctors counseled patients to carry sweet-smelling herbs to ward off contagion, and how the urban population survived the psychological and social difficulties associated with the spread of the disease. A staunch Protestant, Cole also wrote about his religious faith, including an interpretation of Psalm 104, *Paraphrasis . . . van den CIIII Psalm* (1618, 1626) and an exhortation for Christians to make a "good death," *Of death a true description* (1624, 1629).[12]

Cole's interest in natural history, his education, and his connections with learned men throughout Europe were built upon a family foundation, and the cornerstone of that foundation was his famous uncle. Abraham Ortelius (1527–98), a book and antiques dealer in Antwerp, developed a passion for geography and natural history which led to his appointment in 1570 as royal cosmographer for Philip II of Spain.[13] His *Theater of the World* (*Theatrum orbis terrarum*, 1570), the most important atlas printed in the early modern period, catapulted Ortelius to fame, while his business affairs — dealing in antiquities and books and compiling maps — helped him to cultivate close relationships with other scholars in the Republic of Letters. He was a careful and consistent correspondent, never failing to look for a rare imprint or to make an impression of a unique coin that he thought one of his friends might value. They reciprocated by sending him the latest news from voyages of exploration and fragments of manuscripts that touched on his interests in the ancient or natural worlds. Ortelius was Cole's role model, and the young man self-consciously styled himself after his uncle, signing his own budding correspondence within the Republic of Letters "Jacobus Colius Ortelianus" so that the relationship between the Antwerp cartographer and the London merchant would not go unnoticed.

Cole's letters were sent throughout Europe under the auspices of another maternal relative, Emmanuel van Meteren (1535–1612), his mother's first cousin. The van Meteren family, like the Cole family, had fled the Netherlands for religious reasons. Van Meteren's father, Jacob, played a key role in the publication of the first English Bible in the 1530s, and the family established itself as a pillar of the Protestant refugee community in London. As an adult, Emmanuel lived on Lime Street, became the consul for the Dutch merchants in London, and occupied the all-important office of postmaster. Reliable mail service was an essential component of any natural history network, since specimens ranging from tulip bulbs to rhinoceros horns needed to circulate between interested naturalists. Accompanying these specimens, of course, were the letters on which the

European natural history community in particular, and the Republic of Letters more generally, depended in order to thrive. Having the hub of a Europe-wide mail- and package-distribution center right on Lime Street no doubt made Cole's life somewhat more convenient, but it was van Meteren's skillful management of the post that made him indispensable. When the artist Marcus Gheeraerts wanted to send smoked herring to Antwerp, or Ortelius wanted gifts to arrive at his sister's house in London, they inevitably went through Emmanuel van Meteren and his formidable network of middlemen, merchants, sailors, and travelers to ensure that precious messages and gifts reached their destination. Emmanuel van Meteren, like Ortelius, was part merchant and part intellectual, but his interests focused on history. After years of painstaking work and publication delays, van Meteren's study of recent events in the Netherlands, *Belgische ofte Nederlandsche Historie van onzen Tijden*, finally rolled off the presses in 1599. James Cole played an important role in its completion, as he served as van Meteren's proofreader. "I am doing nothing at present except reading Emmanuel's history," Cole wrote to Ortelius in 1590/91, and "correcting its spelling a little." Cole did his work carefully — it took him two months to read through the first six chapters — and van Meteren dedicated the work to him.[14]

His mother's relatives gave Cole an entrée into learned circles, and he was attentive to ways he could further increase his status both in England and abroad. While his first marriage to Maria Theeus, the daughter of a prominent member of the Dutch Church in London, gave him additional presence among the Stranger community, his second marriage to Louisa de L'Obel strengthened his existing position in the Republic of Letters. Cole's father-in-law, Mathias de L'Obel, was known for his pioneering work on botanical taxonomy and his work with Pierre Pena on the *Stirpium adversaria nova* (1571). During his first visit to England, between 1569 and 1571, he met with the French naturalist Charles de L'Écluse to do some fieldwork in Bristol and struck up a friendship with Hugh Morgan, Queen Elizabeth I's apothecary, who had a number of novel West Indian plants in his garden. In 1585 L'Obel returned to London to take up permanent residence at the southern end of Lime Street. Once in London, L'Obel cultivated patrons as well as plants, and in time he won a position as supervisor of Lord Zouche's gardens in Hackney. The L'Obel family continued to seek out, and win, high-placed patrons in the next generation, when Mathias de L'Obel's son, Paul, was appointed apothecary to James I shortly after his coronation in 1604.[15]

Surrounded by family on Lime Street, and tied to his uncle Abraham's extensive network of business associates and correspondents, James Cole occupied a place at the heart of London's most distinguished community of naturalists. In addition to Cole and L'Obel, the neighborhood boasted an internationally re-nowned Flemish apothecary skilled in cultivating and distilling plant materials for use in medical treatments. James Garret (d. 1610) was prosecuted by London's medical establishment, the City's College of Physicians, for practicing medicine without its permission — though this doesn't seem to have made a dent in his considerable reputation or his trade. Garret was well known in London for importing novel drugs and plants from the East and West Indies, and his fellow apothecaries knew that they could rely on his manuscript translation of José de Acosta's useful treatise on these drugs, since no printed translation was then available.[16]

Like Cole and L'Obel, Garret was connected to a wide circle of naturalists, including Jean-Henri Cherler and Charles de L'Écluse (1526–1609), the latter of whom described him as "my dear friend, a man of honour, greatly delighting in the study of herbarism."[17] L'Écluse is remembered today for altering the face of European gardens with bulbs and tulips brought from the East, including irises, hyacinths, lilies, and crown imperials, and for writing important botanical treatises. His affection for Garret may have stemmed from their mutual passion for tulips, which Garret cultivated and propagated in his London garden plot along the City's crumbling wall in Aldgate. When L'Écluse visited England in 1571 and 1579, he stayed on Lime Street with his friends Garret and L'Obel. He was also in London in 1581, when he received some strange roots Sir Francis Drake had gathered on his 1580 voyage through the Straits of Magellan. L'Écluse, in honor of this gift, named the specimen Drake's Root.[18] Because L'Écluse's expertise was matched only by L'Obel's during the sixteenth century, for brief periods the small stretch of Lime Street was the residence of *both* of Europe's most influential plant specialists.

England's premier enthusiast for the study of animals and insects also lived in the Lime Street neighborhood, and was a part of Cole's community of naturalists. Thomas Penny (c. 1530–88), like James Garret, established a popular medical practice despite the best efforts of the College of Physicians to quash it, and also actively engaged in the exchange of specimens with others in London and abroad. But animals and insects were even more fascinating to him, a passion fostered in Zurich, where he studied with the naturalist Konrad Gesner in 1565. When Gesner died a few months later, Penny left the city with some of Gesner's animal studies, then turned his attention to the study of plants. He arrived at the French university of Montpellier, where he became friends with the medical student Mathias de L'Obel, who would one day be his neighbor in London. While still in Europe, Penny also became close to Joachim Camerarius (1534–98), the author of pioneering books in fields of study that would become botany and zoology. Camerarius described Penny in his *Hortus medicus* (1588) as "the eminent London doctor, very skilled in natural history, my particular friend."[19] Camerarius received a number of plants from Penny, including spurges and sedums, which were included in his studies, and he returned this professional favor by sending Penny insect specimens.

Penny returned to London in 1569 and moved into a house near the Cole, Garret, and van Meteren families adjacent to the Leadenhall Market. It was Penny who drew Thomas Moffett (1553–1604), another physician and naturalist, into the orbit of Cole's Lime Street community. Moffett was a London native who had been at Cambridge with Penny before studying abroad at the University of Basel. Moffett, like his friends Garret and Penny, was frequently at odds with the College of Physicians, which disapproved of his foreign notions about medicine, especially his interest in the controversial new chemical medicines of an itinerant German medical practitioner called Paracelsus. Despite the difficulties he experienced with the London medical establishment, Moffett was known to enjoy the friendship of other learned men of his time. "To the knowledge and learning he acquired by his studies," William Oldys wrote in the eighteenth century, "he added the improvements that were to be gathered from conversation with learned and knowing men." A 1580 visit to Italy introduced him to the joys associated with the close study of insects, which remained his passion after his return to England. After Penny's death, Moffett gathered up all his friend's manuscripts concerned with insects and compiled them with his own observations into a monumental work of more than 1,200 pages. No London printer would agree to publish it, especially since Moffett could not resist tinkering with it and adding new entries whenever an exciting specimen came to his attention. It was only after Moffett's death, when his apothecary sold the tome to another immigrant friend of James Garret and the L'Obels, Théodore Turquet de Mayerne, that the volume finally saw the light of day as *The Theatre of Insects, or Lesser Animals* (*Insectorum sive minimorum animalium theatrum*) of 1634. Only Moffett's georgic verses promoting the planting of mulberry trees and the cultivation of silkworms were published in English in his own time.[20]

Cole, L'Obel, van Meteren, Garret, Penny, and Moffett were the key members of the Lime Street naturalist community. A few other men who lived nearby — William Charke (fl. 1581–93), Johannes Thorius, and Johannes Radermacher — were linked to the group, but their interests either were confined to a single discipline (William Charke was an avid student of geography, and Johannes Thorius was a physician who remained committed to medicine) or were so wide ranging (as in the case of the merchant and historian Johannes Radermacher) that they did not regularly participate in the conversations about natural history that are my focus. Other Lime Street residents known to have had an interest in natural history, such as alderman and mayor Sir James Harvey, do not seem to have been included at all in the group's conversations, deliberations, and exchanges. Though Harvey lived in one of the largest houses in the neighborhood and was highly respected in local naturalist circles for his famous Lime Street garden (John Gerard visited Harvey to see his "mad apple," or eggplant, bear fruit in an unusually warm year), he does not appear in the evidence about the community that now survives.[21]

London's small neighborhoods and bustling streets provided ample opportunities for communities like the one on Lime Street to flourish. As a port city, London could also serve as an *entrepôt* to other far-flung destinations and people. Though often hidden behind English-sounding names like Cole and Garret, the composition of the Lime Street neighborhood reminds us of the truly international character of London during the age of Elizabeth. London was home to numerous European religious refugees, as well as continental merchants and craftsmen seeking a more lucrative market for their goods and services. As a result, neither London specifically, nor England more generally, could boast a distinctively "English" science at the time. The international character of Elizabethan science makes it remarkably similar to other cosmopolitan areas of early modern Europe, such as Florence, Prague, and Leiden. Like their European counterparts, men and women interested in the study of nature and living in London in the sixteenth century could find a wealth of new ideas and approaches right on their own street, as well as by communicating with friends and relatives abroad.

London also provided the Lime Street community, and other communities like it, with sufficient economic opportunity that they could engage in the work of the natural sciences without needing to seek out patrons for their financial survival. A striking feature of the Lime Street community was that its individual members were economically self-sufficient. A physician with a good practice like Penny, a busy apothecary like James Garret, and a prosperous silk merchant like James Cole simply did not need to find a noble patron or a position at court to make ends meet. Cole's father's silk business was so successful, for example, that he bore the dubious distinction of paying the highest tax of any of the Strangers living in the wealthy parish of St. Dionysius Backchurch.[22] The profits of a medical practice, an apothecary shop, or a mercantile business freed members of the Lime Street community to explore their intellectual interests, buy books, acquire specimens, and even make extensive trips abroad.

Among the Lime Street community only the L'Obel family had strong court connections, and even those cannot be said to have been entirely beneficial.

Though L'Obel was closely connected to Edward la Zouche and his garden at Hackney, for example, he was banned from the garden in 1600. When L'Obel was charged with stealing plants and broadcasting the fact that it was he — and not his aristocratic patron — who was responsible for the garden's distinctions, Zouche was forced to sever their relationship or suffer damage both to his horti-cultural investment and his reputation. "I am not so out of love with my garden," Zouche fumed in a letter to L'Obel after the naturalist walked out of the garden with several bags of valuable plants and bulbs, "that I would have it made worse than it is." L'Obel may have been drawn to plant theft because Zouche was un-able to pay him a sufficient salary for his work. Zouche hints, however, that Mrs. L'Obel played a role in the affair by entertaining delusions of grandeur about her husband and his position in his patron's world. Zouche complained to Mathias de L'Obel that he "would willingly to my cost [have] had . . . your company ever," but the "greatness of Mrs. L'Obel's mind . . . hindered it." L'Obel's son, Paul, fared no better when he left the safety of Lime Street and the City for the perils of court patronage. Shortly after becoming the king's apothecary in 1604, he was embroiled in controversy and implicated in the murder of Sir Thomas Overbury, one of the king's favorites. Overbury, who was taking drugs supplied to him by Paul de L'Obel, died after ingesting poison in the Tower of London in 1613. In the king's efforts to find a perpetrator of the crime, fingers began to point in L'Obel's direction, and in spite of the fact that he was almost certainly inno-cent, his reputation was ruined during the lengthy inquiry and trial.[23]

Even had they needed the money, the naturalists in the Lime Street commu-nity may have had excellent reasons for eschewing the court whenever possible. These reasons appear to have been shared by a number of their European friends and correspondents. Rudolf II's court physician, Johannes Crato, complained endlessly to Ortelius about the grind of court life, and longed for a quieter life in the city of Antwerp. Johannes Rumler made even more pointed remarks to James Cole in 1609 regarding the elevation of both Paul de L'Obel and Rumler's brother, Wolfgang Rumler, to royal favor at the court of King James. Though Rumler admitted that his brother "deserves great praise for pleasing princes," he urged Cole to warn Wolfgang not to trust in his sudden success. "Court life," Rumler wrote, "is a life of splendid misery." Rumler compared the hollow re-wards of the court to the benefits associated with his own career trajectory: "I am practicing as a physician in my native city," he wrote with satisfaction, "and am independent of everyone."[24] Even if these statements warning against court pa-tronage ring a bit hollow, Rumler's letter suggests that avoiding the perils could be a deliberate strategy for naturalists able to rely on their independent financial status as silk merchants, physicians, and apothecaries to fund their research. Like the English-sounding names that disguise foreign residents of London, however, such occupations can also serve to screen Elizabethan naturalists and other practitioners from the historians of science who want to uncover their activities.

The cosmopolitanism of London and the opportunities that the City provided for financial well-being fostered the Lime Street community's activities and served to engender the spirit of cooperation and collaboration that helped them to study the natural world. Instead of competing for limited resources, the Lime Street naturalists plied their complementary trades and swapped information and specimens without having either to cater to the whims of patrons or to engage in the fierce rivalries of court life. In the case of the Lime Street naturalists, London's urban sensibility helped them to maintain a fruitful tension between English and Stranger, market forces and collegiality, collaboration and competition. These attitudes overlapped in important ways with codes of conduct in the Republic of Letters. Within those complicated European networks, London's Lime Street naturalists secured a reputation for learning, congeniality, and intellectual civility.

MAPS, TULIPS, AND SPIDERS: COLLECTING
AND COLLABORATING ON LIME STREET

On 9 January 1586 Abraham Ortelius wrote from Antwerp to his twenty-three-year-old nephew James Cole in London.[25] England was at war with Spain, with the Netherlands providing a bloody battlefield for the struggle between one waning and one ascendant imperial power. But there was no mention of these tricky matters in Ortelius's letter, which related to natural history and the study of antiquities. Though the two men were separated by geography and political disputes, it is clear that Ortelius and Cole felt that they were part of a seamless community of shared interests and associations. There was nothing strange, therefore, when Ortelius asked to see James's sketch of a rhinoceros horn, because he thought he had seen a man hawking it in an Antwerp market. In an earlier letter James had asked for his uncle's assistance on behalf of his friend and neighbor Thomas Penny, who was trying to gather up his notes on insects and was still looking for rare and unusual specimens to distinguish the work. Ortelius, sadly, had no arthropods to send Penny, save a Neapolitan tarantula which had already been featured in Pier Andrea Mattioli's 1557 discourses on Dioscorides' *materia medica*. Nonetheless, Ortelius praised Penny's project, which he thought would make a unique contribution to learning.

A year later Ortelius again wrote to James Cole, and again the letter centered on matters of common concern to naturalists: this time geography, history, and plants. In the letter Ortelius thanked his nephew for ascertaining the precise location of "Wigandecua," or Virginia, from the newly returned members of the Roanoke expedition. Ortelius also praised James's knowledge of history, while lamenting the fact that his nephew could not take advantage of his well-stocked Antwerp library for his studies. To close the gap between Antwerp and London, Ortelius enclosed some valerian and sunflower seeds (one of the trendiest garden flowers at the time) gathered from his garden by James's aunt Anna.[26] Ortelius reckoned that sending seeds to James was like "sending owls to Athens" because his nephew was such an accomplished gardener. Ortelius hoped that they would serve as an adequate substitute for the African marigold seeds that James wanted, and which the Antwerp cartographer could not procure from his European contacts.

It was through the circulation and collection of these *naturalia* — a packet of seeds, a drawing of a rhinoceros horn, a spider, a snippet of information about Virginia — that the Lime Street community expressed its vitality at home and made its reputation abroad. Though it is easy to dismiss these objects as intellectual bric-a-brac, the fragmentary evidence of an unsystematic interest in the natural world, each item was part of an intricate web of exchange that stretched from Russia to the New World, from Denmark to Africa. Every time a dried plant specimen changed hands it became infused with new cultural and intellectual currency as its provenance became richer, its associations greater. Every gift of a flower bulb or a fossil came with an unspoken understanding that the recipient would take the specimen and not only credit its donor but find some way to repay the donor either directly or indirectly with something of equal value and importance.[27]

Within this circuit of exchanges natural objects led double lives; they were both subjects of study and inquiry, and artifacts cherished for their rarity and beauty. As subjects of study, natural objects provoked commentary and argument as their features and merits were debated and discussed within the community. As material objects, they were hoarded in cabinets, were swapped for other desired items on a naturalist's wish list of specimens, and provided cultural ornamentation that spoke to kings and queens interested in the rare and unusual, as well as to scholars and intellectuals. At a time when most of Europe was locked in war over matters of religion and imperial ambition, the exchange of natural objects prompted an intellectual civility that stood in stark contrast to national disputes. While a naturalist like Cole might have difficulty traveling from Protestant England to the Spanish-occupied Netherlands, his sketch of a rhinoceros horn or a seed packet could cross borders with relative ease and foster friendships that rose above linguistic, religious, and national obstacles.

Tracing natural objects as they traveled between scholars, appeared in learned commentaries, and came to rest, temporarily, on the shelves of a collector's natural history cabinet provides us with a way of seeing the Lime Street community in action and embedded in a larger network of intellectual exchange. Two impulses governed the Lime Street naturalists: the urge to collaborate, and the urge to collect. As with so many aspects of life in early modern London, including the study of nature, this involved a delicate interplay between cooperation and competition. Whether the object in question was a tulip or a map, discovering more about it and how it passed from individual to individual can illuminate the Lime Street community, and help us understand how a cosmopolitan city like London could serve to foster an interest in the natural sciences.

The Lime Street naturalists' urge to collaborate and their compulsion to collect found primary focus in the area of plants — their collection, propagation, and identification. After plants the Lime Street naturalists were most interested in animal and fossil specimens, with books, maps, and antiquities rounding out the list. Though this might seem like a bewildering and incoherent range of interests, in the sixteenth century they represented the classic components of natural history, as exemplified in a cabinet of curiosity. Cabinets of curiosity were temporary resting places for natural and artificial objects of wonder and delight. Frequently and provocatively crossing the ancient divisions between the natural and the artificial, an early modern curiosity cabinet could hold unicorn horns and mummified remains as well as artistic representations of shells, statues honoring the human form, and mechanical gizmos of every description. Monarchs throughout Europe assembled enormous collections of rarities, most notably the Holy Roman Emperor Rudolf II. Elizabeth I also had a cabinet of curiosities that she kept in her private chambers, which housed portrait miniatures and precious gems, along with her correspondence. At Greenwich, the Queen kept a room filled with clocks, globes, a saltcellar modeled to resemble a Native American inlaid with precious stones and ornamented with feathers, and rare tapestries made of peacock feathers. The Swiss traveler Thomas Platter reported that these were gifts from lords she had favored, "for they were aware that her royal Highness took pleasure in such strange and lovely curios." Her chief minister, William Cecil, also had a cabinet full of "rare things of workmanship" that included rocks, coins, and other antiquities, including a strange mineral specimen sent to him from Prague by the notorious alchemist Edward Kelly.[28]

By the end of the century, the collecting bug had bitten the middle class and many London residents had curiosity cabinets. The Barber-Surgeon and naturalist William Martin (fl. 1575–1606), for example, had a collection of curios that included portraits of recent kings and queens of England, the likeness of a "Hebrew physician that spoke twenty-eight languages," an ostrich egg, maps of Jerusalem and Ostend, books, and surgical instruments. One item from Martin's collection, a picture of the ancient plant-hunter Dioscorides, may even have made its way into James Garret's collections on Lime Street.[29] When the Swiss traveler Thomas Platter visited London in 1599, Lime Street's Mathias de L'Obel took him to visit the house of Sir Walter Cope (1553? – 1614) in the nearby suburb of Kensington to talk about Londoners' addiction to tobacco and to see his curiosity cabinet. Nestled among a fearsome assortment of weapons were feathered headdresses, clothes from China, an assortment of shoes, instruments, the horn and tail of a rhinoceros, the far rarer horn that had sprouted out of an English woman's head, an embalmed child, caterpillars, Virginian fireflies, a pelican's beak, a sea mouse, and an entire Native American canoe "with oars and sliding planks, hung from the ceiling." "There are also other people in London interested in curios," Platter wrote, but he was confident that Cope was "superior to them all for strange objects, because of the Indian voyage he carried out with such zeal."[30]

We know that the Lime Street naturalists had similarly varied collections of natural and artificial objects. Thomas Penny maintained a "dried garden," or *hortus siccus*, a collection of preserved plant specimens that he kept safely between sheets of paper. He also collected representations of insects as well as actual specimens, and these later passed into the hands of his friend Thomas Moffett. Moffett treasured his "storehouse of Insects," where he kept his rarest specimens, including an African grasshopper that he received from Pieter Quiccheberg of Antwerp, the son of the famous collector Samuel Quiccheberg. James Cole inherited Abraham Ortelius's large collection of natural objects, maps, and antiquities — including artifacts from the New World, the therapeutically treasured gallstones of Persian goats known as bezoar stones, precious gems, and ancient marble statues — and stored them all on Lime Street within two large curiosity cabinets that had both shelves and drawers.[31] Cole kept the cabinets in his countinghouse, where most merchants locked up their most valuable items as well as business profits and ledgers. While historians often focus on these splendid cabinets and their quirky contents, I want to shift our focus from the objects in their temporary resting place and consider instead how they were acquired, discussed, and traded within the Lime Street community and abroad. To do so we need to explore references to fieldwork, travel, and exchange in their correspondence, manuscripts, and published works.

One of the striking features of life on Lime Street was the collaborative spirit of its residents and their intellectual activities as they worked together on common problems or in pursuit of commonly valued knowledge. The late sixteenth century was a time of enormous excitement and challenge, as naturalists tried to sort through the overwhelming range of new plants and animals that seemed to be coming into their hands from every direction, and working with other students of nature eased the difficult business of comparing and contrasting previously unknown species. Moffett and Penny enjoyed doing fieldwork together, wandering "here and there a-sampling." L'Obel and Penny also went out into the field to examine plant specimens. James Garret helped Moffett study the field cricket by pulling off its wings and rubbing them together "very cunningly" to determine whether their music came, as other naturalists thought, from their wings or from a kind of hollow tube in the insect's stomach. Garret also studied the habits of the worms he observed living in his garden violets for Penny's work on insects, reporting that they were "very small and black, and run very fast." The English physicians Peter Turner and William Brewer joined Thomas Penny on an expedition in Heidelberg, where they were students in the 1570s.[32]

Working with a wide range of European naturalists provided members of the Lime Street community with additional collaborative opportunities and gave them a way to establish their expertise and to acquire additional specimens. Moffett and Penny received wasps, an illustration of a praying mantis located in Greece, and depictions of magnificent butterflies from the onetime Lime Street resident Charles de L'Écluse, who was then living in Vienna. These included an image of a butterfly that was so splendid and unusual Moffett rhapsodized that it was "as if Nature in adorning . . . this had spent her whole painter's shop."[33] Lime Street returned the favor when James Garret sent L'Écluse dried plant specimens from the New World and the fruits from the clove berry tree, while Thomas Penny sent him a blue-flowered gentian.[34] James Cole asked Franciscus Raphelengius, the nephew of the famous Antwerp printer Christopher Plantin, to send him a white tulip if he could find one, and sent Marcus Welser, a family friend living in Augsburg, rose bushes and flower bulbs. Many new plants found their way to Cole's garden thanks to Ortelius, who sent his nephew the seeds of an American chestnut tree and the rare plane tree. The French physician Jean Antoine Sarrasin (1547–98) thought Penny might like a picture of a praying mantis he had seen in Geneva, and Abraham Ortelius offered to send him his Neapolitan tarantula for inspection. Joachim Camerarius (1500–1574), the Leipzig naturalist, also sent Penny illustrations of an unusual ox-horn beetle in the Duke of Saxony's curiosity cabinet, as well as specimens of two kinds of weevils found in German barns. James and Pieter Quiccheberg sent insects from Antwerp and Vienna, including beetles, wall lice, and grasshoppers. And the surgeon Edward Elmer sent rare Russian beetles back to London for examination by the Lime Street naturalists.[35]

The mailbags contained all manner of intriguing objects and ideas as they passed between London and the European continent: hairy caterpillars from Normandy, strange biting caterpillars from Hispaniola, and scorpions from Barbary all found their way to Lime Street through the post. Sometimes, contributions came from closer to home. Thomas Knyvett (1545/6–1622) and his brother Edmund, "famous for his curious search into the knowledge of natural things," frequently sent information from their studies of insects to Moffett and Penny. Edmund Knyvett sent Penny a painted portrait of a small silver fly that sat on flowers, which helped the naturalist to frequently spot them in "hedges, and places with privet," as well as depictions of a beetle and a caterpillar. In addition to information about dragonflies, the physician and naturalist William Brewer kept Moffett and Penny posted on the angling habits of local fishers, carefully watched the generation of a glowworm, and then sent a specimen of the insect he found on the heath to Penny after he had carefully desiccated and preserved it. Lancelot Browne (d. 1605), a London physician, told Penny about a small fly that nestled in flowers "for warmth's sake and feeding." Not all of Lime Street's English informants were learned men, or of high birth, however. A simple "country man" shared his observations about the life cycle of the fly with Penny, for example, and the physician later hired an area resident to study the fen cricket, to "observe as often as might be its condition, and to make relation of it."[36]

Collaborative ventures like these often sprang from fieldwork. Out in the field, the goal was to observe something novel, rare, or peculiar. One of the great benefits associated with the study of these natural wonders was that through them "the minds of the studious may be filled with variety and rarity," as Thomas Moffett explained. L'Obel, in particular, loved fieldwork, a passion that was born during his student days at the University of Montpellier and which continued throughout his life. With Thomas Penny, L'Obel studied plants in Geneva and the Jura Mountains. In England he spotted a yellow "Star of Bethlehem" in a cornfield in Somersetshire, and observed a type of ivy-leaved mustard growing on the rocks in Portland near Plymouth. While traveling on the northwestern border of England and Wales, L'Obel identified two new plant varieties: a yellow pulsatilla and a blue-flowered butterwort. "I am discovering some beautiful plants in the mountains!" L'Obel enthused to his son-in-law James Cole. Moffett was also a tireless fieldworker, scouring "all Helvetia, Germany and England" for firsthand knowledge of a particular species of grasshopper, but he was unable to find his elusive prey. When he spotted an unusual specimen, like the "rare fly, not every where to be seen . . . that feeds on a mud wall made with mud and putrefied materials," Moffett preserved it "though dead, in a box for the rarity of it."[37]

Trips abroad represented working journeys for the Lime Street naturalists rather than opportunities for recreational travel. Fieldwork could be conducted in new environments, specimens previously seen only in books could be examined *in situ* growing wild or in cultivated gardens, and friendships made through correspondence could be strengthened by face-to-face contact. James Cole undertook a natural history tour of Europe in the summer of 1597 after the death of his first wife, Maria Theeus.[38] Accompanied by another "learned Englishman," Cole traveled throughout Europe, visiting the naturalist Joachim Camerarius in Nuremberg, the historian Marcus Welser in Augsburg, and the collector Adolphus Occo in Aachen before sweeping the length and breadth of Italy. In Italy, Cole and his English friend visited Professor Magini at the University of Bologna to see whether he had any interesting ancient maps, met artists in Florence, and visited Fulvio Orsino's unparalleled precious stone collection at the Vatican in Rome (odd as it might seem for two avowed Protestants to gain entrance there).

But it was in Naples, at the house and museum of the della Porta brothers, that Cole encountered a collaborative spirit of science reminiscent of his experiences back home on Lime Street. At the home of Giambattista and Giovanni Vincenzo della Porta, Cole had the opportunity to experience the late-sixteenth-century Neapolitan intellectual scene at its best. Naples was a center for natural history with an active intellectual colony linked to Rome's Lincean Academy, and the presence of well-known naturalists like Niccolò Stelliola and Ferrante Imperato, as well as polymaths like the della Portas, drew in visitors from all over Europe.[39] Cole was dazzled by Giambattista della Porta's mathematical studies on the quadrature of the circle, and by Giovanni Vincenzo della Porta's work on the judicial astrology of Ptolemy, but their museum of rare coins and marble statues made an equally strong impression and may have inspired renewed efforts by Cole to collect curiosities for his Lime Street cabinet.

While Naples was impressively remote from London by early modern standards, even more exotic fieldwork opportunities were emerging as voyages of exploration opened up new worlds. Eager naturalists were quick to acquire examples of novel specimens from North America, South America, and Africa, since the conditions of travel made excursions to these far-flung regions difficult and dangerous. Acquisition of these desirable, trendy natural commodities was not cheap, as Thomas Moffett discovered when he had to purchase a praying mantis "from Barbary that was brought out of Africa with some cost to us." It was far more cost efficient, and more in keeping with the bonds of obligation one cultivated in the Republic of Letters, to procure exotic specimens from friends like Pieter Quiccheberg in Antwerp, who sent Penny a young African grasshopper, which Moffett continued to treasure in his storehouse of insects after his friend's death. Other friends also made valuable contributions to the Lime Street community's efforts to study nature, as when Ludovicus Armacus, "a very diligent surgeon," brought Penny a grasshopper from Guinea and a caterpillar from Africa, and the artist taken on Raleigh's Roanoke expedition, John White, gave him another grasshopper "brought forth from Virginia."[40]

Because information was coming into the hands of the Lime Street naturalists from so many different sources — from their own fieldwork, from reports of fieldwork conducted by friends, and from intelligence gathered by informants as far away as Russia and New Spain — they carefully distinguished between the natural objects they witnessed themselves and those known only secondhand. Moffett noted that he had seen four small dragonflies, but not the thin, gray dragonfly that fed on apples which William Brewer brought to Penny's attention. Moffett made such careful distinctions because reported facts were not always reliable no matter how credible the source was. "The chameleon which some have reported, but falsely, to feed only on air," Moffett wrote with satisfaction, "feeds on flies, which, with his tongue six inches long, putting it forth suddenly and waving it to and fro, he hits [them] unawares, draws [them] to him, and devours them, as I have seen with mine own eyes in the year 1571." He was equally gratified when his own observations confirmed his friend Peter Turner's reports that earthworms fed "most greedily on a piece of white, unleavened bread."[41]

In light of the problems associated with verifying secondhand reports, the Lime Street naturalists preferred receiving actual specimens of plants and animals, or even careful drawings, rather than verbal descriptions. Like a cabinet of curiosities, an image of a plant, insect, or animal froze the specimen in time and allowed the naturalists to share it with other interested parties and study it minutely and at their leisure. An artistic representation of a natural object could impart a greater sense of verisimilitude to the viewer than a description, especially when the descriptive power of words seemed unequal to the task of accurately capturing an especially rare or magnificent specimen. Moffett was glad that he received an elegant drawing of a colorful butterfly from his friend Charles de L'Écluse because it was "easier to wonder at and admire, than with fit expressions to describe." Even so, representing a specimen — in words or visually — was recognized as an imperfect process of translation from the real object. "How hard and uncertain it is to describe in words the true proportion of plants (having no other guide than skilful, but yet deceitful, forms of them, sent from friends . . .) they best do know who have most deeply waded in this sea of simples," wrote Mathias de L'Obel.[42]

Much of the reliability associated with drawings and other representations of plants and insects depended on the quality of artists one could call upon to make the drawings. Though Ortelius praised Cole's drawing abilities, he was surely an amateur when compared to artists known to the Lime Street community like Marcus Gheeraerts the Elder, Marcus Gheeraerts the Younger, John White, and Joris Hoefnagel.[43] Even when a careful line drawing was made for publication or to send to a friend, the absence of color could limit its usefulness to other students of nature. Moffett carefully instructed his readers how to modify the line drawings that were to be included in his published *Theater of insects* to make them better reflect live specimens. "We have here set down exactly the form and magnitude of the Cranesbill-Eater," Moffett wrote next to a drawing of a hairy caterpillar, but in order to fully appreciate what it looked like, the reader "must make the white spots that adorn its black girdles of an iron color; and paint the belly and feet, and the white space between the girdles, with a leek-green color." Moffett gave similar instructions for hand-coloring the line drawing of a tarantula: "If you paint the white places with a light brown, and the black with a dark brown, you have the true spotted Tarantula."[44]

Whether from a live example, a preserved specimen, or a representation, simple observation did not always lead to perfect understanding of a natural object. Sometimes, closer study was required. In 1587, for example, the physician Thomas Moffett traveled to the west of England. He found an entire wasps' nest in a small village and decided to verify written reports that the females of the species lacked a stinger. Undaunted by the prospect of studying an insect capable of injuring him, Moffett poured hot water on the nest and killed all of the wasps. Closer examination of the dead wasps led Moffett to conclude, "I think they all in general are armed with stings," since he "could find none that had not a sting, either within their bodies, or sticking out." Mathias de L'Obel and Pierre Pena poisoned dogs in Paris, Louvain, and Heidelberg with arsenic and mercury in order to test the efficacy of its antidote, the herb "True Love." And Penny and William Brewer experimentally verified Dioscorides' theory that the salamander, contrary to legend, was not capable of living in fire when they incinerated one they found in the hills around Heidelberg.[45]

The cultivation of gardens gave naturalists like those on Lime Street another important opportunity to make hands-on inquiries into nature and to display their collecting prowess. Ortelius was jealous of his nephew's London garden, and the rare muscari, tulip, narcissus, and lily specimens he had there, some of which were completely unknown in Antwerp.[46] Thomas Penny's garden offered Mathias de L'Obel his first glimpse of water betony. L'Obel sent Gaspard Bauhin plants from his London garden to help him complete his *Pinax theatri botanici* (1623). Plants came into Lime Street from all corners of the globe, with the expectation that their health and welfare would be carefully tracked as they became acclimatized to English conditions. Ortelius sent to Lime Street South

American sunflower seeds and North American chestnuts that he had acquired from friends with New World connections. The French naturalist Nicholas Fabri Claude de Peiresc sent fragrant, colorful styrax shrubs to Mathias de L'Obel and James Cole, along with some double daffodils fresh from Argiers. "I hope that you will find within some curiosity worthy of your beautiful little garden," Peiresc wrote to Cole in the note that accompanied the package, since they "grow here in France" but are not "very common where you live."[47]

Despite Cole's splashy horticultural successes, James Garret was the most skilled practical gardener among the Lime Street naturalists, internationally renowned both for his ability to propagate new and ever more colorful varieties of tulip as he unwittingly exploited the virus that was becoming rampant in the flower stock, and for his ability to procure new plant specimens from all over the world. John Gerard, who was to become Garret's enemy, explained in *The herball* how the apothecary had "undertaken to find out, if it were possible, the infinite sorts [of tulip], by diligent sowing of their seeds, and by planting those of his own propagation, and by others received from his friends beyond the seas for the space of twenty years." Gifts of plants poured through the door of his drugstore on Lime Street: West Indian grasses, seeds from the Peruvian balsam tree obtained by Lord Hunsdon, new species of beans, a potted *Herba mimosa* brought out of Puerto Rico by the Earl of Cumberland, and a Virginian version of the china root.[48]

Not all plants were as hardy as Garret's tulips, however, and transplanting new or unfamiliar plants into cold, English gardens involved real risk, for the specimens did not always survive their first winters in the ground. Mathias de L'Obel's precious ginger specimen obtained from William of Nassau's garden "perished through the hardness of the winter." While visiting his uncle in Antwerp in 1588, James Cole planted some new specimens sent from Nuremberg by Camerarius, including foxglove, pink lily of the valley, fritillaries, and purple anemone. None of the plants fared well, and Ortelius wrote the following spring that most of them had failed to emerge from the earth after the winter passed. Despite these gardening setbacks, Ortelius kept Cole posted as to the more promising fate of a red and white tulip that the two had planted in Antwerp.[49]

Years of observation, fieldwork, travel, correspondence, collaboration, and collection informed the work of the Lime Street naturalists. Thomas Moffett drew attention to the benefits associated with their hard-won expertise when he discussed the work of older bees in a beehive: they should be ranked above "the rest in industry and experience, for years have taught them skill." As time passed and their knowledge grew, the Lime Street naturalists became wary of hasty conclusions about the intricacies of the natural world. Moffett was reluctant to include anything in his treatise on insects that was not "confirmed by long experience," especially if those insights might be challenged by readers. After finding several accounts of how glowworms, when combined with quicksilver in glass beakers, could create special lighting effects, Moffett asserted, "I will not believe [it] until such time as the experiment be made before mine own eyes."[50]

This discernment helped the Lime Street naturalists extract the choicest morsels of natural knowledge from the bewildering variety of the created world. Only after making specific observations could general theorizing take place. An example of just such a step-by-step process occurred between Ortelius and James Cole in 1589 on the subject of fossils. Ortelius had been discussing and receiving fossils from his European contacts for years before Cole began to put forward his own ideas about these rare, inorganic "jokes of nature," which God had deliberately sprinkled about to delight and distract students of nature. In 1579, for example, a Flemish admirer sent Ortelius a stone in which the cartographer could "observe the wonderful ingenuity of nature." The stone was hollow and contained smaller stones inside which caused it to rattle. Also inside were "various shells of snails, mussels, and cockles turned into stone." Cole's friends on Lime Street were also intrigued by the puzzling objects. In the winter of 1579 Thomas Penny sent a fellow naturalist a large tooth unearthed from a Cambridge field, and he also observed small worms "with six feet in old rotten stones," which amazed his neighbor Moffett, who had always understood "that all things . . . though they may corrupt in time, yet they will breed no worms."[51]

Fossils were a great mystery to early modern naturalists, and theories abounded as to when and how the animal and plant remains became trapped in the earth. While most scholars believed that fossils were inorganic, some were perplexed at the structural similarities between fossil remains and now-living creatures. One obstacle to drawing a clear connection between, for instance, an enormous fossilized bone and the bone of a much smaller living creature, was partially explained away when a theory emerged that these off-scale remains found in the earth actually grew there. When reports reached Lime Street from abroad of living animals found trapped in stones, these occurrences only made the debates about whether fossilized specimens were organic or inorganic more pressing, and more confusing. The Swiss physician Felix Platter was clearly veering toward the notion that fossils were somehow organic when he reported to Moffett that "he found a great live toad in the middle of a hard stone that was sawed in sunder." Platter concluded that the toad "was bred there" in the midst of the stone. Moffett found himself agreeing with Platter, for the German reports that would once have seemed "incredible and monstrous" could now be juxtaposed with accounts coming from a quarry in Leicestershire that noted the very same occurrence.[52]

Cole reached a daringly different conclusion. He believed that the shells and the animals within them were indeed the vestiges of organic creatures that had once inhabited the earth. Neither their inordinate size nor their placement could sway his belief. As for the fossilized shells that came to rest on the mountains where so many collectors found them, Cole explained that the earth had hardened into stone around the organisms before or around the time of Noah's Flood.[53] Ortelius, while claiming to have no ideas of his own on the subject, was troubled by the sheer impossibility of Cole's theories, especially the notion that fossilized shells were not inorganic jokes of nature but organic remains of living creatures, perfectly preserved. Ortelius felt that the other alternative to the inorganic theory of fossils — that the fossilized remains of animals had somehow "grown" in the earth from normally sized organic remains — had not been entirely ruled out, either. Prompted by his examination of an enormous petrified snail in Antwerp, Ortelius admitted that nothing that large could possibly have lived *on* the earth, so if fossilized shells were organic they must have grown *in* the earth.

Cole and Ortelius were not the only members of their extended community to weigh in on the subject of fossils. Another of Ortelius's correspondents, the French antiquary and naturalist Dionysius de Villers, argued with him over his notion that fossils could grow in the earth, pointing out that while individual bones left in the earth might grow to enormous size through some mysterious nutritive quality in the soil, how could one explain enormous skeletons found in their entirety? Villers believed that the answer could be only that once a race of giants had roamed the earth, a conclusion he reached after studying a huge tooth (probably a mammoth tusk) in his curiosity cabinet. Moffett, musing over the same troubling evidence of the fossil record over on Lime Street, and perplexed by Penny's belief that worms had grown inside crumbling stones, concluded that rocks could indeed breed organic life forms, despite the contrary opinions of ancient philosophers. "I began to weigh the matter narrowly, and to put into an equal balance, without fraud, all of their opinions," Moffett wrote, and "at last I found that our Ancestors were here and there most foully deceived, and I ascribe more to my eyes and the eyes of Penny, than to all their [ancient] words."[54]

These are not the amateur ramblings of casual collectors — these are state-of-the-science deliberations about one of the most pressing natural history debates of the period. It was not until 1616 that Colonna, in his *Dissertation on Tongue-stones*, publicly suggested something akin to Cole's theory that fossils were neither the enlarged remains of once organic creatures, nor inorganic jokes of nature, but exact imprints of otherwise unknown organic creatures which had once lived on earth.[55] With two other Italian naturalists, Agostino Steno (1629–1700) and Nicolaus Steno Scilla (1638–1686), Colonna was able to articulate and publish a theory of fossil remains that sounded remarkably like Cole's much earlier, private arguments. After comparing living creatures with fossilized remains, the three men set forth a convincing case as to why fossils were organic, setting off a firestorm of controversy that continued to consume the attention of naturalists well into the nineteenth century.

Fossils were not the only confusing aspect of the natural world, and the Lime Street naturalists and their friends devoted similar attention to two other zoological mysteries: the barnacle goose and the tarantula. Abraham Ortelius was a key figure in these exchanges, for he had a high reputation throughout Europe for his knowledge of animals, and naturalists from the Netherlands to Naples consulted him about rare and exotic species. It was the naturalist Niccolò Stelliola, for example, who asked Ortelius to share everything he might discover about another joke of nature: the barnacle goose (Figure 1.2). The barnacle goose was reputed to be born from a rare tree in northern Scotland that sprouted barnacles rather than fruit. From these barnacles hatched a goose that wandered away from the tree on four, rather than two, feet. The fact that no barnacle tree, nor any barnacle goose, had ever been found did not deter naturalists from looking for them. Stelliola felt sure that Ortelius would know what there was to know about the rare species, reminding him that anything he could divulge would represent "a great service to students of nature" and might incidentally prove his theory that two of the goose's four reputed feet were actually wings. Thomas Penny, on the other hand, asked Ortelius to confirm James Cole's reports that the cartographer had a tarantula specimen that had four eyes. This revelation surprised Penny, since he could not remember any other naturalist mentioning the fact, and sketches made for him in Italy showed only two eyes. Despite Ortelius's reluctance to style himself as an expert on animals and insects, Penny urgently requested that he have the tarantula — and any other rare specimen that he might know of in Antwerp — sketched immediately. Penny assured Ortelius that he would pay any price for the drawings through the agency of their local postman, Emmanuel van Meteren.[56]

Despite the Lime Street naturalists' interest in careful observation, experimentation, and the framing of hypotheses and conclusions, they were not eager to leap into the thickets of natural philosophy and make statements about God's overall plan for the cosmos, or to engage in debates about first causes or final movers. Students of nature had to avoid claims that they completely understood the natural world. Thomas Moffett warned "how foolishly and vainly man's wisdom doth many times vaunt itself . . . if not founded upon right Reason, the mistress of all Arts and Sciences." "There are many productions of nature," Moffett counseled, "the causes whereof it is impossible for any man to know, much less

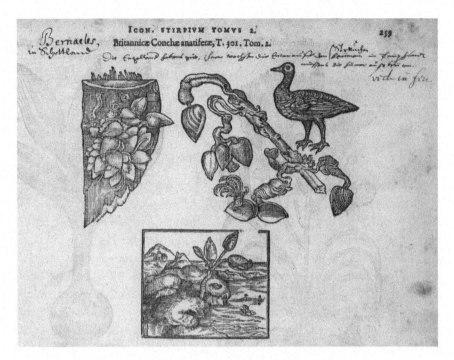

Figure 1.2. Scotland's famed "barnacle goose," from Mathias de L'Obel's *Plantarum, seu stirpium icones* (1581). The Lime Street naturalists were contacted by other students of nature for their expertise with native British flora and fauna. Reproduced with the permission of the Henry E. Huntington Library.

to show to others." This, Moffett argued, was really God's plan: to keep mankind admiring the power of God while acknowledging human blindness and ignorance. God created natural marvels like fossils and tarantulas "only for his glory, that he might both confound the shallow understandings of men, and also teach them to acquiesce in his wisdom." When "searching out the natural causes of things," Moffett concluded, "it is impossible to go any farther."[57]

As the Lime Street naturalists collaborated on their projects and shared the results of their inquiries with one another, they became more aware of the value of personal experience and the importance of verifiable knowledge about the natural world. The evaluative urban sensibility that already helped them to cope with the interplay that existed in London between collaboration and competition, English and Stranger, private and public also helped them to make distinctions between the experiences of their many informants and the information that they discovered themselves and witnessed with their own eyes. As all Londoners knew, you could not believe everything you were told — even if the one doing the telling was a famous naturalist or a learned friend. Underlying Lime Street's collaborative efforts there was a fine thread of competition, as individuals within the community strove to put forward more exact and reliable information about the natural world. But it was competition from without, rather than within, that ultimately obscured Lime Street from historical view.

LOSING LIME STREET

The Lime Street naturalists kept busy collecting and collaborating, and Elizabethans knew that an active, important natural history community existed in the heart of Elizabethan London, one that was equal to any in Europe in terms of interest and abilities. Yet the names of James Cole, Thomas Penny, James Garret, and even Mathias de L'Obel, are strangely absent from the standard histories of botany and the history of science. The main reason for their absence, I believe, is the English Barber-Surgeon John Gerard. John Gerard represents everything that the Lime Street naturalists were not: he was not part of the economically self-sufficient Lime Street community, he was not able to keep collaboration and competition in balance, and he did not obey the rules of appropriate collegial behavior established in the Republic of Letters. Instead, he relied on patronage, individual ambition, and his willingness to enter the publishing market for his success.

As we have seen from the complicated circle of exchanges that bound the Lime Street community together and linked them to important figures abroad, the work of natural history depended upon collaboration and cooperation as much as it did competition. These collaborative enterprises in turn depended on strict adherence to codes of civil, learned conduct to keep disputes in check. Adopting proper forms of address on letters, acknowledging learned assistance, and promptly repaying obligations were ways of expressing respect and collegiality. The economy of obligation that underpinned the Lime Street community could escape from the taint of mercantile buying and selling only if it was conducted in a way that was polite, impeccably honest, and transparently cooperative. Expressions of civility abounded in the letters, manuscripts, and printed treatises written by the Lime Street naturalists, whether they were addressing internationally renowned authors and scholars, aristocrats, or local luminaries. Moffett described Joachim Camerarius as "that most learned and courteous gentleman" and William Brewer as "a learned man and an excellent naturalist" and "a learned man and my good friend." Moffett also prized Brewer's "integrity of conversation," which, along with the physician Peter Turner, was "second to none." Among the nobility Moffett praised Edmund Knyvett for being "a Knight that is very courteous to learned men, and singularly noble both by descent and virtue."[58]

Like curiosity cabinets, the popular friendship albums (*alba amicorum*) of the period froze these carefully cultivated cordial relationships on the page.[59] Like the cabinets, however, their seemingly static appearance belies the ways in which they, too, were mobile objects of exchange. Today they are museums of friendship, but in their own time they were shipped around Europe either in their entirety or in loose sheets to be bound later into a treasured volume. It seems, for example, as though everyone active in the European Republic of Letters inscribed something in Abraham Ortelius's friendship album, which passed into James Cole's hands after his uncle's death in 1598. Notable English intellectuals, such as the historian William Camden and the mathematician John Dee, carefully inscribed mottoes, epigrams, and illustrations in Ortelius's album. Like an early modern version of a modern photograph, your page in a friendship album was meant to conjure up an image in the owner's mind of shared experiences, happy memories, and ongoing amity. Sometimes, the pages even included carefully trimmed engraved portraits to better conjure a friend's face.

Emmanuel van Meteren and Mathias de L'Obel's son, Paul, owned friendship albums that have survived. Other Lime Street residents inscribed mementos into these albums, as well as into other European and English collections. Emmanuel van Meteren's album includes entries by James Cole, Abraham Ortelius, and Charles de L'Écluse, as well as beautiful still-life miniatures by artists like Joris Hoefnagel and Lucas d'Heere. Nicholas Fabri Claude de Peiresc also ornamented his pages, expressing his undying friendship for the Flemish historian. Paul de L'Obel's friendship book was similarly star-studded, including the autographs of the classical scholar Joseph Scaliger, France's royal gardener Jean Robin, his father's old friend and neighbor Charles de L'Écluse, and King James I's two apothecaries, Wolfgang Rumler and Lewis de Myre, as well as entries by his brother Matthew and brother-in-law James Cole (Figure 1.3).[60]

In the albums and correspondence of the period, scholars and intellectuals had to negotiate a fine line between modesty and self-promotion. They wanted their entries to be noted and admired by others in the Republic of Letters, but they also wanted to be appropriately self-deprecating. After William Camden inscribed his entry into Abraham Ortelius's friendship album, illustrated with the head of Hermes and astrological symbols, he wrote to his friend, "I gladly wrote my name in your album and have added my symbol, though it deserves no place among so many clever ones, being a specimen neither of fine thought nor of

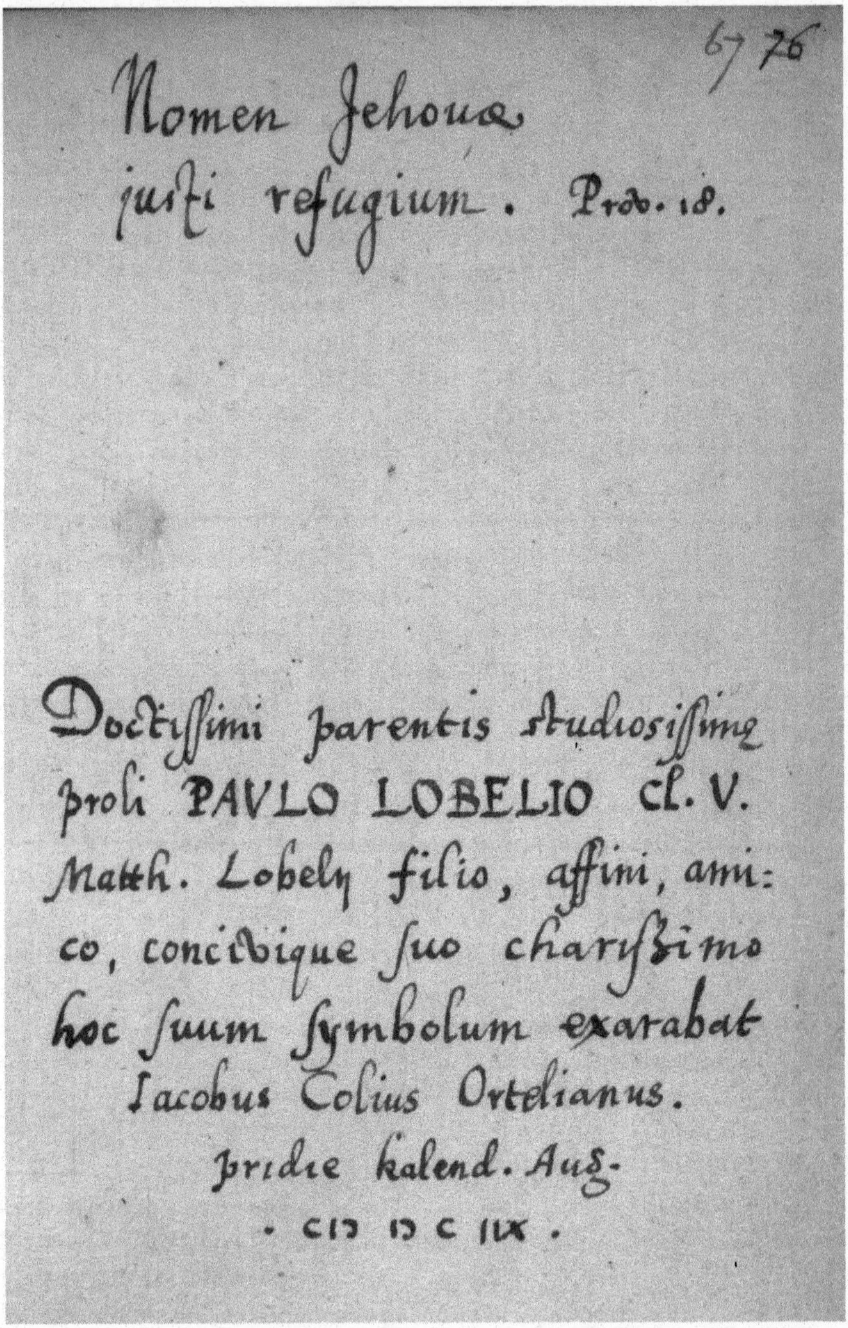

67 76

Nomen Jehoua
justi refugium. Prov. 18.

Doctissimi parentis studiosissimæ
proli PAVLO LOBELIO Cl. V.
Matth. Lobely filio, affini, ami-
co, concivique suo charissimo
hoc suum symbolum exarabat
Iacobus Colius Ortelianus.
pridie kalend. Aug.
· CIƆ IƆ C IIX ·

Figure 1.3. James Cole's entry in the *album amicorum* or friendship album
of his brother-in-law Paul de L'Obel. Tangible signs of friendship, like those here,
were valued in the European Republic of Letters. Harley MS 6467,
reproduced with the permission of the British Library.

beautiful expression." Despite Camden's protestations, it is clear that a great deal of time — and often money — was lavished on most album entries. The English secretary of state Sir Thomas Wilson followed the example set by many who were artistically challenged: he hired a painter to illuminate his page in Ortelius's album. Each expenditure of time, effort, and money was worth it, however, if it added luster to one's reputation and elevated one in another's esteem. After James Cole returned the classicist Johannes Woverius's album to him, for example, he received hearty congratulations for his well-worded entry and for bestowing on Woverius tangible proof of their friendship.[61] Cole's learning was now a matter of public record within Woverius's highly influential Flemish circle of philosophers and artists.

Every letter and every exchange between individuals in the Republic of Letters was padded out with compliments and gifts — but each came with strings attached to bind the giver and the recipient into further complicated obligations and exchanges. "Really, Abraham, your nephews overwhelm me with presents," Johannes Faviola chided Ortelius in a letter brimming with insincerity that included a postage-due receipt for a few gifts he received in Middleburg from the Cole family in London and for which he now wanted reimbursement. Like the expressions of modesty in a friendship album, expressions of thanks, reluctance, and gratitude had to be taken with a grain of salt and carefully sifted through for double meanings. When Rudolf II's court artist, Joris Hoefnagel, wrote to Abraham Ortelius in 1593, his letter was a tour de force of the delicate steps required to make, and fulfill, promises of friendship and cooperation in a way that kept the wheels of exchange turning. Ortelius had done some business for Hoefnagel, so the artist sent him "a little drawing from my hand, hoping that it will not displease you," along with an old drawing of a flowerpot he had long ago dedicated to the cartographer as a token of thanks. It turns out that having paid off one debt, Hoefnagel now wanted Ortelius's assistance completing his "book of art" that already had examples of "about three hundred good and notable masters." "This study requires the help of friends," Hoefnagel earnestly explained, "and for the little flowerpot I desire nothing but art for art." Later in the letter, after sharing the latest news, Hoefnagel dropped broad hints as to the artists whose works he would like to acquire: "I possess nothing by Henry Bles, nor of Joos van Cleve, Frans Floris, or the Pourbuses."[62]

Despite Herculean efforts to avoid offense and embody the spirit of generosity at every opportunity, relationships in the Republic of Letters and on Lime Street were sometimes peppered with competition, tension, and conflict. James Cole was especially trying to his uncle, who often felt slighted by his nephew's hasty arrivals and departures in Antwerp and the cavalier way the young man treated the

offers of advice he received from Ortelius's friends. These were the elder states-
men of the Republic of Letters, and had all been carefully cultivated by Ortelius
for decades, so the geographer was easily exasperated by reports about Cole's
lapses of civility. The Augsburger Marcus Welser sniffed that he was glad to meet
Cole during his 1597 visit, but "he made no use" of the advice given to him about
his future travels. The Heidelberg professor Jan Gruter reported to Ortelius that
Cole was with him so briefly they had barely spoken three words. He could only
hope that Cole would make up for his hasty departure by writing to him. Later,
Cole did repair their relationship through correspondence and gift giving,
thereby proving to Gruter that he was still "retained in memory even though
absent."[63]

Civility was all the more important given the serious confessional differences
and political struggles that divided so many countries in the early modern period. Within the Republic of Letters, whose members lived in regions like the
Netherlands and France that were rocked by serious religious disputes, religion
was bound to be an issue. Cole sometimes explained his reluctance to visit Ortelius in Antwerp by referencing how troubling he found it to contemplate entering the Spanish-occupied Netherlands, given his firmly Protestant outlook.
Cole's attitude was prudent, but his uncle was still hurt by his refusals. "I invited
you to stay with us, but I will excuse you," Ortelius wrote Cole in 1593. "I suppose
that religion, which binds all good persons, binds you also." But, as in any community based on bonds of family, friendship, and obligation, Cole's decision had
consequences, as his uncle was quick to remind him. "I would willingly have en-
trusted all that I posses into your hands," Ortelius commented dourly in the final
lines of his letter; "now I will think of another."[64]

Given the deeply ingrained practices of civility in the Republic of Letters, it
was a particularly delicate business to criticize the work of a friend, as James
Cole discovered when he notified his uncle of the errors he found in Charles de
L'Écluse's edition of Garcia da Orta's work on the plants of southeast Asia. Ortelius, proud of his nephew's perspicacity, sent the remarks on to his old friend,
who responded somewhat tartly that he was grateful that Cole had pointed out
those parts of his work that "were not clear" so he could amend his new edition.
Though L'Écluse remained on warm terms with other Lime Street naturalists
like Garret, the atmosphere between him and Cole remained chilly. Years later,
Cole tried to treat with delicacy L'Écluse's subsequent brush-off of his own work
in praise of the study of plants, the *Syntagma herbarum* (1606). "I doubt whether
I could, as you advise, offer the work to him [L'Écluse] without blushing," Cole
wrote to Ortelius. "Four years ago in Frankfurt I asked him to read it, but he professed to have no time . . . [and] I fear to intrude upon him again." Still, Cole

sent a copy of the whole work to his uncle so that he could send it on to L'Écluse in apparent ignorance of their squabble.[65]

To avoid public embarrassment, it was often prudent to take a proactive stance and correct one's own mistakes before a colleague was obliged to step into the breach and do it for you. Mathias de L'Obel, upon discovering certain inaccuracies in his published illustrations of ginger, took John Gerard a new illustration while he was writing *The herball.* L'Obel explained to Gerard that he had relied for the illustration on an "honest and expert apothecary, William Dries . . . [who] sent me from Antwerp to London the picture of ginger, which he held to be truly and lively drawn." Later, L'Obel realized it was simply copied from a Flemish herbal rather than sketched from life, and was full of misleading inaccuracies, especially with regard to its foliage. Rather than "suffering this error any further to [be] spread abroad," L'Obel told Gerard about the mishap so that he could be spared the embarrassment of perpetuating the mistake.[66]

Given L'Obel's efforts, it is all the more surprising that Gerard abandoned the clear rules of conduct governing relationships in both London and the European Republic of Letters and ran afoul of the Lime Street community, committing what the historian Charles Raven has described as "almost all of the sins of which a man of letters or of science can be guilty."[67] Gerard's *Herball* was a monumental project, constructed of 1,400-plus folio pages of woodcuts and descriptions. First printed in 1597, the work was reprinted repeatedly in the seventeenth century and used by everyone from apothecaries who made medicines to fine ladies who embroidered twirling tendrils of peas and fanciful cucumbers on their bed hangings and petticoats. Scattered references within the work to the naturalists who lived on Lime Street and a letter of praise from Mathias de L'Obel suggest that Gerard's massive achievement was known and admired by the Lime Street community. If we consult the evidence outside the text, however, we reach a very different conclusion.

Most of the external evidence demonstrates that the careful balancing act between competition and collaboration that was a crucial component of the urban sensibility of London's communities of science was proving difficult to maintain with respect to the Lime Street naturalists and John Gerard. First, he was literally an outsider who resided on the far side of London in the suburbs of Holborn. There was more space for gardens there, but there was not the vital intellectual community that could be enjoyed in the City's center on Lime Street. What Gerard lacked in terms of companionship he more than made up for by seeking out patrons and positions at court, where he was not only appointed surgeon to the queen but also made superintendent of William Cecil's gardens on the Strand and at his country house, Theobalds. Gerard spent most of his life shuttling between the court, his own gardens at Holborn, and his duties for Cecil. He gained access to a number of rare specimens through Cecil, many of which came into England in diplomatic pouches, such as a double white daffodil from Constantinople, which flowered once and never again, despite careful tending.[68] He was solicitous, too, of the many aristocrats who sent him specimens from their gardens, including Mathias de L'Obel's patron Edward la Zouche, Lord Zouche. Gerard received a number of seed varieties, including Italian mustard seeds, from Zouche, "which do flourish in my garden, for which I think my self much bound unto his Lordship."[69]

Gerard was also connected to London's two chief medical organizations, the College of Physicians and the Barber-Surgeons' Company. While there was often tension between the company and the college, Gerard petitioned the college to appoint him superintendent of a special physic garden that would be devoted entirely to the study of the medical uses of plants. In 1586 the college granted his request, although there is no evidence that the garden was ever established. Still, Gerard enjoyed good relations with members of the college. Physician Stephen Bredwell joined him in plant-hunting expeditions around London, including one on which they inspected a veronica growing in the village of Barnes. When *The herball* was published, Bredwell wrote one of the many letters published at the front of the volume in praise of the work, mentioning both Gerard's skill as a gardener and his avidity for fieldwork.[70]

Like the Lime Street naturalists, Gerard carefully cultivated his relationships with other scholars at home and abroad. Among European naturalists, Gerard counted a number of important figures his friends, including Charles de L'Écluse and Joachim Camerarius, who gave Gerard a German iris and the seeds of a goat's thorn bush, which grew successfully for two years before they perished "by some mischance." Gerard regularly exchanged plants and seeds with Jean Robin, the French royal gardener. Robin sent him cress seeds, hyacinths, a "rare and strange" *epimedium* (which Gerard christened "Barrenwort"), a crocus, and double yellow daffodils from Paris. In England, Thomas Hesketh and the apothecary Thomas Edwards both sent him plants found on their plant-hunting excursions. A preacher living in Hatfield, Robert Abbot, told Gerard of native orchids that grew near the Queen's childhood home. And Nicholas Lyte (1517–85), a Somerset merchant and active naturalist, gave Gerard cabbage seeds from France, a yellow carnation from Poland, and even rarer specimens from Syria he had acquired through a business associate in Aleppo.[71]

Gerard engaged in frequent fieldwork during the course of his studies. He observed plants growing in Kent and Essex, went hunting for plants in a village outside London with Robert Wilbraham, and was the first to identify a type of daffodil he saw in an old woman's garden in Wiltshire. Fully utilizing the riches of his own backyard, Gerard made frequent trips into the fields around London, noting the presence of wall pennywort on the door between Chaucer's tomb in Westminster Abbey and the old palace, and finding garlic and orchid specimens in Islington, creeping pinks in Deptford, whitlow grass growing in tufts on the brick wall in Chancery Lane, bugle in the village of Charleton, bittersweet in a ditch by the Earl of Sussex's house in Bermondsey Street, speedwell near the churchyard in Chiswick, and mustard in Edmonton. In the outer reaches of Southwark, where criminals and Catholic sympathizers were executed, Gerard found the water plant called "Great Arrow Head." Hampstead Heath and the surrounding woodlands also provided him with excellent opportunities to study native species, including orchids, yellow pimpernel, goldenrod, hyssop, and lily of the valley.[72]

An avid gardener, Gerard published the first edition of his catalogue of garden plants in 1596, giving us a glimpse of the specimens that grew behind his garden walls.[73] At least part of his garden was devoted to the medicinal herbs and plants he knew so well, noting that he grew several kinds of dock in his garden for his "use in physic and surgery." In addition to English natives, Gerard was devoted to collecting exotic rarities. He planted sugar cane in his garden, and found that the "coldness of our climate made an end of mine." Gerard had similar problems with other West Indian natives like rice, but had more success with tobacco, with which he had "experimented every way to cause it quickly to grow." Mediterranean plants, like the "Sea Onion" he obtained from sources in Constantinople, were typically more reliable transplants than those from the New World, though Gerard did not succeed in making Syrian cotton seeds flourish. The unexpected growing habits of some transplanted specimens were quick to get his attention. After Henry Lyte gave him Roman beet seeds, Gerard planted them in his garden, where they grew to enormous proportions and put forth a good quantity of seeds in 1596. Mother Nature had some tricks up her sleeve, however, for when Gerard planted the gathered seed "taken from that plant, which was altogether of one color" they brought "forth plants of many and variable colors," he wrote. "Nature doth seem to play and sport herself . . . as the worshipful gentleman master John Norden can very well attest, unto whom I gave some of the seeds aforesaid, which in his garden brought forth many others of beautiful colors."[74]

Given Gerard's interests in collecting plant specimens, doing fieldwork, cultivating his garden, and corresponding with naturalists at home and abroad, the Lime Street community should have provided a congenial set of friends with whom he could consult about matters of common concern. In fact, the Lime

Street naturalists sometimes appear in *The herball,* amid the references Gerard makes to his own garden, to his own fieldwork, and to the plant specimens he acquired from friends. References to members of the Lime Street community demonstrate that they did indeed extend their collaborative spirit to Gerard and his studies — at least at first. James Garret showed him a rare garlic bulb that grew in his garden, and gave him specimens including an unusual lily and a European "Lady's Slipper" orchid, for example.[75] James Cole, described by Gerard as "a learned merchant . . . exceedingly well-experienced in the knowledge of simples," told him of his discovery of a native orchid called "Lacy Traces" in the fields around Stepney. Cole also shared the precise location in the village of Hogsdon, where he found the rare rose ribwort growing wild. Gerard was also friendly with Mathias de L'Obel before the debacle surrounding *The herball's* manuscript. The two men traveled together into Kent to "discover some strange plants not hitherto written of" that he included in *The herball.* Gerard was forced to cite L'Obel repeatedly because of the many new species he had named in the course of his studies, and Gerard graciously admitted that L'Obel "very properly" Latinized the name of an English grass. Though the two men chatted companionably about how residents of the Netherlands made an oatmeal-like gruel from the seeds of dew grass, Gerard's competitive stance toward his illustrious friend was evident in his dismissal of L'Obel's description of the eight anemones known to him in favor of the "twelve different sorts" growing in his own garden.[76]

How can we account for Gerard's reluctance to fully admit how much he was indebted to the Lime Street naturalists? One possible answer lies in his marked ambivalence toward foreigners and foreignness. Running through *The herball,* and Gerard's life, was a pronounced fascination with, and anxiety about, the presence of European immigrants and their status in England. Tensions between the European Strangers and proper English citizens were common in Elizabethan London. When medieval and early modern Londoners searched for someone to blame for crime, poverty, epidemic disease, or economic hard times, they often scapegoated the immigrants who lived in their midst. Sometimes the tension turned into violence, as on the first day of May in 1517, when a riot broke out between disgruntled London apprentices and wealthy Strangers out to celebrate spring at the maypole erected in the Lime Street neighborhood, in front of St. Andrew Undershaft.[77] Despite the undeniable friction that from time to time divided people in the City — no matter where they were born or what language they spoke — the immigrants were a resource for new technologies, new ideas, and new cultural practices. Strangers introduced industries like pinmaking into England and established schools that offered foreign-language instruction to London citizens; and artists like Hans Holbein and Marcus Gheeraerts made significant contributions to the development of English art.

Nonetheless, Gerard spoke for many when he mocked the interest that some English citizens showed in foreign novelties, seemingly unaware that he shared the tendency with them. In his discussion of goldenrod and its use in medicinal preparations, he heaped scorn on the unsophisticated London consumers who spent half a crown on an ounce of imported, dried goldenrod while turning their noses up at fresh specimens gathered in Hampstead. "This verifies our English proverb, *far fetched and dear bought is best for ladies*," Gerard wrote dismissively of the willingness of Londoners to pay any price for the rare, the imported, and the strange.[78] But such predilections were symptomatic of the ambivalence that the society at large had toward the immigrant population: they found the new, exotic, and strange attractive provided they could own it, control it, and ultimately subjugate it.

It is revealing that Gerard collected nonnative rarities eagerly, and cultivated them carefully, but seemed almost triumphant when they failed to thrive in the hardy English soil. When William Cecil's double white daffodil from Constantinople failed to flower a second time, Gerard noted that "it should appear, when they were discharged of that birth or burden which they had begotten in their own country, and not finding that matter, soil, or climate to beget more flowers, they remain ever since barren and fruitless." While he barely touched on the considerable number of nonnative daffodils that easily naturalized in English gardens, Gerard spent an inordinate time dwelling on the foreign failures.[79]

But some flowers, after careful cultivation, could be naturalized, or, in Gerard's words, "made denizen." Here Gerard's phrase illuminates the complicated relationship that he and other Elizabethans had with the foreign presence in England, for denizens were immigrants granted the status of semicitizen, with slightly reduced taxes and slightly more privileges. To be made denizen was to begin the process of cultural assimilation, blending into English society until, in a few generations, none of your neighbors was likely to know that your family had once been Strangers. Of a species of woundwort native to the French region of Narbonne, Gerard wrote, "These plants grow naturally about the borders or brinks of rivers . . . whence they were brought into England, and are contented to be made denizens in my garden." Gerard reserved the status of denizen for plants that neither propagated recklessly nor perished quickly, but (under his firm guidance) predictably, and safely, grew. Gerard was annoyed at the "French Mercury" he brought home from a bishop's house in Rochester, "since which time I cannot rid my garden from it." Sometimes simply passing through an English middle man was enough to transform a plant from feeble foreign specimen to robust English denizen. Gerard received jimsonweed seeds from Jean Robin in Paris that "did grow and bear flowers, but perished before the fruit came to ripeness," but the jimsonweed seeds he received from Constantinople from Lord Zouche "bear fruit and ripe seed." Noble and splashy flowers, like the Persian Lily or the Crown Imperial, were given the privileged status of "denizen" because, once transplanted, they continued to benefit the English landscape. "This Persian Lily grows naturally in Persia and those places adjacent, whereof it took his name," Gerard wrote, "and is now (by the industry of travelers into those countries, lovers of plants) made a denizen in some few of our London gardens."[80]

Given Gerard's anxieties about immigrants and foreign plant specimens, it is not surprising that he guarded *The herball* so jealously from the interference of Garret and L'Obel. Perhaps it was the knowledge that he had roused the indignation of the Lime Street naturalists that made Gerard especially eager to shield his work further with staunch English supporters. Gerard dedicated the work to his patron, William Cecil. With the exception of L'Obel's (forged) letter, all the remaining prefatory materials were penned by English physicians, surgeons, and naturalists who highlighted the enormous contribution Gerard had made to *English* science. Latin poems touting Gerard as a "botanologian," and a "surgeon and herbarist" poured from the pens of medical men like Frances Herring and Thomas Newton. Gerard included not the usual one letter to the reader but three: one by the physician Stephen Bredwell, one by the surgeon George Baker, and one written by Gerard himself. The letter from Baker is especially illuminating, since it mentions an intellectual duel with a foreign naturalist. "I do not think for the knowledge of plants, that he is inferior to any," Baker wrote. "I did once see him tried with one of the best Strangers [in matters of natural history] that ever came into England," Baker continued with satisfaction, "and one whole day we spent therein, searching the rarest simples. But when it came to the trial, my French man did not know one [plant] to his four."[81]

Despite this English bulwark, subsequent naturalists expressed frustration regarding the quality of Gerard's *Herball*. When Thomas Johnson decided to edit and update the work in 1633, he attempted to sort out some of its problems, including mismatched illustrations and incomprehensible names. Johnson found it more difficult to manage Gerard's inconsistent use of sources and his tendency to conflate different plant species and to draw conclusions from those conflations. "Our author in this chapter was of many minds," Johnson finally wrote, throwing up his hands in exasperation at Gerard's muddled treatment of the saffron crocus. When Gerard turned to varieties of gentian, Johnson wrote, "Our author in this chapter so confounded all [of the plants], that I know not well how, handsomely to set all right." Johnson found Gerard's descriptions of some specimens "so barren, that little might be gathered by them" by students of nature. Johnson also called some of Gerard's fieldwork into question after visiting two sites where Gerard claimed he had seen a ragwort that was thought to be an alpine native. "I have been at the former and latter of these places to find out plants, yet could I not see this plant," wrote Johnson. While it grew "in the garden of Mr. Ralph Tuggy . . . I fear [it is] hardly wild in this kingdom."[82]

Gerard may be considered the father of English botany because of *The herball*, but he was by no means the best, or even the best known, naturalist in Elizabethan London. That honor fell instead upon a community of men in the heart of the City who worked tirelessly to increase natural knowledge and kept their competition in check with collaboration. But the Lime Street naturalists, like so many other groups that we will meet in the following pages, made a fatal miscalculation: their community was so vital and exuberant they could not imagine that it would ever cease to make notable contributions to the study of nature. As a result, they did not publish their findings but relied on face-to-face interaction and manuscript records. This might not have been important in their own time, since they clearly found all sorts of ways to communicate their ideas and progress to their friends and associates. But the failure to publish proved fatal after the Lime Street naturalists died and faded from memory. Even seventeenth-century accounts that were published of Gerard's egregious misuse of the community failed to resurrect the Lime Street naturalists in the popular memory, for there was nothing to resurrect. Today they lie buried, like the parish church of St. Dionysius Backchurch where so many of them once worshiped, under many layers of later activity.

The legacy of Gerard's *Herball* is really a message of survival: publish or perish. Urban communities like the one on Lime Street could be impressive and internationally renowned in their own time, but such communities were not impervious to the passage of time. Intellectual exchange, when it was based on manuscripts and proximity, was dynamic but it was also fragile. This was a lesson that subsequent English naturalists, as well as other students of nature, learned well. Though there were always a few reluctant holdouts like Isaac Newton, whose work had to be pried out of his hands by friends who insisted that others were about to publish similar findings and scoop his great discoveries, most rushed their findings into print. English science became inextricably linked to authoritative publications, but a surprising amount of face-to-face interaction and community negotiation was still required to get any manuscript through the press.[83] Print was a valuable weapon in the arsenal of any student of nature, but it is a weapon that has historically cut both ways. Here it cut out London's most prestigious and active community of naturalists from the historical record. In the next chapter, we will see how it gave coherence to a community in search of a public identity.

2

The Contest over Medical Authority

*Valentine Russwurin
and the Barber-Surgeons*

Valentine Russwurin, an itinerant medical practitioner from the central European town of Schmalkalden, carved out a place for himself in London's bustling medical market sometime in the late spring of 1573, when he set up a temporary market stall just outside the Royal Exchange, at the commercial heart of the City (Figure 2.1). There he displayed his collection of extracted bladder stones removed from reportedly happy and healthy patients, presented testimonials related to his ability to surgically treat cataracts, and held up samples of a powerful ointment that he had developed to combat skin diseases — all the while keeping up a steady patter to amuse and entertain his audience. Russwurin's stall was one of hundreds of structures that sprang up every year in London and filled the gap between the limited number of shopfronts available for rent and the growing market of urban consumers interested in buying books, clothes, food, and other goods and services. In a city like London, where business and trade were regulated, foreigners like Russwurin, along with women who could not fully belong to trade guilds or companies, and scores of other people who just wanted to make a quick profit, resorted to selling their goods and services directly to the crowd. Often no more than slightly raised platforms that provided a higher vantage point for hailing shoppers, or small tables for displaying a selection of wares, such stalls were popular among unlicensed medical practitioners like Russwurin in part because they could be taken down quickly when officials were notified of their unlicensed, and therefore illegal, activities.[1]

Figure 2.1. An itinerant medical practitioner displays his instruments at a stall,
in front of banners studded with bladder stones purportedly removed from satisfied
clients. A well-dressed man, possibly a surgeon, warns a potential patient that the
testimonials are counterfeit, and the patient hobbles off happy to have been warned
of the danger he faced. That the illustration is placed facing a description of Valentine
Russwurin's cures indicates that it may well be intended to show the German empiric
at work. From William Clowes, *A briefe and necessarie treatise, touching the cure
of the disease called morbus Gallicus* (London, 1585). Reproduced with
the permission of the Henry E. Huntington Library.

While Russwurin's German accent and exotic cures were useful in catching
the attention of Londoners strolling to or from the Exchange, he was only one of
a multitude of unlicensed medical practitioners who were plying their trade and
trying to keep officials from the City, the College of Physicians, and the Barber-
Surgeons' Company from shutting them down. An alewife by the Smithfield
Bars sold purging medicines and administered pregnancy tests to anxious Londoners, for example. A Barber-Surgeon's son-in-law, Richard Hottofte, treated
head wounds at the King's Head and Castle pub in New Fish Street. The lord admiral, Charles Howard, hired the popular unlicensed practitioner Paul Buck to
treat his medical complaints, as did the queen's favorite, the Earl of Essex. When
Mrs. Barker of Newgate Street was afflicted with a chronic cough, she sought out
treatment with Alice Skeres, the wife of a merchant, and when a midwife experienced heart pain, she went to Matthew Desilar, a French silk weaver who was at

the same time treating another woman for a debilitating wasting disease. Clerics like Henry Holland treated patients "for friendship's sake," limiting their clients to family members and the poor. Even lawyers were known to dabble in medicine, like Richard Scot of Clifford's Inn, who relied on the apothecary Andrew Field to make his medicines (a contribution duly noted and rewarded in Scot's will, where he bequeathed forty pounds to him).[2]

As this fistful of examples shows, London had a practitioner to suit every patient's purse and preferences. While the City's medical market was not the most tightly regulated in Europe, there can be no doubt that it was thriving. Parish records, wills, and guild proceedings provide us with details about scores of similar medical cases and a wide range of practitioners. At the same time, these records illuminate an important struggle over medical authority that was taking place throughout the City between patients and caregivers surrounding who could — and who should — be permitted to supply the chronically sick urban population with much-needed medical services. But the contest did not take place primarily within the privileged enclaves of the College of Physicians or even the far more populous, but still restricted, Barber-Surgeons' Hall: it unfolded on London's streets, and within her shops and markets.

Valentine Russwurin was cast into a central role in Elizabethan London's contest for medical authority when two young and ambitious Barber-Surgeons, William Clowes and George Baker, singled him out from the throngs of unlicensed medical practitioners in the City and lodged a complaint against him for illicit and improper medical practice.[3] Russwurin had been at work in London for more than nine months when Clowes and Baker, in a remarkable maneuver, ignored precedent and the regulations of their guild and leapfrogged over the Barber-Surgeons court, where disputes between its members and unlicensed practitioners were traditionally sorted out, and into London's Guildhall. There Clowes and Baker confronted the aldermen with a long list of Russwurin's unhappy patients and details about the medical catastrophes that resulted from his purported cures. As is often the case, we do not know how this dispute ended. There are no gossipy reports in the City's records, the Barber-Surgeons' Company never mentions the affair, nor does the College of Physicians. And we never hear from Russwurin again.

Despite the ensuing silence in civic records following the investigation, Russwurin became a notorious figure in the printed medical literature produced by Clowes, Baker, and several other members of London's large Barber-Surgeons' Company. These printed books reveal the reason why Clowes and Baker had singled Russwurin out for censure: he had unwittingly threatened the agenda shared within a community of young Barber-Surgeons who were interested in expanding the scope and increasing the status of surgery in the City by promoting new medical therapies associated with a Swiss practitioner named Philip Theophrastus Bombast von Hohenheim. Born to an impoverished nobleman and physician at the end of the fifteenth century, von Hohenheim followed in his father's professional footsteps despite his deep reservations regarding the medical education of the time, which emphasized mastery of classical medical texts rather than experience with sick bodies. By the time of his death in 1541, he had earned himself the nickname Paracelsus and had traveled throughout Europe and the Mediterranean region in pursuit of medical and natural knowledge in the woods, fields, workshops, and mines, all the while offering his growing therapeutic skills to those that needed them. Through printed works published during his lifetime, and manuscripts collected and published by his followers after his death, he introduced different conceptions of disease into western Europe and altered how diseases were treated by using novel chemical medicines, rather than purely botanical ones. The most popular of his works was a treatise on syphilis that advocated the therapeutic use of mercury, a cure still in use in the twentieth century before the discovery of antibiotics as one of the few efficacious ways to treat this sexually transmitted disease — although it had serious possible side effects, including madness and death.[4]

Though barber-surgery had always been medicine's inferior stepsister, Paracelsian therapies such as the use of mercury in ointments and in orally administered potions offered a new range of medical options to men like Clowes and Baker that might help to bolster the status of their craft. Traditionally, Barber-Surgeons like Clowes and Baker were trained through the hands-on apprenticeships of the craft tradition and licensed to do all medical procedures that required the use of tools — such as tooth extraction, operations, and sewing up wounds — as well as those that benefited from the topical application of medicines to cure the body's superficial ailments. Physicians, on the other hand, were learned, Latinate scholars who held university degrees and were allowed to diagnose and treat the whole body with internal medicines. Physicians were concerned with assessing and regulating overall health, while the surgeon was seen as a simple therapist applying external remedies like poultices or performing surgical procedures. These practical boundaries between physician and surgeon were quickly crossed, however, when it came to preoperative and postoperative treatment of patients. Could a surgeon supply an opiate to a patient about to be cut for bladder stones? If a patient became nauseous after surgery, was it the surgeon's responsibility to provide a draught to calm the patient's stomach? And in the case of a disease such as syphilis, which exhibited painful skin symptoms that normally fell under a surgeon's responsibilities, who should be permitted to treat the disease, and how? In most of these boundary disputes the Barber-Surgeons were not only trying to give better treatment to their patients; they were also struggling to inch their practice toward "physic" and the administration of internal medicines.

While physicians and surgeons worried over these community boundary disputes, on the streets of London it was becoming clear that Paracelsian therapies were wildly popular among consumers. Despite their popularity — or perhaps because of it — there was considerable debate both in the City and throughout Europe about what exactly constituted Paracelsian therapies, how they should be made, who should administer them, and what danger they posed to the population. Most of the university-trained members of the College of Physicians advocated caution and had a continued respect for time-honored therapies. They preferred using bloodletting to bring a body's "humours" into proper balance and advised patients to ingest botanic simples to purge the body. Much of the College, though not all, distrusted the potent, distilled chemical preparations that constituted the Paracelsian pharmacopoeia and disapproved of the Barber-Surgeons who wanted to employ them. Itinerant practitioners, unlicensed medical empirics, whose knowledge was based on treating illnesses and not solely on book learning, and apothecaries, on the other hand, argued for free trade in Paracelsian remedies and reveled in the plethora of cures on offer in the market, from idiosyncratic nostrums to toxic chemical potions purportedly devised by the great Paracelsus himself. Clowes, Baker, and like-minded Barber-Surgeons in pursuit of a wider medical practice elected to walk a fine line between the cautious members of the College of Physicians on the one side and, on the other, the enthusiastic itinerant practitioners and unlicensed empirics who all too often gave Paracelsianism a bad name. Clowes and his friends needed to distance themselves in the medical marketplace from popular Paracelsians like Russwurin — many of whom were simply slapping a popular brand name on vaguely chemical concoctions — while at the same time embracing a "proper" form of Paracelsianism in their own practices and crafting a community identity that included dispensing the new medicines.

The surgeons' quest to be seen as a more strictly defined community who were acknowledged as experts in the practice of surgery and permitted to administer oral medicines got its unlikely start when a traveling Paracelsian empiric began to ply his trade in the Elizabethan city of London and ran headlong into this ambitious subset of the Barber-Surgeons' Company. Just as Clowes, Baker, and their friends were beginning the hard task of situating themselves between the physicians and the other medical practitioners in London, Valentine Russwurin came to town claiming to be one of Paracelsus's students and disciples. As a result, for more than two decades the unfortunate Russwurin's name cropped up whenever and wherever these men felt it could be useful in their ongoing efforts to expand the scope of surgery in the City, tighten their grip on Paracelsian therapies, and define their community of surgeons with respect to their competitors.

The Barber-Surgeons' dispute with Russwurin was not primarily about the well-being of patients or even the presence of foreign medical practitioners, as vexing as those problems might have been. Twin concerns about community and authority lie at the heart of the Russwurin affair, pushed onto center stage by ambitions surgeons intent on enlarging both the status and scope of their craft. These Barber-Surgeons practiced and promoted a learned surgery that harked back to ancient traditions but was resolutely modern in its emphasis on anatomy, its familiarity with key medical and surgical texts, and its commitment to sound training and practical hands-on experience. Clowes, Baker, and their friends were deeply engaged in transforming the art and craft of surgery into a science and a profession that blended theory and practice for the benefit of patients. To do so they needed to purge the medical marketplace of undesirable elements and ill-trained practitioners and establish a community identity for those who remained that was medically authoritative and technically expert.

In this chapter I trace what *Paracelsian* meant to the different constituencies involved in the Russwurin affair, and how those definitions shifted in the face of new imperatives as certain practitioners entered, became well known, and left London's medical market. By tracing connections between key figures, we can produce a highly localized and particular account of medicine as it was practiced on the streets of London and see how the medical market functioned as a complex economic and intellectual organism.[5] If we examine the Russwurin case as a boundary dispute between different constituencies and communities in the medical marketplace, it reveals how permeable the walls were between physician and surgeon, elite and empiric practitioner. Layers of overlapping obligations and the labyrinthine twists and turns required to navigate London politics made the case all the more difficult to mount and, ultimately, resolve. Yet the case also reveals the ways in which this seemingly confused palimpsest of guild regulations, City laws, networks of friends and enemies, and consumer pressures of London's medical market did work when it came to arbitrating, investigating, and resolving medical disputes.

I also show that the burgeoning Elizabethan print culture could be a crucial weapon not only in the contest over medical authority but also in the pursuit of a coherent public identity. Print provided Londoners with an arena in which new voices were heard and grievances could be aired without the need to seek traditional forms of permission from a guild or the City. In the Russwurin case, what looks like a chaotic conflict in which individuals deliberately flouted regulation and convention in their pursuit of self-advancement is, on closer inspection, a carefully launched community campaign against a common enemy using the power of print. While the Barber-Surgeons might squabble in private among themselves, and Baker and Clowes had to be reprimanded for beating each other up in the fields outside the City's walls in an effort to settle some other minor dispute, in public they were a force united against bad medicine and itinerant practitioners like Russwurin.[6] To understand how this group of Barber-Surgeons seized on print culture and used it to their advantage, we must first understand how London's medical marketplace could let a man like Russwurin achieve such notoriety. Valentine Russwurin's meteoric rise also merits some attention, since his success is tied to his ability to cloak himself with the authority of Paracelsus. Finally, we will examine the Barber-Surgeons and how their "battle of the books" developed into an attempt not simply to unseat Russwurin but to expand their market share and frame a coherent community identity within the City.

MINDING THE GAPS: HOW LONDON'S MEDICAL MARKET WORKED

As with Valentine Russwurin, the case of Paul Fairfax illustrates the challenges and opportunities medical practitioners faced in Elizabethan London.[7] Fairfax appeared like an ominous blazing star in London's medical market in 1588 before fading into obscurity again in early 1589. A manufacturer of his own miraculous medicine, "aqua coelestis," Fairfax handed out pamphlets describing his distilled water in the markets, boasting that he had cured Mr. Treen's son in Southwark who suffered from dropsy, and Mr. Spagman's daughter who was plagued with head pains. Fairfax was caught plying his trade by London's chief medical regulatory body, the College of Physicians, and charged with the illicit practice of medicine.[8] The college's records are fulsome on the subject of Mr. Fairfax. Though a "traveled man," his pamphlets displayed both "arrogance and ostentation," and with his reputedly miraculous water Fairfax "cheated the people of their money." After the college threatened him with imprisonment if he didn't cease trafficking in cures, Fairfax was required to pay a stiff fine of five pounds and promised to behave in future.

Though the college no doubt hoped this was the last they would see or hear of Paul Fairfax, the physicians knew enough about London's medical market to anticipate that he would crop up again in connection with similar complaints. Worse, offenders hauled up on recurring charges frequently reappeared brandishing testaments to their medical skills from people too important to ignore. The college records registered no surprise when, a few months later, Fairfax was again charged with the illicit practice of medicine. This time he produced evidence that he had taken a medical degree at the University of Frankfurt, as well as a letter from Henry Carey, Lord Hunsdon — Elizabeth I's Lord Chamberlain and a close relative, and thus someone whom the college was not eager to slight. Nevertheless, Fairfax was clapped into prison for his continued practice of medicine, and an explanation was fired off to the queen's cousin that outlined Fairfax's many deficiencies in training and knowledge, his dubious university degree, and his propensity for making up medical terms to "entertain the simple hearer and to delude the unlearned multitude."[9]

Fairfax, like most medical practitioners in London, understood all too well how the medical market worked and how, if necessary, someone without a license might evade the multiple, overlapping regulatory bodies that vainly attempted to oversee the practice of medicine in the City.[10] The system was riddled with gaps in coverage, and a clever use of print culture, patient testimonials, a false university degree, good connections, and the laws of supply and demand enabled Fairfax and other practitioners to work for months and even years before running afoul of a patient, a rival practitioner, or a regulatory body like the College of Physicians. Even if, like Fairfax, you were brought up before the authorities on charges of improperly practicing medicine the game was not over: allegiances and alliances continued to come in handy to reduce fines, spring you from prison, or keep detractors at bay.

Crisscrossing lines of influence, allegiance, antagonism, and affiliation between practitioners and the corporations that sought to oversee their activities can make it difficult for historians to understand how cases were brought against practitioners, because each patient had many options when it came to prosecuting a medical practitioner. This complexity does not necessarily imply chaos, however. Just as a patient today has the option of taking her grievances to one (or more) of a number of regulatory bodies to seek advice and compensation, so too did the men and women of Elizabethan London. The City had a system of medical care that included multiple overseers of medical practice; many different agencies had the potential to shut down medical practitioners for malpractice or improper treatment; and patients played a role through petition and complaint in establishing a set of expectations regarding quality of service. London's medical market functioned because of the constant pressure placed on it by an enormous potential clientele of chronically sick people who were well-versed in their legal rights and not squeamish about taking their providers to law for malpractice.

Because of the multiplicity of practitioners and the overlapping regulatory bodies, trying to map the contours of London's sixteenth-century medical market from the bird's-eye perspective provided by the intervening centuries and the accidents of historical preservation results not in a neat organizational flowchart but in something reminiscent of a guide to the London underground, with a strange mix of possible departure points marked by people, institutions, and places. While some of these features of the medical marketplace can be linked through easily spotted lines of influence and antagonism, fragmentary records and incomplete accounts leave odd lacunae as well. At the contemporary Elizabethan street level, however, a very different picture would have emerged of medical practice in the City: there would have been medical practitioners everywhere with connections to one another, their patients, and City officials that are not preserved in the records. Each neighborhood in London had an apothecary, empiric, physician, surgeon, midwife, herb woman, dentist, urologist, bonesetter, or cataract specialist occupying a storefront or setting up a temporary stall next to the neighborhood's parish church — and all of them had patients. If we take to the streets and alleys of the City, we can better examine this landscape and investigate how the medical market actually worked during the reign of Elizabeth. Patient and practitioner stories illustrate that London's medical market functioned well precisely because of its loose regulation, which provided varied opportunities for individual practitioners and permitted both collaboration and competition. In the interstices between patients, practitioners, and London's medical regulations, those with persistence and imagination could make a living, craft a public identity, and even begin constituting a community.

In response to the complicated world of Elizabethan medicine, some medical men like John Hall, John Securis, William Clowes, and George Baker tried to divide their world neatly into good practitioners like themselves, who brought order to the market, and bad practitioners like Russwurin, who actively promulgated disorder. This was, in all likelihood, a deliberately false dichotomy evoked to get the attention of readers and civic officials. While early modern people found it comforting to draw sharp contrasts between order and disorder to help them cope with the overwhelming experience of living in urban environments, the historian Peter Lake argues that the two competing forces were in fact "connected by a finely graded chain of incremental moral infractions and failures" rather than standing at great, unspannable distances from each other.[11] This was especially true for London, since it contained so much diversity and variety, as Donald Lupton noted in the 1630s. Rather than accepting at face value the neat categories so clearly set forth by men like Baker and Clowes, I am fascinated by what Lake describes as an important third category: the "underside of order": the

murky middle ground of good and bad most familiar to City residents.[12] Order
and disorder were interdependent in early modern London, as were the regular
and irregular practitioners that Clowes and his friends tried vainly to put in opposition. In Elizabethan medical writing, as in many other types of literature
from the period, the authors were employing deliberately strong and divisive
rhetoric in an effort to draw public attention away from the confusing middling
ground where so many practitioners and their patients could be found.

Valentine Russwurin's experiences lend weight to this perspective. When he
entered London's medical marketplace in 1573, he was able to figure out fairly
quickly its key features and the way that it functioned. Elizabethan London's
medical market had three constituencies: the corporate institutions that sought
to regulate an orderly medical practice, like the college, the Barber-Surgeons'
Company, the church, and the City; the practitioners who plied their trades,
some with corporate backing and others without; and the patients who consumed medical treatments and advice without much regard for the question of
who was licensed to practice and who was not. The individuals who kept the
market lively and resisted the organizing tendencies of the regulatory bodies
were the patients, the chronically ill population of London. Londoners demanded that medicine be available to them at all times and at all price points,
and the market responded to that demand. Age, gender, class, and status did not
necessarily determine who you saw for medical treatment, however. Courtiers
saw empirics for their aches and pains, and those same practitioners were also
sought out by ordinary citizens, while elite physicians and well-known Barber-
Surgeons served the poor in hospitals like St. Bartholomew's as a form of charity.

Despite the importance of patients to the workings of London's medical marketplace, it is difficult for historians to re-create that world from the patient's
point of view, in large part because few patient narratives have survived.[13] Most
patients who speak to historians do so through the records of court proceedings
or official inquiries. As a result, we have a better sense of when the medical market failed to live up to patient expectations than we do about the normal course
of patient care. Evidence from patient complaints suggests that most Elizabethans had a wide choice in practitioners and cures, and felt that they had avenues of recourse if their chosen healer failed to live up to his or her promises.[14]
In London patients were quick to seek redress from the City's two chief medical
regulatory bodies, the College of Physicians and the Barber-Surgeons' Company. The shoemaker Christopher Hardy, a servant from Southwark, knew that
he could cross the Thames and complain to the college about the empiric Paul
Buck's failure to administer medicines according to their agreement — even
though Buck was not a member of the college. When her husband died after re-

ceiving treatments from Widow Austen, Mrs. Bate seems to have been more put out at the loss of kitchen utensils and a tin dish she had given to the healer as payment than at the loss of her husband. Mrs. Bate sought the assistance of the college to get her property back. The Barber-Surgeons' Company saw its fair share of patient complaints, too, as when Mr. Bulleyn appealed for a refund after a company member's wife, Mrs. Newnam, undertook to cure his scrofulis and failed to do so. And the ironmonger Richard Selby sent his wife Agnes to the Barber-Surgeons' Company to complain against member William Wyse, who failed to cure his injured leg.[15]

The College of Physicians and the Barber-Surgeons' Company were the newest and most active regulatory bodies in Elizabethan London. It was only during the reign of Henry VIII (r. 1509–47) that Thomas Linacre and his associates had founded the College of Physicians, and this elite university-trained group was given wide discretionary powers over all medical practitioners within a seven-mile radius of the City. The college represented the educated medical establishment and gathered its authority from its association with Oxford and Cambridge Universities, where the texts of the Roman physician Galen provided the backbone of the medical curricula. The college increasingly frowned upon graduates of continental universities like Padua and Basel, where students learned newfangled ideas about anatomy and the Paracelsian chemical therapies not included in Galen's canonical works. In theory, only practitioners approved by the College of Physicians were supposed to prescribe the orally administered, plant-based simples that were the therapeutic focus of Galenic medicine. These "inward medicines," as they were called, worked to restore humoural balance to the body, recalibrating each patient's unique mixture of elemental essences. The Barber-Surgeons' Company's role in medical regulation was similarly novel when Elizabeth came to the throne, as the group had been formally consolidated in 1520 from the Barbers' Company and a defunct corporation of surgeons.[16] The Barber-Surgeons' Company was charged with examining and certifying all surgeons in London and within a one-mile radius of the City — essentially the walled city and its immediate suburbs.

No regulatory system was entirely successful in early modern Europe, whether it was centralized, as in Spain and parts of Italy under the protomedico, or decentralized and multifaceted, as in England.[17] There was simply too much patient demand for services — and too many people willing to provide nursing care, medicines, and other therapies. Old worries about the low status of manual labor further complicated the situation for elite physicians. While members of the College of Physicians tried to maintain a tight grip on the administration of medicinal simples made from plants and other organic substances, for example, they had to depend on the labor and expertise of other practitioners when doing business. Physicians relied on herbwives to gather plants, on apothecaries to transform the plants into a pill or potion, and on surgeons for treating superficial skin ailments, setting bones, or making incisions. The dynamics of the market thus made any theoretical division between elite physicians and the rest of the medical practitioners untenable.

Despite the difficulties associated with distinguishing between practitioners and their activities, the College of Physicians and the Barber-Surgeons' Company, in an effort to live up to their obligations to regulate London's medical marketplace, actively pursued both malpractice on behalf of disgruntled patients and illegal practice. While practitioners could not insure themselves against the former, some protection from the latter could be had by acquiring a license to practice from the church. The church was the original regulatory body for medicine and had licensed medical practitioners (including physicians, surgeons, midwives, and empirics) since the Middle Ages, and the church's regulatory function had been confirmed by Henry VIII in 1511. The licensing powers of the Bishop of London, the Dean of St. Paul's, and the Archbishop of Canterbury persisted throughout the Elizabethan period despite the establishment of new regulatory bodies. The church could exact a fearsome penalty — excommunication — from those who practiced without a license, and this practice continued throughout the Elizabethan period. Thomas Woodhouse was excommunicated from his parish church of St. Botolph Aldgate for practicing surgery without a license in 1595, for example. While the charters of the college and the company gave them the ability to oversee medical practices in the City and license practitioners, these charters did not negate earlier statutes that made the church responsible for the care of bodies as well as the care of souls. Not until 1948 were the Henrician statutes confirming the medical authority of the church repealed.[18]

Surgeons and physicians who wanted the additional credential of an ecclesiastical license typically appeared before church officials with members of the Barber-Surgeons' Company or with their university degrees. William Clowes, Russwurin's fiercest opponent, did not receive his ecclesiastical license to practice in London until 1580, although he appears in Barber-Surgeons' records as early as 1569. Even after a decade of association with the company, and though he was a published medical author and member of the illustrious St. Bartholomew's Hospital staff, Clowes went to the bishop accompanied by five members of the Barber-Surgeons' Company who vouched for his skills.[19] Empirics and midwives, who had no corporate body like the Barber-Surgeons or the college to appear on their behalf, also brought witnesses. Elizabeth Alee of St. Mary

Woolchurch received her midwifery license in 1591/92 after presenting six women who were satisfied with her medical proficiency.[20] Wives of other medical practitioners, such as Barber-Surgeons and apothecaries, gave additional luster to a midwife's application for a license, and Dorothy Evans brought Anna Bovey, the wife of the Barber-Surgeon William Bovey, to the bishop as one of her witnesses, while Rosa Priest (herself the wife of a Barber-Surgeon) had three Barber-Surgeons' wives speak on her behalf.[21]

Though the church, the college, and the Barber-Surgeons' Company officially oversaw London's medical marketplace, the City's aldermen also played a role. Occasionally the Court of Aldermen gave itinerant and foreign practitioners the go-ahead to treat London's citizens, and they also mediated disputes over medical practice. Peter "Pickleherring" van Duran (fl. 1559–84), a brewer from the parish of St. Olave's in Southwark, satisfied the aldermen that he "professe[d] the knowledge and science of surgery" and was given permission in 1563 to "set up bills [of advertisement] on posts in such parts of the City as to him shall seem good to give the people knowledge of his said science." John Smyth (fl. 1556–73), a member of the Barber-Surgeons' Company, was chastised by the aldermen for posing a nuisance to City residents by roaming about the streets and plying his trade. Such activities led to the formation of crowds, put the public health at risk from evil smells and bad air, and upset the sensibilities of London's finer sort with the display of human blood and other gore. Smyth was instructed that in future he should "make open show" of his surgical skills "against his own house and door and not elsewhere." And the entire Barber-Surgeons' Company came under civic censure when they began "interrupting the surgeon of the poor house when he [was] at work," conduct that the aldermen immediately stopped.[22]

Elizabethan Londoners inherited these multiple, overlapping regulatory bodies from the Middle Ages and the Henrician period, when the king began chartering new regulatory bodies without dismantling earlier regulatory processes. All the evidence suggests that Londoners — patients and practitioners alike — were familiar with the regulatory and supervisory role of these four institutions and made full use of them in their pursuit of medical care while still seeking out practitioners who occupied the spaces between them.[23] Patients and practitioners also appear to have recognized that what the crown had established it could just as easily ignore, so even Elizabeth I played a part in the way medical business was done in London. Ultimately, the queen had the authority to rescind any license or overturn any decision made by her church, her City of London, her College of Physicians, or her Barber-Surgeons' Company, but such reversals were not a common occurrence. Elizabeth, like most of her predecessors, knew that London's support was necessary for political survival, and she was not eager to flaunt any privilege unless absolutely necessary. One of the rare cases in which Elizabeth did intervene involved the Dutch empiric Margaret Kennix (fl. 1576–83), who, following persistent efforts of the college to close down her medical practice, successfully appealed to the queen for protection. Like Paul Fairfax, Kennix knew the importance of a highly placed connection when it came to disputes with the college. In 1581, Elizabeth instructed Sir Francis Walsingham to remind the college that "it was her highness's pleasure that the poor woman should be permitted . . . to quietly practice and minister for the curing of diseases and wounds by the means of certain simples." Walsingham cited two reasons why the queen wanted Kennix to be allowed to practice medicine undisturbed. First, "God has given her a special knowledge [of simples] to the benefit of the poorer sort." Second, since Kennix's husband was unable to work, her entire family "wholly depend on the exercise of her skill" for its economic survival. Unable to let Walsingham (or the queen) have the last word on such an important matter as their corporate rights and responsibilities, despite the fact that the crown had the ultimate say in the matter, the affronted college responded that Kennix's "weakness and insufficiency is such as is rather to be pitied of all, then either envied of us or maintained of others."[24]

Kennix and other unlicensed practitioners could draw further rationale for practicing medicine from the words of the queen's father, Henry VIII. Ignoring his own decision to grant regulatory powers to the College of Physicians and the Barber-Surgeons' Company, and still failing to snip off the dangling threads of ecclesiastical licensing, Henry VIII made the crazy quilt of medical relations in the City even more confusing with his 1542/43 charter that ensured a place in the medical marketplace for "honest men and women" to whom God had imparted knowledge of herbs, roots, and waters. Lambasting the Barber-Surgeons' Company in particular for trying to monopolize medical practices in the City and driving experienced healers from the marketplace, Henry VIII's "Quack's Charter," as historians have dubbed it, gave every citizen the right to concoct ointments, baths, poultices, and plasters to treat diseases such as a sore breast, burns, sore mouths, and bladder stones.[25] The only forbidden medical activities were surgical operations (the province of the Barber-Surgeons' Company) and the prescription of inward medicines (the province of the College of Physicians). Elizabeth and Walsingham were clearly acquainted with the language of her father's statute, and used it when arguing on behalf of Kennix's right to practice medicine.

Healers intent on practicing medicine and surgery, like Margaret Kennix, Valentine Russwurin, and Paul Fairfax, found themselves wedged between eager consumers clamoring for their cures on the one side, and the regulatory powers of the College of Physicians and the Barber-Surgeons' Company on the other. But the conflicting pressures exerted by patients and corporate institutions, which were regularly vented by City, church, and royal officials, were still not sufficient to drive practitioners from the medical market. As long as patients sought their services, practitioners remained interested in offering pills, potions, and purges to Londoners. Some came from the provinces, but more often than not the itinerants were foreign medical practitioners with exotic names and no less exotic cures. George Baker, who would go on to write scathingly about foreigners in his letter praising Gerard's *Herball*, had already grumbled in print about the Strangers in their midst, especially those like Russwurin, who "have such a great name at their first coming. But after . . . their work is tried and then the proof of them seen: the people for the most part are weary of them." English men and women, eager to sample the new treatments, were partly to blame for the influx of foreign practitioners according to Baker. "Such is the foolish fantasy of our English nation that if he be a Stranger, he shall have more favorers than an English man, though the English man's knowledge do far pass the others," he wrote. One physician charged that a few of London's native medical practitioners went so far as to adopt a foreign dress and accent in order to gain a clientele in the City's competitive marketplace.[26]

While John Hall warned readers of his *Most excellent and learned woorke of chirurgerie, called Chirurgia parva Lanfranci* (1565) to exercise caution given that the number of truly skilled surgeons and physicians was minuscule in comparison with the numbers of "smiths, cutlers, carters, cobblers, coopers, curriers of leather, carpenters, and a great rabble of women" who offered their cures and threatened the health and well-being of patients, such warnings typically fell on deaf ears. Male and female practitioners like Russwurin, Fairfax, and Kennix still flocked to the City with their novel treatments, waving letters testifying to their skills from foreign princes, nobles, and towns, and even displaying physical proof of their expertise in the form of banners studded with bladder stones extracted from satisfied clients. And so, while William Clowes did "not mean to speak" of the bevy of unlicensed practitioners who disturbed his sense of medical propriety in *A prooved practise for all young chirurgians* (1588), he could nevertheless readily conjure up a vivid list of some of London's most popular medical practitioners, including "the old woman at Newington, beyond St. George's Fields, unto whom the people do resort, as unto an Oracle," "the woman on the Bankside, who is as cunning as the horse at the Crosse Keys," and "the cunning woman in Sea Coal Lane, who has more skill in her coal-basket than judgment in urine, or knowledge in physic or surgery."[27] Like the fabled nine-day wonder, these unlicensed practitioners often disappeared as quickly as they reappeared, leaving only their dissatisfied patients behind. The physician John Securis heaped blame for this shameful diversity of practitioners not only on the failure of the college and the company to regulate medicine but also on the endless gullibility of the populace, who willingly gave everyone benefit of the doubt when it came to a cure, even if the practitioner were a "Sir John Lack-Latin, a peddler, a weaver, [or] . . . a presumptuous woman."[28]

To fully understand Russwurin's rise to such heights in London's medical world, as well as his steep fall, we must remember that one man's quack might be another man's preferred provider. In the competitive climate of Elizabethan medicine, no practitioner was immune from charges of quackery, impropriety, and malpractice. Indeed, some of the most hostile contemporary attacks on medical practitioners were leveled against members of the College of Physicians. Early modern Londoners evidently found the university-trained physicians as quirky and objectionable as any empiric. A poetic late Elizabethan or Jacobean send-up of the college and its members featured their lechery, bad diagnoses, and incompetence. Of the esteemed Dr. Paddy, a long-term officer of the college, the anonymous poet wrote:

> Fair ladies, all glad ye, here comes Doctor Paddy,
>> The best at women's clyster.
> What ever be her grief, he gave her relief,
>> If once he but kissed her,
> And kiss her he might, as he is a true Knight,
>> And a valiant man at Arms.
> He never drew blood, but for the party's good,
> And was well paid for his harms.[29]

Giving a female patient a clyster, or enema, had obvious sexual connotations, as it involved the insertion of a greased pipe into the patient's anus, and the reference to blood-drawing might have had similarly bawdy connotations relating to the bloody evidence when a virgin was first penetrated sexually. Sexual indiscretions were not the only topic covered in the poem, however. Peter Turner, the Lime Street naturalist and physician interested in Paracelsian medicines and alchemy, was accused of killing his patients with chemistry. College members who ate, drank, or bought too much were also criticized: John Argent's medicine was dubious because he was "too fat to have any skill," while Matthew Gwinn was suspiciously thin for a member of such a profligate body, and Daniel Sellin was more fond of drinking wine than of performing his medical duties and delivering the City's public lecture in medicine.

How these ostensibly learned members of the medical profession came by their knowledge — as well as the thin line that divided physician from empiric and surgeon — was covered extensively in the poem, as well. Dr. Sawell, the son of an immigrant surgeon, was charged with spending too much time with the lowly members of the Barber-Surgeons' Company and with gleaning his knowledge of medicine from his father's old, foreign formularies. Thomas Rawlins's background as an apothecary's apprentice was highlighted, as was his rambunctious behavior in the City:

> Doctor Rawlins, for all your brawlings,
>> You are but a scurvie leach.
> For 'til it was your chance, your self to advance
>> By getting a widow by the breech
> You were but an Apothecary, or rather but his man,
>> But men rise by degrees, as well as trees,
> And I pray you, Sir, what then?[30]

This poem, ribald and satirical as its author intended, makes it clear that the university-trained physician was not guaranteed acceptance or immunity from criticism in the medical marketplace.

The success of all practitioners, regardless of their background or status, depended on providing a good service to their clients. But no medical practitioner could ensure that all his or her patients would be cured. In Elizabethan London the best way for medical practitioners to survive the challenges posed by patients who failed to improve after weeks of attention, clients who died despite their best efforts, and the scrutiny of the authorities, was to have as many local ties as possible. Education, marriage, neighbors, patients, corporate associations, and friends provided practitioners with potential alliances when cures went badly. Despite the differences that divided English citizen from immigrant Stranger, physician from surgeon, and apothecary from empiric, the looseness of London's medical market fostered close relationships between practitioners through shared patients, borrowed remedies, common institutional affiliations, and neighborhood ties. While a Stranger like Russwurin might enjoy a short-term advantage because he cut an exotic figure, this exoticism could present a long-term disadvantage if strong relationships with other practitioners and clients failed to develop.

Tracing the relationships between patients and practitioners sheds additional light on how well London's medical marketplace functioned. When Mrs. Sharde (fl. 1589–90), an unrepentant empiric, was once again brought before the College of Physicians for practicing medicine without a license, a French immigrant physician, William Delaune (fl. 1582–1610), asked "for favor and goodwill towards Mrs. Sharde" because he "had a patient in her house for a month on a purgative diet to cure a fever." The apothecary Edward Barlow (fl. 1581–94) worked regularly with the immigrant physicians Johann Vulpe (fl. 1581–89) and Hector Nunes (fl. 1553–92), as well as the English physicians Thomas Penny (fl. 1569–89), Richard Forster (c. 1545–1616), and Walter Bayley (fl. 1580–91/92). Barlow's regular contact with members of the College of Physicians continued despite his being reprimanded in 1581 for practicing medicine without their consent. His book of prescriptions, a manuscript formulary to guide his hand when he prepared medicines, also reveals how physicians shared responsibility for difficult cases. Some Barber-Surgeons, like William Ferrat and Thomas Symons from the St. Bartholomew's Hospital neighborhood, opened up joint practices. When the parish of St. Lawrence Pountney paid for the cure of a young boy's head injury in 1591–92, it hired a team of caregivers, including a female nurse, a female surgeon, and a male barber:

> Item to Goodwife Goodgame for healing his head the 5 of September, 13s 4d
> Item to the barber for shaving his head, 2d
> Item to the barber for healing his arm when he hurt it September 7, 2s 6d
> Item more for necessaries for his head, 1s 4d
> Item more for cloths for the boy's head, 8d
> Item to Goodwife Snoden for the nursing of Robert Matthews from the 5 of September 1591 until the 22 of December 1592, 68 weeks at 12 d the week, £3 8s[31]

London practitioners knew they had to cooperate in the medical market to survive.

The most lasting cooperation often resulted from marriage. Medicine became a family business in Elizabethan London, with practitioners spanning the generations and marrying into other medically inclined families. William Baxter (fl. 1578–1602), a member of the Barber-Surgeons' Company, married the already established empiric Emma Philips (fl. 1571–1603), whose brother, Edward Philips (fl. 1583–1602), was an up-and-coming apothecary. When Emma received the unwanted attentions of the College of Physicians, who described her as "an ignorant and bold woman" and committed her to prison for practicing without a license, she was released only four days later because of her husband's assurances that she would keep out of the medical market in future. Midwives often married medical practitioners: Lieven Alaertes of Ghent (fl. 1586–1602), for example, was the wife of the unlicensed surgeon Guillaume Alaertes (fl. 1588–92).[32]

While family connections could help you if you ran afoul of the regulatory bodies, happy clients were any practitioner's best defense. Satisfied patients tended not to complain to the authorities, and while the college and the Barber-

Surgeons' Company did launch proceedings against practitioners whose illicit practices they discovered themselves, most practitioners came before these corporate bodies because patients brought them there for failure to cure, failure to return money or goods when the cure was not successful, or malpractice. But there are also records that refer to medical practitioners who made extraordinary efforts to relieve the suffering of their patients, even though death and complications were commonplace in Elizabethan medicine. When Mrs. Querings, the wife of a gun stock maker, gave birth in St. Botolph Aldgate in 1589 to a stillborn girl, described by the parish clerk as "deformed" and "being but a [hand's] span long," it was her midwife, Mrs. Pullett (fl. 1584–89), who paid the sexton for the burial. Though the sexton often buried children of the poor at no cost, it is possible that the irregularities surrounding the baby's appearance might have made him reluctant to perform his duty. Similarly, a surgeon in the nearby parish of St. Helen's Bishopsgate paid "for a pit in the body of the church for a child that died at Barbadine Bonsiniors" house in 1567.[33]

London's loosely regulated medical marketplace, with its imperfectly pieced-together regulatory bodies and potent mixture of practitioners and patients, provided the ideal environment for an itinerant practitioner like Valentine Russwurin to flourish and prosper. Like Margaret Kennix and Paul Fairfax, Russwurin succeeded by exploiting the City's regulatory gaps, making powerful alliances, and satisfying his patients' needs for new and more powerful therapies. This combined strategy was often successful — at least for a time — and was embraced by both licensed and unlicensed practitioners. While some physicians and Barber-Surgeons bristled at the public's inability to discern the distinguishing features of the well-trained practitioner with impeccable credentials and official stamps of approval, on London's streets it was usually impossible for most people to tell the difference between an unlicensed practitioner such as Paul Buck and a college official like William Paddy. By returning to the case of Valentine Russwurin and examining it in greater detail, we will discover that his great misfortune was neither his treatment of patients nor his exceptional status in the City's medical market. Instead, it was his expertise in Paracelsian medicine that put his London career on a collision course with the ambitions of the Barber-Surgeons William Clowes and George Baker.

VALENTINE RUSSWURIN'S LONDON CAREER

If Paul Fairfax was a blazing star in London's medical cosmos, Valentine Russwurin's arrival deserved the status of a grand conjunction, a celestial event caused when powerful and sometimes contradictory planetary forces were brought together, foretelling the end of one age and the beginning of another. In the case of Russwurin, no political regimes rose or fell, no new religions were founded. Just as the practice of natural history was altered by the events surrounding the publication of John Gerard's *Herball*, so, too, was London's medical market altered by the inquiries launched into the activities of this one itinerant German practitioner.

Russwurin's presence was the catalyst that polarized London's medical marketplace into camps marked with Paracelsian, anti-Paracelsian, and "proper Paracelsian" signs.[34] In London, Russwurin employed three public personas — the skilled Paracelsian physician, the unlicensed empiric, and the itinerant quack — a mixture which, like a many-headed mythical beast, ensured that his reputation lived on long after he was officially defeated and driven out of town. The diverse appeal of Russwurin's three personas was reflected in the diversity of reactions to him. When Russwurin entered the medical marketplace, he was praised by London's citizens for his "felicity" and "facility" in the practice of medicine, and his advice was sought by court luminaries like William Cecil. From these auspicious beginnings, however, Clowes and Baker transformed him into the City's most dangerous and fraudulent practitioner, a threat to good order, the English commonwealth, and each of Elizabeth's cherished citizens.

No evidence explains how and when Russwurin entered England. The first mention of him is in a report about an "Allmaigne surgeon" drafted by one of William Cecil's servants on 29 August 1573. William Herle, one of Cecil's most active intelligencers, wrote enthusiastically to his master about an unnamed surgeon's skill at preserving and restoring eyesight, and Russwurin's subsequent long letter to Cecil counseling him on the root causes of his lifelong enemy the gout, and advising him on the best way to correct his mother's blindness, makes it clear that Herle's "Allmaigne surgeon" was Russwurin.[35] Cecil's family suffered from a host of medical complaints, and he searched ceaselessly for new, more effective methods to treat their ailments. With the appearance of a new surgeon in the medical marketplace — one from abroad, who styled himself an "Opthalamista" — Cecil had new hope that at least some of these conditions might be relieved.

Herle was deeply impressed by accounts of Russwurin's activities in London and abroad, which he obtained from "men of good knowledge and credit," including such notable continental figures as the Duke of Königsberg. Russwurin's English patients, Herle reported, were no less delighted with the man and his treatments, and found that Russwurin dealt "very honestly and speedily with them." Among his triumphs was the case of Joan Winter, a sixty-six-year-old widow blind for nearly a decade, whose sight Russwurin restored through the use of both surgical instruments and medicinal ointments. Good postprocedural care was one of Russwurin's strong points, and another patient, Alice Burton, praised him for keeping her in his own home "in a dark parlor . . . until her sight was confirmed." Further endorsements of Russwurin's skill came from the naturalist and physician Peter Turner. Fresh from medical school at the progressive University of Heidelberg, where he had disputed with Sigismund Melanchthon on purging medicines and debated the anatomy of the kidneys with Thomas Erastus, Turner was present at Burton's bedside when Russwurin "expounded sundry things out of Galen, making . . . Doctor Turner . . . perceive . . . sundry defects of the eyes . . . not able to be discerned . . . until the books of Galen and . . . [her] eyes were confronted together."[36]

All was not rosy, however. Herle noted that disgruntled voices could be heard in the halls of both the College of Physicians and the Barber-Surgeons regarding Russwurin's activities. These ominous rumblings Herle dismissed as envy; he did not consider them serious complaints. The "physicians and surgeons . . . have used many . . . speeches to deface him," wrote Herle, "which the Lord Mayor has now forbidden by open proclamation." Despite the carping of the medical establishment, Herle was certain that Russwurin "has done and does things . . . in my opinion of rare effect," an opinion confirmed by Dr. Turner's sense that the German had "a singular knowledge of the eyes, and an experience therein above other men."

Russwurin emerges from Herle's account as a sober, dedicated, and experienced medical practitioner, blessed with happy patients and troubled only by the envious tongues of some English practitioners. The adept practitioner is equally in evidence in the single surviving piece of evidence written by Russwurin himself, in which he dons the garb of a skilled Paracelsian physician. In the opening of the letter, Russwurin relates that his success in medicine resulted from his experience making and administering chemical medicines and a hands-on knowledge of the human body. Instead of reading a few texts by Galen like the university-trained doctors, Russwurin achieved his knowledge of medicine with "great labor and pains, not only of the head . . . but of the hands and body." Medicine, as Russwurin practiced it, depended on searching out the secrets of nature in the "fields, woods, and mountains" like his mentor Paracelsus, rather than in the pages of books.[37]

Chemical medicines like those Russwurin advocated were often attributed to Paracelsus in the medical marketplace, and arguments surrounding their manufacture and use were dividing physicians, surgeons, empirics, and apothecaries all across Europe. In this contentious atmosphere Russwurin proudly claimed to be a "painful Paracelsian" rather than a "coop-crammed Galenian," and he allied himself with other humble medical practitioners struggling to win a place for themselves, and for chemical therapies, within the marketplace. In England, however, Russwurin noticed that many physicians and surgeons displayed a disturbing tendency to dismiss chemical medicines as poisonous compounds that compromised patient health. Russwurin attributed the negative reaction of the English physicians and surgeons to intellectual snobbery as well as ignorance and envy. Russwurin wrote: "From my first coming into this land . . . there has not escaped . . . a meal or meeting where any of the university doctors have been present, wherein I have not been backbitten, slandered, and also impudently . . . belied."[38] Russwurin claimed to be unused to such disrespect; his medical opinions were esteemed by such highly regarded continental physicians as Pier Andrea Mattioli and Rembert Dodoens, "who being . . . present when I have been dealing with my patients have not been too lofty or ashamed to . . . have learned . . . at my hand such things as . . . they should never have found . . . in Galen nor Avicenna." The English doctors' dismissiveness, Russwurin noted, had not put a dent in the sale of chemical medicines in London — and some of the most active consumers of his products were the same physicians and surgeons who publicly criticized him. "If . . . chemical . . . preparations be venomous or . . . shorten a man's life within 3 or 4 years after the use of them and [are] therefore abominable . . . in a good commonwealth," Russwurin chuckled, "I marvel . . . [at] what makes [the physicians and surgeons] . . . resort to the preparers of such medicines . . . to buy all the chemical medicines they can get."

Had England's physicians and surgeons done as Peter Turner did and followed Russwurin as he visited his patients, they, too, might have learned a great deal about Paracelsian medicines and methods of treatment. For it is clear from Russwurin's letter to Cecil that he was not an ignorant or illiterate practitioner — he had mastered Paracelsian theories about the body, was knowledgeable about chemical distillation, and had extensive hands-on medical experience. To diagnose and treat Cecil's gout, for example, Russwurin relied on the Paracelsian technique of not simply examining his patient's urine for color and opacity but also weighing it and separating it into its constituent elements. This analysis revealed that Cecil's body was functioning as its own distillation apparatus, as Russwurin vividly explained. Warm vapors rose from Cecil's lower digestive system and cooled in the brain before descending as watery humours into his eyes and ears, causing headaches and eye complaints. Heavier and colder humours gravitated toward Cecil's extremities, where they hardened into "tartar" and caused his painful gout. The remedy, Russwurin pronounced, was to cleanse the blood with chemical medicines and to apply additional chemical medicines to the extremities to warm them and break up the hard tartar that had settled there.

Russwurin's explanations of bodily processes and his chosen therapeutic strategies contrasted sharply with those of Cecil's Galenic physicians. Instead of a distillation apparatus, Cecil's physicians saw his body as a closed vessel that registered health only when the fluid humours it contained were kept in balance. In their opinion it was the head, rather than the digestive system, that was the root cause of Cecil's gout. His doctors quoted from Galen that the brain was like a sponge full of water that, when disturbed by foods or external conditions, released fluids into the body, causing swelling and pain in the extremities. The remedy, according to Cecil's Galenic physicians, was to engage in an active program of enemas and purges to keep the fluid from building up in his brain in the first place, accompanied by the application of topical medicines to the head to further intercept the watery humours.[39]

After Russwurin counseled Cecil on his gout and promised to cure the cataracts that were causing his mother's blindness, Elizabeth I made this self-styled "physician, spagyrist, and opthalamist" a denizen in the winter of 1574.[40] By that spring Russwurin was famous in the City for his treatment of eye complaints, bladder stones, and skin ailments. While Russwurin treated ophthalmologic patients like Alice Burton in his home in Bishopsgate Street, he preferred to advertise his ability to treat painful (and prevalent) bladder conditions at a stall erected outside the gates of London's Royal Exchange, where he could display his awe-inspiring collection of extracted bladder stones to curious passersby. Such advertising was common among itinerant and unlicensed practitioners, though less so for medical men and women with established reputations and court connections. Some put up alluring signs, like the surgeon Edward Parke (fl. 1564–88), whose shop sign described him inaccurately as "the scholar of St. Thomas of Wallingford." Parke was competing with three other surgeons in the parish of St. Dunstan's in the West, and though he cannot have been surprised when the Barber-Surgeons' Company ordered him to remove the sign, the opportunity to set himself apart from his commercial competition (even fleetingly) was too great to resist.[41] Still more practitioners had single-page broadsheet advertisements printed up that could be handed out or pasted onto walls and doors. Charles Cornet (fl. 1555–98), described by the College of Physicians as "an ignorant Fleming and a most shameless buffoon," put up his bills of advertisement "on all the Corners of the City" despite the fact he was not licensed to practice medicine. Russwurin, on the other hand, decided to emphasize his status as an itinerant practitioner from a far-off land by erecting the market stall that was the hallmark of the traveling empiric or charlatan.[42]

Russwurin's market stall drew patients by the score, but it also attracted the attention of an irascible neighbor: the Barber-Surgeon George Baker. Baker lived on Bartholomew Lane adjacent to the Royal Exchange, and he would have had to pass by Russwurin's stall regularly in the course of doing business, seeing patients, and visiting friends and neighbors. Baker may have been one of the bystanders who witnessed Russwurin's botched attempt to cure Helen Currance, a musician's wife suffering from urinary complaints. On 3 April 1574 Russwurin "did attempt with his instruments to have taken out of her bladder a stone." Witnesses later alleged that "finding none there . . . he took a stone out of the pocket of his hose . . . conveyed it into a sponge . . . [and] forced it in [her] *pudendo*." When this procedure failed to relieve her discomfort, Russwurin sent her a powder that made it impossible for her to urinate. Uncomfortable side effects from the powder included blisters in her mouth, nose, face, and "inward parts of her body." Despite Mrs. Currance's troubles and Russwurin's demonstrated inability to cure them, he continued to treat patients for urinary complaints and eye ailments. Mr. Castleton, a scholar from Cambridge, was most egregiously affected by his ophthalmologic efforts. Though Castleton still retained some vision when he first consulted Russwurin, shortly after contracting a cure, Valentine, "by his rustical dealings, put out his eyes clean, and so deprived him of all his sight." Castleton had Russwurin arrested at the Royal Exchange, "where he did display his banners and wares," while he was in the middle of a sales pitch to the crowd.[43]

Enraged at this interloper who was practicing surgical procedures in his own backyard, Baker enlisted the help of his friend and fellow Barber-Surgeon William Clowes. Together they brought "certain complaints and objections" from Russwurin's patients to the attention of the Court of Aldermen on 22 April 1574—a course of action guaranteed to cause offense and raise eyebrows.[44] Clowes and Baker were publicly highlighting the failure of London's medical regulations (and the officers who enforced them) to keep empirics like Russwurin in check and to monitor the administration of Paracelsian therapies to the City's population. More-experienced members of the Barber-Surgeons' Company might have advised using familiar channels to make their objections known to civic officials and the college. But Clowes and Baker (both of whom were in their thirties) were not inclined to act according to precedence and prudence. London's wheels of change ground slowly, and while their elders were content to make measured progress to their objective, young men like Clowes and Baker often preferred a more gymnastic approach to problem solving, dancing around obstacles and swooping down on adversaries. Just as they bypassed their own company's court to settle their initial dispute with Russwurin, they resorted to the even higher court of public opinion to advance their suit on behalf of surgery's right to supervise the administration of Paracelsian therapeutics.

According to their later account of the proceedings, Russwurin left behind him a very long list of dead patients — twenty-three in all — from every walk of life, including Master Mace, a grocer; the servant of goldsmith Master Dummers; and two recent immigrants to the City. The aldermen, forced to act upon news of medical malpractice on this vast a scale, put together a subcommittee that was instructed to call upon the expertise "of the most discreet and best skilled surgeons of this city" to judge Russwurin's "knowledge and skill in surgery." Two eminent physicians also served on the subcommittee: John Symings had been president of the College of Physicians, and Giulio Borgarucci was connected to the queen and to Cecil. Russwurin was safely incarcerated in Newgate Prison while the committee conducted its inquiries, and a new committee judged the case on 10 May 1574.[45] This time the committee included aldermen, clerics, and five physicians — including one of the College of Physicians' censors, Roger Gifford; Richard Smith of Oxford; Richard Smith of Cambridge; and Hector Nunes. John Symings was dropped from the committee, but Dr. Borgarucci remained.

No account survives of the subcommittee's findings or of the final judgment, so we must rely on one of Russwurin's fiercest critics, William Clowes, for some sense of what their inquiries revealed. According to Clowes, Russwurin had only one defender: a "proud bragger . . . a man of little skill and less honesty . . . [who] practices surgery without any order or authority." A single voice whispers the defender's name in an anonymous note in the margins of the Huntington Library's copy of Clowes's work: he was "John Hester Alchemist at Paul's Wharf," a known supporter of Paracelsian ideas and an active manufacturer of chemical medicines (Figure 2.2).[46] Hester claimed that "Valentine Russwurin was a wise alchemist" and that Clowes and his cronies were "ignorant fools and asses." Russwurin impressed John Hester as a "wise Alchemist" because he incorporated chemical ideas into his medical practices. He had chemically analyzed Cecil's urine, weighing it carefully to find that it was "eight ounces and a little more, wherein it hath no difference from a sound man his water at all." When he discussed the problems Cecil's mother was having with her eyes, he focused on the hard "tartar" of her cataracts." These preoccupations put Russwurin's practices well within the concerns of Paracelsian therapeutics, and anyone who had read Paracelsus's works would have considered Valentine a genuine disciple. Clowes confessed that he was not competent to judge Russwurin's alchemical skills but did "know wise alchemists, of my opinion, that account him indeed an arch cozener, and interloper, and quacksalver."[47]

What became of Russwurin after the aldermen's inquiries were concluded is a mystery. Was he arrested or expelled from England, or did he flee of his own vo-

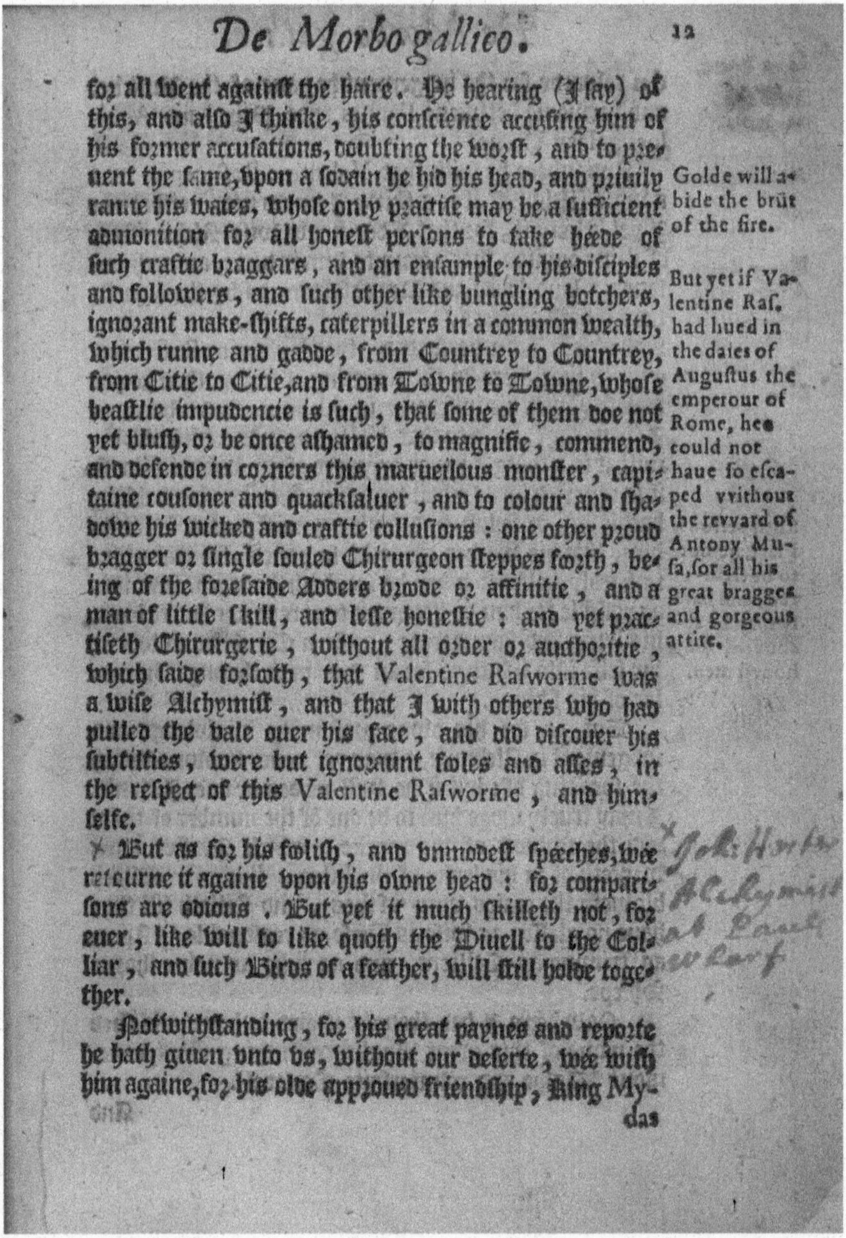

fo2 all went againſt the haire. He hearing (I ſay) of
this, and alſo I thinke, his conſcience accuſing him of
his fo2mer accuſations, doubting the wo2ſt, and to p2e-
uent the ſame, vpon a ſodain he hid his head, and p2iuily
ranne his waies, whoſe only p2actiſe may be a ſufficient
admonition fo2 all honeſt perſons to take héede of
ſuch craftie b2aggars, and an enſample to his diſciples
and follovvers, and ſuch other like bungling botchers,
igno2ant make-ſhifts, caterpillers in a common wealth,
which runne and gadde, from Countrey to Countrey,
from Citie to Citie, and from Towne to Towne, whoſe
beaſtlie impudencie is ſuch, that ſome of them doe not
yet bluſh, o2 be once aſhamed, to magnifie, commend,
and defende in co2ners this marueilous monſter, capi-
taine couſoner and quackſaluer, and to colour and ſha-
dowe his wicked and craftie colluſions : one other p2oud
b2agger o2 ſingle ſouled Chirurgeon ſteppes fo2th, being of the fo2eſaide Adders b2ode o2 affinitie, and a
man of little ſkill, and leſſe honeſtie : and yet p2ac-
tiſeth Chirurgerie, without all o2der o2 autho2itie,
which ſaide fo2ſooth, that Valentine Raſworme was
a wiſe Alchymiſt, and that I with others who had
pulled the vale ouer his face, and did diſcouer his
ſubtilties, were but igno2aunt fooles and aſſes, in
the reſpect of this Valentine Raſworme, and him-
ſelfe.

But as fo2 his fooliſh, and vnmodeſt ſpéeches, wée
returne it againe vpon his owne head : fo2 compari-
ſons are odious. But yet it much ſkilleth not, fo2
euer, like will to like quoth the Diuell to the Col-
liar, and ſuch Birds of a feather, will ſtill holde together.

Notwithſtanding, fo2 his great paynes and repo2te
he hath giuen vnto vs, without our deſerte, wée wiſh
him againe, fo2 his olde app2oued friendſhip, King My-
das

Golde will a-
bide the brūt
of the fire.

But yet if Va-
lentine Raſ.
had hued in
the daies of
Auguſtus the
emperour of
Rome, hee
could not
haue ſo eſca-
ped vvithout
the revvard of
Antony Mu-
ſa, ſor all his
great bragges
and gorgeous
attire.

John Hester
Alchymist
at Pauls
Wharf

Figure 2.2. A marginal note reveals that one of Valentine Russwurin's defenders was
John Hester, a well-known chemist who operated a shop on Paul's Wharf in London.
From the Henry E. Huntington Library's copy of William Clowes, *A briefe and
necessarie treatise, touching the cure of the disease called morbus Gallicus* (London,
1585). Reproduced with the permission of the Henry E. Huntington Library.

lition after discovering that London's medical marketplace no longer had a place for him? He disappeared from the historical records as quickly as he appeared. But Russwurin continued to occupy a place in the London imagination for well over a decade, kept alive in the printed books produced by William Clowes, George Baker, and their friends. Though Russwurin was not unique in London's medical market — there were other itinerant practitioners, Paracelsians, and unlicensed empirics when he arrived, and their work continued long after he was gone — he rose to prominence at a moment when his activities threatened a group of Barber-Surgeons including Baker and Clowes. Russwurin's successful London career, which was based on his familiarity with contested chemical and Paracelsian medicines, was further proof to ambitious practitioners that there was money and a reputation to be made in this area of medical practice. The controversies sparked by Russwurin's case continued to actively shape the dynamics of London's medical market until the end of the Elizabethan period. More important, they helped to define a community of like-minded surgeons and set the stage for a reconfiguration of the traditional boundaries between physic and surgery.

THE BATTLE OF THE BARBER-SURGEONS

When Russwurin rose to notoriety in London in 1573 and 1574, William Clowes was already in an agitated, anxious state. He was frustrated by his relatively low status in London's medical market, convinced that his surgical skills and medical knowledge were greater than most of his fellows in the Barber-Surgeons' Company, and intent on advancing his career as a surgeon (though he did not yet know how). Clowes reached this sad state of affairs after serving an apprenticeship with one of his company's most distinguished surgeons, George Keble. Years of military service in the Low Countries under the Earl of Leicester won him some noble supporters and did nothing to dampen his ambitions. While in the Low Countries, Clowes served alongside a leading surgeon of the time, John Banister, and the two men gained important experience in what we would call emergency or trauma surgery. Banister went on to become one of the few men in England to receive licenses to practice both physic and surgery, and was highly regarded for his knowledge of medical theory as well as for his practical expertise.[48] During his time with Elizabeth's armies Clowes also came into contact with foreign surgeons from Italy, Spain, Holland, Flanders, and France who shared new techniques and remedies with him. Some of those techniques and remedies were undoubtedly chemical, if not Paracelsian, in inspiration and composition. In the chaotic world of the battlefield, no one raised an eyebrow if a surgeon like Clowes employed novel and controversial medicines or crossed the line into physic by giving his patients oral preparations to help them with their pain or to speed healing.

Back in London, however, Clowes was expected to toe the line, respect the restrictions placed on surgeons, advance the good reputation of the Barber-Surgeons' Company, and thus help to foster urban civility. Clowes found it difficult to meet these expectations, and was often hauled in front of company officials who reprimanded him for his medical failures and his appalling public behavior. In the fall of 1573, just a few months after Russwurin's arrival, one of Clowes's patients accused him of taking his money and not curing him as promised. This charge would be repeated later, when William Goodnep complained that Clowes had promised to cure his wife's syphilis, but had not done so. While these complaints were unfortunate, they were also routine in the Barber-Surgeons' Hall and were dealt with quickly. It was Clowes's behavior toward his fellow medical practitioners that was more troubling to the company. In addition to Clowes's brawl with his good friend George Baker, the immigrant empiric John Goodrich (fl. 1555/56–86) reported that Clowes had insulted him and "most of the masters of the Company with scoffing words and jests." Clowes promised once again to do better, but he still found it difficult to guard his tongue and check his temper and so was reprimanded for similar offenses a few years later when he insulted his fellow surgeon Richard Carrington in the collegial confines of Barber-Surgeons' Hall.[49]

Clowes had a quick temper and tongue but was carefully monitored by his company's officials to make sure that he didn't bring disrepute to the corporation. Men like Russwurin faced no restrictions, however, and could be checked only by disaster or discovery by regulatory officials. Every day Clowes faced empirics and apothecaries in the market who took up the new chemical medicines and became famous for administering them, while he was expected to behave and do business as usual for Barber-Surgeons: sewing up minor wounds, applying ointments to skin conditions, and setting broken bones. Paracelsian therapies, while wildly popular with Londoners, were virtually ignored by the College of Physicians' Galenic medical establishment. Slipping through gaps between physic and surgery, chemical medicines provided new opportunities for practitioners like Russwurin, who were unhampered by guild regulations and untroubled by the threat of steep fines or short prison sentences. For men like Clowes, however, they were further evidence of the problems associated with London's restrictive medical regulations.

Clowes was not the only frustrated Barber-Surgeon in London. Several other men in the company also recognized that they faced the unpleasant prospect of being squeezed into an ever smaller segment of the market by the opposing forces of the college on one side and the apothecaries and chemical medical practitioners on the other. Clowes's friend John Banister also wanted the craft of surgery to expand, rather than contract, and the two men were joined in their efforts by the Barber-Surgeons George Baker, William Pickering, William Crowe, John Read, and John Gerard. Rather than focusing exclusively on the broad and unsolvable problem of unlicensed practitioners, these men turned their attentions specifically to Paracelsian remedies and the practitioners who dispensed them. From the arrival of Russwurin in 1574 until the late 1580s they tirelessly fought for the right to administer chemical medicines and for the responsibility of determining what was, and what was not, a proper Paracelsian cure.

Before Russwurin, Clowes and his friends had already struggled to increase their status in the medical market and gain recognition from the authorities for their surgical skills; after Russwurin they launched a concerted program of publication to win over popular opinion. Russwurin and his English supporter John Hester were increasingly invoked in their printed treatises on surgery and medicine to illustrate the perils that could ensue if the Barber-Surgeons were not recognized by London's authorities as the appropriate dispensers of Paracelsian medicines. Between 1574 and 1590, Baker, Clowes, and Banister released a steady stream of vitriolic rhetoric into the press. Russwurin was mentioned briefly in a single paragraph of Clowes's *Short and profitable treatise touching the cure of the disease called Morbus Gallicus by unctions* (1579), but he devoted several pages to the events of 1574 when the book was issued in a second, expanded edition in 1585. We might ask why Russwurin demanded more attention rather than less a decade after he had disappeared in apparent disgrace from London's medical market? The answer can be found by examining the books on both surgery and chemical medicines that were published during the intervening years. Studying these works reveals that the battle of the books for this group of Barber-Surgeons reached a crisis point in 1585, when apothecaries like John Hester began a publishing effort of their own to educate consumers on their full range of options when it came to chemical medicines.

It was in 1585 that the battle over chemical medicines also finally received serious attention from the College of Physicians, which did what all organizations seem to do when facing a challenge to their authority: it formed a committee. A group of physicians was charged with looking into the establishment of a London pharmacopoeia of accepted medicines that would include Paracelsian treatments.[50] Once the physicians entered the fray, Russwurin's catalytic role in the crisis was forgotten, and Clowes's friends among the Barber-Surgeons switched their attention to demonstrating how skilled and learned their company was in an effort to ensure that they were at least consulted regarding the pharma-copoeia. As with so many grand early modern plans to better regulate and control the medical market, the college's goal of drawing together London's ultimate drug formulary and medicine list never reached fruition, and by 1590 Clowes's cohort felt secure that they were winning the larger war over chemical medicines, even if they might lose a battle to rivals every now and again.

These Barber-Surgeons' weapon of choice, the printed book, proved effective in their long-term strategy to increase the status of surgery as well as in their short-term rhetorical objective of being London's most audible voice on behalf of proper Paracelsian medicines.[51] While most medical practitioners of the time were devoted to the manuscript transmission of medical formulas and cures, William Clowes was equally drawn to the power of the printed word when it came to advertising and advancing the credentials and skills of English Barber-Surgeons. During his career, Clowes wrote and published original treatises on curing gunshot wounds and syphilis, and supervised editorial and translation projects that made classical works of surgery available in the vernacular. He also helped to establish a canon of English surgical classics written by such august members of the Barber-Surgeons' Company as his former master, George Keble, and Thomas Gale. All of Clowes's works were highly collaborative ventures, with dedications to the Barber-Surgeons' Company, testimonial letters and poems in praise of the author written by prominent members, medical formulas gathered from practicing surgeons, and appended treatises and epilogues written by such high-profile practitioners as Baker and Banister. Living on Lime Street defined London's natural history community, but the printed page defined this surgical community, transforming a group of like-minded Barber-Surgeons into a visible intellectual group dedicated to ensuring the public health, promoting the worth of surgery, and increasing the prestige of English surgeons.

But several obstacles stood in the way of their efforts. The College of Physicians was one such obstacle, but I believe chemical practitioners like Russwurin were even more threatening — and far easier to attack publicly. As a result, a very different Russwurin emerges from the pens of the Barber-Surgeons, one who stands in stark contrast to the skilled and prosperous medical practitioner we saw in Herle's report to Cecil. Herle, on the lookout for novel medicines for his master, saw Russwurin as a potentially productive and useful presence in the English medical market. The English Barber-Surgeons led by Clowes, eager to expand their role in that same market by appropriating chemical medicines, saw him as a threat. Using a medicalized language of contagion, poison, and infection, they emphasized Russwurin's foreignness, his violation of English bodies, and his insidious effects on the City of London. Like all contagions, Russwurin's negative influence spread, infecting English men like Hester, who also became venomous to civic order and the commonwealth.[52]

The Barber-Surgeons were called to arms in the battle over chemical medicines by George Baker in 1574. While Russwurin was drawing clients to his stall at the Exchange that spring, Baker was putting the finishing touches on a compendium of useful medical knowledge, *The composition or making of the moste excellent and pretious oil, called oleum magistrale.*[53] The work not only contained a discussion of a sensationally popular Spanish oil that was used to treat everything from lethal wounds to hemorrhoids, it also included a translation of parts of Galen's *De compositione medicamentorum per genera* (On the composition of medicines) relating to wound treatment and joint complaints, and a rehearsal of surgical errors common in London. The book's title page assured readers that it was "Very profitable and necessary for all surgeons and all others which are desirous to know the right method of curing." Baker did not enter the realm of print without adequate defenses against potential critics: the work is bracketed at the beginning by Baker's dedication to his patron, Edward de Vere, the Earl of Oxford, and at the end with poems in praise of the author from the illustrious Barber-Surgeons John Banister and William Clowes.

Baker covered all possible bases in his compendium: novel medicinal oils from a Morisco empiric active in Spain; ancient wisdom from the father of physic and surgery, Galen; and even an assessment of contemporary faults in surgical practice in need of rectification. But he was not yet an open advocate of chemical medicines. In the preface, Baker warned consumers to be wary of practitioners who relied too heavily on the "rude observations" of Paracelsus's followers. Without proper knowledge of the theories of medical greats like Galen, Paracelsian practitioners were unable to heal their patients in a way that was "orderly, . . . [artful], and sure." "Although the common practicers do heal many diseases," Baker admitted, "they themselves must needs confess that the end of their labor depends upon the pleasure of fortune."[54] His translation of Galen into English was intended to integrate ancient theory into the practice of surgery, as was his discussion of how inexperienced "common surgeons" might correctly look after patients seeking cures for syphilis.

One therapeutic option for syphilis that Baker did not discuss was the use of chemical medicines based on mercury. The new chemical medicines traveled hand in hand through the medical marketplace with the century's new disease, known variously as the "French pox," the "Italian disease," and "lues venerea."[55] More vexing to patients and practitioners than the issue of which foreign country could be blamed for its origins were questions about what treatment the disease required. This answer would, in turn, dictate which subset within the medical marketplace would have proprietary rights over the disease and be able to treat the many patients seeking cures. Barber-Surgeons were quick to point out that Galen, who lived in late antiquity, offered little help. The fact that most symptoms of syphilis manifested themselves on the body's exterior in the form of sores and blisters meant that the treatment was not the concern of physicians, but surgeons. Still, the disease was resistant to traditional medicines, whether they were administered inwardly by physicians or topically by Barber-Surgeons. The new chemical and Paracelsian therapeutics offered a host of more effective treatment options in the form of mercury-based concoctions taken orally or transformed into ointments and lotions. Barber-Surgeons argued that since Paracelsus was not taught in English universities, and because chemically based medicines were so effective in treating many skin ailments in addition to syphilis, the use of mercury and other chemically based therapeutics, no matter how they were administered, should be within the legal purview of the craft of surgery.

Before the Russwurin affair erupted, however, Baker could afford to play down the importance of chemical medicines and play up his responsible adherence to the writings of medical authorities like Galen. That stance was not viable following the aldermen's inquiries into Russwurin's activities and the publication of numerous works praising chemical medicine. In 1575 the apothecary John Hester and Francis Coxe, a notorious magician once pilloried for sorcery, both published treatises defending the manufacture and use of chemical medicines, for instance.[56] Only three works on medical distillation and chemistry were published in England in the fifteen years between 1558 and 1573, but it had taken just under two years for Baker, Hester, and Coxe to double that number.[57] These works competed at the booksellers' shops with other books produced by Barber-Surgeons in the years following the Russwurin affair, including John Banister's anatomical textbook, *The historie of man* (1578), and a reissue of Thomas Vicary's *A profitable treatise of the anatomy of mans body* (1577), newly edited by a team of four surgeons at St. Bartholomew's Hospital led by William Clowes.

There was also a demand for books on chemical medicines, and so the printers were eager to issue George Baker's translation of Konrad Gesner's illustrated handbook to medical distillation, *The newe jewell of health* (1576). Baker received an incomplete draft translation of *The newe jewell* from Thomas Hill, a collector of books and manuscripts and an avid student of nature who died before completing his project. Like most enthusiastic scholars, Hill had more than one unfinished manuscript on his desk. His papers also included a treatise on the interpretation of dreams, and a translation of the Italian empiric Leonardo Fioravanti's treatise on plague remedies, including chemical medicines. When dividing the manuscripts among his friends before his death, Hill gave Fioravanti's treatise to John Hester, instead of George Baker, to polish for publication. Baker and Hester, it seems, had shared friends as well as a shared interest in chemical medicines, and in 1576 the two were getting along rather well despite their opposing points of view on the recent Russwurin affair. In his dedication of *The newe jewell* to Anne de Vere, Countess of Oxford, Baker was as enthusiastic about Hester's skills as he was about the power of chemical medicines. "We see plainly before our eyes," Baker wrote, "that the virtues of medicines by chemical distillation are made more available, better, and of more efficacy than those medicines which are in customary use. The trial of which [chemical medicines] we do daily prove to our great credit, and our patients' comfort." Baker advised his readers looking for the distilled preparations to visit one of a select group of apothecary shops, including John Hester's establishment on St. Paul's Wharf by the Thames, for he was "a painful traveler in those matters, as I by proof have seen."[58]

This happy division of labor between the leading advocates of chemical medicine continued for several years. In 1579 Clowes, Baker, and Hester all published works on chemical medicines, and a group of Barber-Surgeons completed their mammoth undertaking of an English translation of Guy de Chauliac's *Chirurgia magna* (Great surgery), accompanied by still more of Baker's translations of Galen and a collection of medical formulas gathered up by Clowes.[59] There seemed to be enough interest in chemical and surgical medicines to keep everyone's client list and pocketbooks full. Perhaps that is why, when Clowes published his first account of the Russwurin affair, all the evidence suggests that he bore no lingering animosity toward Hester. In the first edition of *A short and profitable treatise touching the cure of the disease called Morbus Gallicus by unctions* (1579), Clowes restricted his invective to Russwurin and the "runagates and vagabonds" like him who "do disorderly and unadvisedly intrude themselves into other men's possessions . . . [and] take upon them to practice in that art wherein they have never been trained, or had any experience." Though he had vowed "to have touched on no man particularly," Clowes couldn't resist mentioning Russwurin, "the stonecutter, an impudent deceiver," "who in most shameful sort abused the Queen's good subjects, under the title of a physician, a surgeon, ophthalmologist, [and] stone cutter."[60] Though Clowes went on to discuss the use of chemical preparations of mercury in the text, no mention was made of Hester, either in Clowes's treatise or in George Baker's appended work on the nature and properties of mercury. The competition over chemical medicines had not yet come to the boil.

Though Baker, Clowes and their friends among the Barber-Surgeons produced no further medical, surgical, or chemical works for the next five years, there was no comparable period of inactivity for John Hester. This may have been what transformed amity into enmity, for Hester continued to churn out translations of European texts on chemical medicines, and the medicines themselves, until he earned himself the nickname "the great alchemist of London."[61] A careful look at the books issued under Hester's name indicates that he was no longer limiting himself to the subject of chemical medicines, however. Like the Barber-Surgeons who found it difficult to draw clear lines between physic and surgery, Hester was finding it impossible to stick to distillation and keep free of medicine. In the winter of 1580/81 Hester published his translation of a treatise by Leonardo Fioravanti on surgery, followed in 1582 by Fioravanti's book of surgical and medical secrets.[62] Both contained chemical medicines. Finally, in 1583, Hester put the name Paracelsus onto an English-language title page for the first time when he published *A hundred and fourtene experiments and cures of the famous phisition Philippus Aureolus Theophrastus Paracelsus* (1583). These three works not only discussed the manufacture of chemical medicines, they covered their administration to patients. Though he claimed that he translated the works only "for their medicines, which I usually make," and "not for their methods which I meddle not with," it would be difficult for any reader to pay attention only to the medical formulas.[63] Hester protested in advance against those who might think that he was contributing to London's problems by putting such information in the hands of vulgar readers, writing that if "some runagate varlets carrying all their cunning in a surgeon's box in their pocket, and their learning in a cap case at their back, abuse these or the like medicines . . . that is not the fault of the matter, but of the men."[64]

Hester's translation of Paracelsus's work included a preface by the French physician and alchemical enthusiast Bernard Georges Penot (1519–1617).[65] Though not written by Hester, Penot's remarks contained ample matter to fuel the anger of Clowes and his friends among the Barber-Surgeons. In his preface Penot heaped scorn on critics who, "when reports are spread of the strange cures of sundry grievous diseases, which are wrought by the benefit of tinctures and vegetable and mineral spirits," then turn on the chemists and say "that they ought to be driven out of the commonwealth, and that they are deceivers, and that their extractions and preparations . . . will profit nothing, and that the spirit of vitriol is poison, the essence of antimony and mercury is nothing." These same critics, Penot charged, "like cunning and crafty thieves, privately and with fair promises, pick out from the poor chemists the secrets of physic, and secretly learn those things, which they forbid the common people as poisons." Good physicians should not "abhor the chemical arts" but should instead embrace distillation and learn to "separate the pure from the impure" and "make . . . medicines pure and clean." This was the true power of chemical medicines, argued Penot: they were pure essences of vegetable and mineral substances, with all their impurities removed by distillation.[66]

Following Hester's encroachment into matters of physic and surgery, and his inclusion of the Penot preface, members of Clowes's community took off their gloves and began to fight in earnest against a far more serious and entrenched enemy than an itinerant medical practitioner like Russwurin. The Russwurin affair was resurrected in 1585 not because he had returned to England, nor because empirics were any more a threat than they had been in 1574 or 1579. Clowes revived Russwurin because doing so enabled him to attack Hester and other authors who were publishing books on Paracelsian and chemical medicines. One of his targets was Richard Bostocke, a gentleman lawyer who asserted that Paracelsian medicines restored the ancient medicine of Adam, purifying centuries of heathen additions by pagan philosophers and physicians like Aristotle and Galen. In *The difference between the auncient phisicke . . . and the latter phisicke* (1585), Bostocke linked the early modern practice of alchemy to the "true and ancient physic which consists in the searching out of the secrets of nature." Built on two therapeutic pillars, one consisting of universal medicines that restored health and the other of particular medicines that treated specific diseases, Paracelsian medicine required its practitioners to master the art of drawing out impurities. Through chemical separation, all of nature could be beneficial to the body of man, even substances that were venomous or unwholesome.[67] Bostocke's recommendation, to medical consumer as well as practitioner, was to throw aside the heathen beliefs of Galen and Aristotle, who understood neither the nature nor treatment of the human body, and embrace the medicine God intended for use, now restored by Paracelsus.

In the expanded 1585 edition of his syphilis treatise, *A briefe and necessarie treatise, touching the cure of the disease called morbus Gallicus,* Clowes went to work on Hester, Bostocke, and the proponents of chemical medicine. Russwurin became the chief representative of the "rude rabble of obscure and imperfect experimenters . . . prating proud peasants, and ignorant asses" who labored over their distillation apparatus and then haunted the streets of London foisting harmful chemical medicines on her citizens. Clowes described how this group of undesirable medical practitioners circulated through the City like a poison, proclaiming their talent for cures "in open streets and market places . . . prating, bragging and lying, with their libels, banners, and wares, hanging them out abroad."[68] A crude woodcut accompanies this passage, providing a vivid illustration of an itinerant practitioner showing the tools of his trade — knives, pliers, and urine flasks — to potential patients at an impromptu stand, while above him hang banners studded with bladder stones and sworn testimonies regarding his expertise (see Figure 2.1). Latin phrases alert the educated reader to the counterfeit nature of the testimonials, and an aged patient hobbles off proclaiming his relief at having been prevented from contracting a cure with such a dangerous practitioner.

In the accompanying text, Clowes reminded his readers of Russwurin's notorious activities in London and divulged fresh details, including his patients' names. Of the deaths placed at Russwurin's door, most resulted from complications following surgery for bladder stones. Russwurin's malfeasance was not limited to malpractice, however. Clowes also accused him of violating the bodies of virtuous English men and women with foreign objects in an attempt to bolster his prestige and gain more patients. This "lewd craft," Clowes wrote, was exercised against Helen Currance, a musician's wife, as well as Wilfry Joye, a draper. In both cases, Clowes claimed, Russwurin began surgical procedures to remove suspected bladder stones. An autopsy revealed that Currance had no obstructions in her bladder — the stone rested in her kidney and was inoperable by early modern methods.[69]

Here Clowes was walking an increasingly fine line: while he wanted practitioners like Russwurin discredited, he did not want to turn people away from all chemical medicines or from the Barber-Surgeons who might use them. It was clear that London's appetite for chemical medicines was undiminished and that Clowes's community wanted its company to corner the market. Men like Hester and Bostocke complicated this situation and changed what could have been a simple tug-of-war between physicians and surgeons into a far more difficult game. How could Clowes successfully link an English apothecary and a German itinerant practitioner in the minds of Londoners? He did so by rhetorically transforming Hester and other English supporters of Paracelsian medicine into poisons even more venomous than Russwurin's remedies. English advocates of chemical medicines who were not in Clowes's charmed circle of Barber-Surgeons were transformed into an "Adder's brood," a host of venomous beasts capable of infecting still more English men and women with their foreign cures. Hester was the chief "viper" in this nest of snakes, a man who whispered "into the ears of some of his own [patients] . . . certain rude and lewd speeches" against upstanding medical practitioners. Clowes was determined to stand firm in the face of Hester and his "Scorpions" who "daily cast out their venom and poison against us, and the rest of our loving brethren, the true professors of this art and mystery [of surgery]."[70]

To advance their community's identity and increase the status of surgery as a profession, Clowes and his fellow Barber-Surgeons needed to appropriate novel, foreign cures like chemical medicines even as they tried to distinguish themselves from itinerant foreign practitioners like Russwurin.[71] Exacerbated by public health concerns about the contagion of syphilis, anxieties about contamination and poison often accompanied the general English anxiety about foreignness — or "Strangeness," to use the contemporary English terminology. These anxieties permeated London's civic relations, influenced the reception of medical ideas that came from abroad, and ultimately assisted Clowes in his efforts to construct a coherent public image for surgeons. But London's animosity toward foreigners and foreignness was always problematic, tangled up as it was in the desire for exotic, foreign wares. What was true for London more generally was also true for Clowes and his friends among the Barber-Surgeons who wanted to use the chemical medicines attributed to Paracelsus. When William Pickering wrote his epistle for the end of Clowes's treatise, we see how difficult it was to cast Russwurin as a foreign poison while trying to retain Paracelsian medicine as a viable therapeutic option. Pickering described Russwurin as "a proud prattling Paracelsian — yet indeed no Paracelsian, but a usurper of that name." "I speak not," Pickering further equivocated, "against the good works of the right Paracelsian . . . but [only of] those empirics who . . . have been whipped and banished from city to city."[72]

On London's streets and in her markets Paracelsian therapeutics had taken on many guises and functions. The word *Paracelsian* had been used as a market strategy by Baker, as a marker of identity by Hester, and as a derogatory term by Clowes, and it had provided the itinerant practitioner Russwurin with an entrée to noble patronage. In London, to renounce or embrace a Paracelsian medicine or idea was no longer just a medical judgment, it was a public stance that put you on a particular side in the battle over who should be treating a host of common complaints. Like espousing a modern political affiliation, to be seen as a Paracelsian or anti-Paracelsian after 1585 meant something in the streets of London. The problem for historians is that it meant something different depending on who was doing the talking and who was listening. That is why Paracelsus and his cures were such valuable, and dangerous, weapons in the Barber-Surgeons' struggle to achieve greater medical authority and their quest for a more clearly defined community of practitioners.

After Clowes's friends switched their target from Russwurin to his English supporters in 1585, their struggles with Hester's admirers continued to simmer, boiling over at regular intervals in angry book prefaces penned by members of the Barber-Surgeons' Company. When George Baker edited Thomas Gale's translation of Giovanni da Vigo's surgical works in 1586, he scoffed at Hester's use of

Latin in the preface to the experiments of Paracelsus and claimed that the apothecary did not understand the contents of his own translation. "Now he has set down certain compositions of Paracelsus," Baker wrote, "that the good man himself understands not, for it is not one year since he inveighed against him in my presence." A true understanding of Paracelsus took years of difficult study as well as hands-on experience; translating a few texts into English was not enough to make someone expert. "I for my part have read and used his medicines for eighteen years," Baker stated, "and yet dare not avouch to understand him . . . much less put any part [of his works] in print." Once, he and Hester had been friends, Baker explained, and "when I first knew the man, he was glad to learn those things [from me] that he still uses, and are his best helps."[73]

When Hester took to publishing, however, he trod on the toes of Baker and his friends, and their once congenial relationship turned sour. London was a small city, and there is evidence that people at the time knew that the battle over chemical medicines was fueled by complicated motives that went far beyond a concern for the City's health. An anonymous letter from a learned physician "I. W." to a friend offered a counteroffensive to Baker's attack on Hester and other proponents of Paracelsian medicine. "I. W." confessed that he was "addicted to this new sect of physicians, called by the name and title of Paracelsians," and was distressed that others did not share his passion for chemical medicines. Perhaps critics should read more works actually written by the great German doctor, or even "that book lately set forth by Master B[ostocke] in our mother tongue?"[74]

Clowes ignored the suggestion of "I. W." and returned instead to his favorite object of scorn, John Hester, in A *prooved practise for all young chirurgians, concerning burnings with gunpowder* (1588). He attacked the Paracelsian Richard Bostocke as well. Clowes, like Pickering, wanted to distinguish between proper Paracelsians (such as the Barber-Surgeons) and the "counterfeit, stale Paracelsian quacksalver." These ignorant fellows, Clowes explained, "misinterpret Paracelsus whom truly they understand not, and condemn Hippocrates and Galen whom they [have] never read." Even more alarming, they suggested that the medicines in common use by Barber-Surgeons to help heal wounds were superfluous and contained impurities and should be replaced by chemically purified medicines. To give up the entire traditional pharmacopeia, Clowes felt, was going too far in favor of Paracelsian cure-alls and distilled medicines, which promised to heal any hurt.[75] Clowes's book, stuffed with time-tested medicines proven effective on his own patients in London and on the battlefield, made it possible for the conscientious surgeon to use approved chemical medicines without casting aside useful traditional remedies. Each medicinal formula

Clowes included was accompanied by anecdotal case histories describing its use and its history of successful cures.

For nearly fifteen years Clowes and his fellow Barber-Surgeons George Baker and John Banister had led the charge to strengthen the position of surgeons in London's medical market and to increase the volume of patients who sought them out for treatment. They did so by doggedly pursuing practitioners like Russwurin and Hester who got in the way of their plan to appropriate Paracelsian medicines, and by enthusiastically using print culture to establish a coherent public identity that definitely stood *against* something. In 1588 a new ally joined their ranks, directing their efforts away from old and largely vanquished enemies like Russwurin and Hester and toward a far more impressive foe: the College of Physicians. John Read was the first Barber-Surgeon to publicly assert the surgeon's right to prescribe inward chemical medicines to patients suffering from wounds. In Read's translation of Franciscus Arcaeus's *A most excellent and compendious method of curing woundes in the head* (1588), he described the surgeon's pressing need to be conversant with the "prescribing of inward medicines, and convenient diet." Some wounds, Read argued, did not respond at all to topical treatments, so surgeons needed to be formally educated in physic. This training would necessitate making a clear distinction not between physicians and surgeons, or surgeons and apothecaries, but between *barbers* and surgeons, for "they ought not to be called surgeons who have learned nothing but the composition of two or three emplasters out of barbers' shops." If a surgeon dedicated himself to studying the theory and practice of surgery, nothing should impede him from administering inward medicines, provided that he was not a barber, an unskillful woman, or one of the "blind empirics" mentioned in his friend William Clowes's book on the *morbus gallicus*. In case readers wavered from the point, Reade stated firmly that "all surgeons ought to be seen in physic, and the barbers' craft ought to be a distinct mystery from surgery."[76]

Ultimately, there were winners and losers in the struggle over medical authority in Elizabethan London — and the winners were those who used print to establish a lasting public image that included a specific set of skills, a learned pedigree extending back to antiquity, and a concern for the public health of the City's residents. Historians have long speculated about the role that print played in the Scientific Revolution. Practitioners like Valentine Russwurin would surely suggest that print culture helped to establish a canon of recognizable figures that, because they wrote books, could be easily studied by later scholars who saw published books as the best means available for spreading knowledge to interested parties. Yet knowledge could spread just as easily on the City's streets and in the markets — and it spread far more quickly and cheaply, too. Books were valuable tools in the hands of William Clowes and George Baker, but they were not the only tools that mattered in Elizabethan London.

It took centuries for Read's successors in the Barber-Surgeons' Company to successfully lobby for, and win, a professional association devoted entirely to the practice of surgery. When the College of Surgeons was established in 1745, it was over the bitter opposition of both physicians and barbers, for both groups were justifiably anxious about being squeezed out of the lucrative medical market. Despite these changes to the organizational landscape of medicine in the City, London continued to have a relatively decentralized medical market well into the nineteenth century. Surgeons following in the footsteps of William Clowes and George Baker had become acknowledged and successful medical experts, but there was still room in the medical marketplace for men and women like Russwurin to sell their knowledge and their services.

Clowes, Baker, and the other Barber-Surgeons who joined their cause were thus only partially successful in their efforts to win the contest over medical authority in Elizabethan London. While the lines between surgeon, physician, and apothecary were more clearly drawn, the teeming and vital "underside" of this orderly medical system — the empirics, midwives, and other unlicensed practitioners — continued to exert a pull on potential patients. The line between an itinerant empiric like Russwurin and a recognized surgical expert like Clowes continued to be difficult to discern on the streets of the City. One man's quack continued to be another's preferred medical practitioner, and the network of social relationships that fostered and sustained London's medical market remained resistant to the pressure of print culture for centuries to come. The power of print was not as important to survival within the humming, thrumming markets and among the face-to-face world of London's medical practitioners and patients as it proved to be within the Republic of Letters and the residential enclaves of Lime Street. While the case of the Lime Street naturalists showed how fragile the social foundations of science could be in Elizabethan London if not shored up with an adept use of print culture, the case of Valentine Russwurin and the Barber-Surgeons reminds us that life in the City was a lived experience that could never be fully controlled — not by regulatory officials, and not by the books that some elite practitioners were beginning to use to craft public, community identities.

————◄●►————

3

EDUCATING ICARUS
AND DISPLAYING DAEDALUS

Mathematics and Instrumentation
in Elizabethan London

Sometime around 1590 the mathematical author and educator Humfrey Baker (fl. 1557–90) picked up a stack of broadsides from a London printer and began tacking them up on every conceivable surface of the City and distributing them at the booksellers' stalls along with his popular arithmetic textbook, *The well spryng of sciences* (1568), and his translation of Oronce Fine's work on astrology, *The rules and righte ample documentes, touchinge the use and practise of the common almanackes* (1558). The broadside was divided into three sections. One focused on the theoretical lessons Baker was prepared to teach in subjects such as arithmetic and algebra; another set out the application of geometry to practical problems for "the aid of all mechanical workmen"; and the final, largest section was devoted to his special forte and the bane of many mathematics students, the clever word problem. In his word problem Baker made the most enticing possible pitch to a broad City clientele by using mathematics to untangle the complicated financial arrangements among four business partners trading in London, Rouen, Middleborough, and Elbing. The dizzying twists and turns of their dealings, which involved lengths of cloth, foreign currencies, and customs charges, could be made straight and easy to follow, of course, with the mathematical knowledge that Baker possessed and was willing to impart (for a fee).[1]

Baker encouraged adults, children, servants, and apprentices to visit his house on the northern side of Gresham's Royal Exchange, the bustling commercial venue where the Barber-Surgeon George Baker resided and the Paracelsian Valentine Russwurin plied his trade. There, next to the Sign of the Ship, pupils could make arrangements for private classes, or could even board with Humfrey Baker and his wife, Elizabeth, "for the speedier expedition of their learning." An illustration from Robert Record's *Grounde of arts* (1558) shows how Baker's classroom might have been set up, with the master seated in front of a large, blackboardlike surface where sums could be worked out and his pupils arranged around the table pointing at aspects of the problem that still confused them. High above their heads a shelf held a few precious books (Figure 3.1). In his house, Baker taught arithmetic and geometry, emphasizing how this knowledge was useful in a host of situations and would improve measurements of land, timber, and stone. He also taught accounting, instructing merchants and their children "after a more plain manner than has heretofore been usually taught by any man within this City." Baker even made sure his students could use (and in some cases construct) such mathematical instruments as the quadrant, square, astronomer's staff, and astrolabe.

During the reign of Elizabeth, London was the center of vernacular mathematical education in England, as well as the center of mathematics publishing and instrument making. Londoners bought books on mathematics, frequented the shops of instrument makers, and flocked to demonstrations of new mechanical objects in the City. In London men and women were using mathematics everywhere: shopkeepers making change, an apprentice using an abacus, a carpenter cutting a length of wood, a clockmaker repairing one of the perpetually broken mechanisms that graced London's churches, foreign merchants working out rates of currency exchange, and surveyors measuring out a plot of land for a new house. Work that required some knowledge of mathematics, including navigation, surveying, engineering, architecture, ballistics, carpentry, and masonry, was studied by both elites and craftsmen struck by the potent mixture of mathematical theory and practice required of a skilled practitioner. As Record explained in his arithmetic book, mathematics was "the ground of all men's affairs." Without mathematical literacy, "no tale can be long continued, no bargaining without it can be duly ended, nor no business that man has justly completed."[2]

A growing population, facing stiff competition in the City's markets and workshops, actively sought ways to make their work more accurate, speedy, and efficient. Mathematical teachers, authors, and instrument makers responded to the needs of their urban clientele by promoting the notion that mathematical and instrumental literacy fostered inventiveness, cunning, and quick problem solving. Knowledge of mathematics and profit often went hand-in-hand in the City,

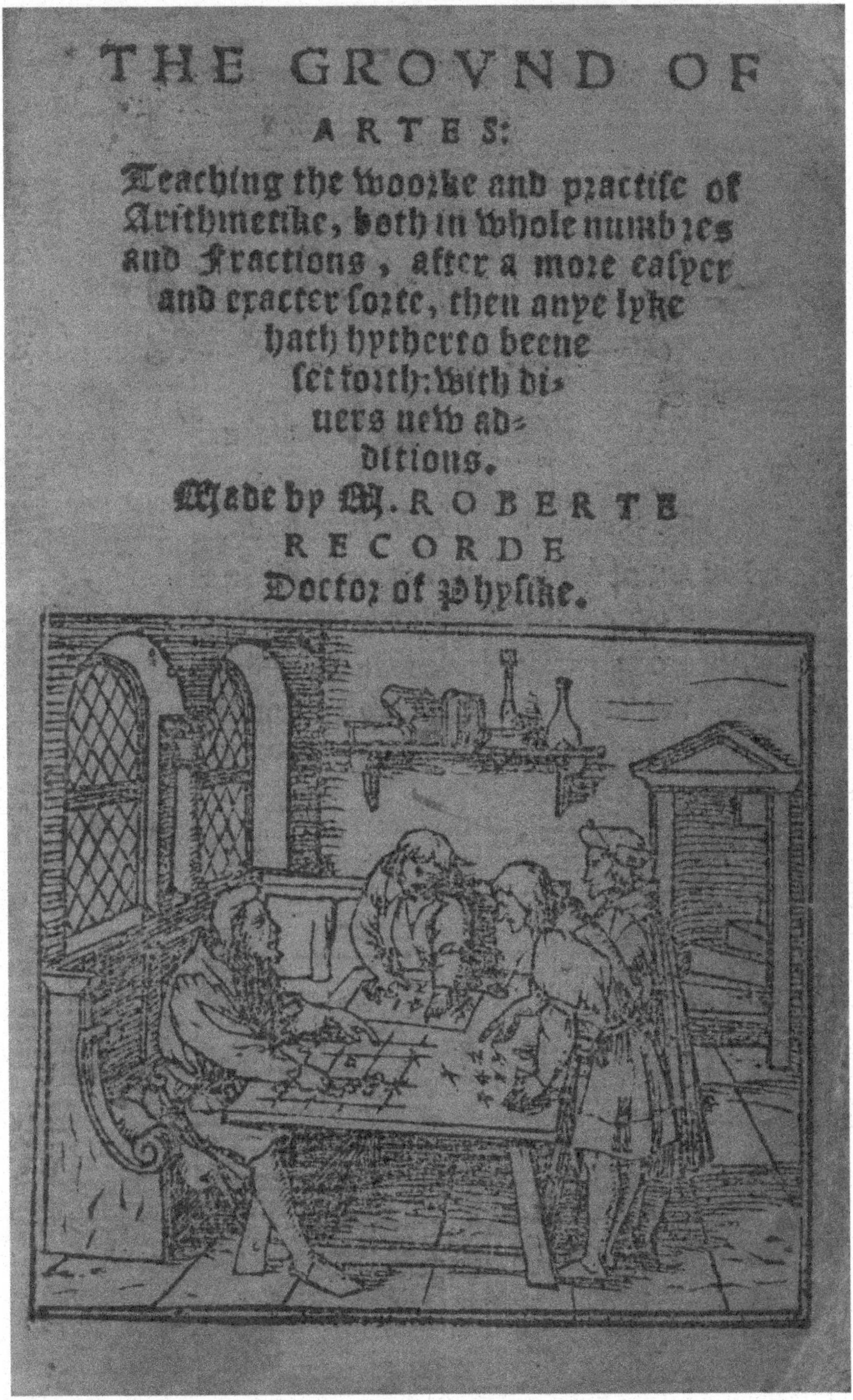

THE GROVND OF
ARTES:
Teaching the woorke and practife of
Arithmetike, both in whole numbres
and Fractions , after a more eafyer
and eracter forte, then anye lyke
hath hytherto beene
fet forth: with di-
uers new ad-
ditions.
Made by M.ROBERTE
RECORDE
Doctor of Phyfike.

Figure 3.1. Here, in an illustration from Robert Record's *The ground of artes* (1558),
a physician works out sums (possibly relating to the formulation of medicines) in a
setting that reflects how mathematics was probably taught in Elizabethan London.
Reproduced with the permission of the Plimpton Collection,
Rare Book and Manuscript Library, Columbia University.

and numeracy was a marker of success for those who made enough money that they needed account books to keep track of it. But mathematics, according to its promoters, need not be all about greed and profit. The use of mathematics in the discipline of astronomy, Record argued, could "encourage men to honor God." "The ingenious, learned, and well-experienced circumspect student mathematical," wrote Leonard Digges, "receives daily in his witty practices more pleasant joy of mind, than your goods (how rich so ever they be) can at any time purchase."[3] Good Protestants could be absolved of some of the guilt associated with prosperity by mathematically contemplating their riches.

Whether for greed or more exalted aims, many Londoners eagerly embraced mathematics during the reign of Elizabeth.[4] Not everyone applauded these efforts to become mathematically literate, however. Some still thought the figuring of sums little more than magic, pointing to the use of calculations in astrology and other forms of divination, and deploring the spread of such an ungodly discipline. Others took an opposing, though equally restrictive view: they thought that mathematics was the language of God, and therefore too divine to be placed in the hands of the vulgar mechanics who plied their trades in the City's streets. Mathematics, these critics argued, belonged in the hands of theologians, not tinsmiths. One of Humfrey Baker's contemporaries, the Cambridge-educated lawyer Francis Bacon, was particularly ambivalent about the growing partnership between the mathematical sciences and the mechanical and technical arts, and about the spread of mathematical and instrumental literacy. Alternatively attracted by the subtle intricacy of mechanical marvels and repulsed by the occult scribbling of "mathematicians and finger-flingers," Bacon warned his readers in the early seventeenth century against dabbling too much in these bewitching sciences.

Bacon's favorite cautionary tale for those passionate about mathematics was the myth of Daedalus, "a man ingenious but execrable." Daedalus contrived a set of wings from feathers and wax only to see his son Icarus fall to his death trying to use them because he failed to understand their limitations. Daedalus and Icarus were emblematic of the dangers that lurked in applied mathematical knowledge, which were "of ambiguous use, serving as well for hurt as for remedy."[5] Despite the potential dangers, most Elizabethans were interested in applying mathematics to the development of new mills and engines, and to the refining of existing techniques in shipbuilding and gunmaking. But the fate of Icarus, whose confidence in mechanical inventions overran his knowledge of the mathematical principles underlying them, was still cause for concern. How could London continue to develop mathematical knowledge and apply it to mechanical problems and projects without inviting similar catastrophes?

Mathematical educators provided an answer: Londoners must be better educated in the nuances of mathematical theory to judge the appropriate limits of
mechanical ingenuity and to ground practical applications in what they called
"the mathematicals." But like so many seemingly straightforward answers to difficult questions, the push for greater mathematical literacy among Londoners
only raised new issues. Who should teach mathematics, who were appropriate
pupils, and who should pay for the classes were the most immediate and pressing
questions, but others quickly sprang up. What should the role of print be in this
new world of urban mathematical literacy? Who would buy the books, what
level of mathematical texts were most in demand among consumers, and what
subjects should be covered? And how much should Londoners rely on the fash-
ionable, expensive mathematical instruments imported from abroad and now
made in the City?

Two Londoners, one a Cambridge-educated mathematician and one a Cambridge-educated merchant, provided early answers to some of these questions
when they promoted mathematical literacy as a way for England to become
more productive and profitable. When the merchant and Haberdasher Henry
Billingsley (d. 1606) decided to devote his spare time translating the great Greek
geometry textbook Euclid's *Elements* into English, he asked his fellow Londoner
and mathematician John Dee (1527–1608/9) to write a preface that would describe the mathematical disciplines, shed light on dimly understood and sometimes feared applications of such disciplines as geomancy and astrology, and elu-
cidate the many practical benefits that would come to England if her citizens
commanded greater mathematical literacy. To avoid the fate of Icarus and the
grief of Daedalus, Dee and Billingsley argued, England must dispel the aura of
mystery that so often surrounded mathematics and must educate a broader range
of people in the theories that provided a foundation for such diverse applications
as instrument making and astrology.

Billingsley and Dee suggested a mathematical compromise, one self-consciously designed to promote mathematics among the widest possible range of
citizens by emphasizing theory and practice in equal measure. Yet by and large,
most London pupils were interested in the pragmatic levels of mathematical literacy. For many, the primary goal was to acquire a basic understanding of mathematics that would lead to more sophisticated mercantile transactions and
record keeping, and an ability to use (though perhaps not fully understand) the
new instrumental technologies developed in City workshops. Most Londoners
felt that hands-on experiences and practical considerations should dictate the
theories that were taught in classrooms, studies, and lecture halls across the City.
Their teachers felt differently, arguing that students could not gain true literacy

until they were thoroughly grounded in the theoretical foundations of arithmetic. After that, students could move, step-by-step, through more advanced mathematical theories and on to the use of sophisticated instruments.[6]

As in the battle between the Barber-Surgeons and the supporters of Valentine Russwurin, proponents of practical competency and advocates of theoretical knowledge took their dispute to the bookstores and the streets. But this was not a fundamental debate about the merits of mathematical literacy. While Oxford and Cambridge mathematicians might deplore the practical side of mathematics, in London few valued an approach to mathematics that was either purely theoretical or purely practical. In the years following the publication of the Billingsley edition of Euclid, after a generation of London students was taught the links between theory and practice by their teachers, the authors of mathematical texts increasingly knit together these two aspects of mathematical knowledge in their work. In doing so, late Elizabethan mathematical authors were able to move to a new argument: that mathematical literacy led to quick and reliable problem solving for merchants, craftsmen, and even officers of the state.

Here we will explore how authors and educators promoted mathematics, how publishers and instrument makers instantiated mathematics in books and instruments, and how London's students and consumers perceived the results. After the publication of the Billingsley-Dee edition of Euclid's *Elements* in 1570 put theory and practice side by side, mathematics and its applications appeared to be inseparable partners. The Billingsley-Dee approach led to a resurgent interest in mathematical education in the City. During the 1580s and 1590s, mathematical educators and authors increasingly emphasized the problem-solving potential of mathematics, and mathematical literacy became crucial to swift and sure solutions in accounting, exchanging money, technical dilemmas, and craft puzzles. Londoners able to analyze a problem through numbers, rather than through trial and error, were a boon to their families and employers. But working out problems with pen and paper was not always practical; complicated situations often required lengthy calculations, and a single simple mistake could lead to disaster. Like Icarus, Londoners were increasingly drawn to mechanical objects and mathematical instruments to help them solve applied mathematical problems. Instrumental literacy fast became a popular goal in its own right.

In all these stories, London played an important role. From the proclamations of their City roots in the Billingsley-Dee edition of Euclid's *Elements*, to the streets where students and merchants shopped for the newest mathematical textbooks and the most innovative instruments, and into the classrooms and houses where teachers argued over the proper way to make their pupils mathematically literate, London provided the social, cultural, and intellectual context for changing popular attitudes toward numeracy and instrumental knowledge. The value of mathematical literacy to both state and citizen was discussed and negotiated in the City as much as in the universities at Oxford and Cambridge. And London could boast the critical mass of expert teachers, skilled technicians, willing pupils, and eager readers that were necessary to transform the aspirations of a few elite mathematicians into a broadly based, vernacular and mathematical way of thinking about the world.

MATHEMATICS FOR MERCHANTS:
VERNACULAR MATHEMATICAL PUBLISHING

Elizabethan London's interest in mathematical education and literacy can be traced back to a most unlikely point of origin: a flying beetle made for a Cambridge University undergraduate theatrical. In 1545 a young Londoner, John Dee, was studying Greek and mathematics there when he agreed to put his mathematical knowledge to work by constructing a mechanical prop for a student production of a classical play that was meant to divert students during the spring vacation. Aristophanes' *Pax* is an earthy Greek comedy about a farmer's audience with Zeus, king of the gods, and the plot calls for a dung beetle that flies toward the sun. The beetle, Dee later claimed, was so realistic that the learned audience fled in terror. Few put any faith in Dee's regular protestations that the gizmo was a simple mechanical device whose inner workings could be understood mathematically and explained to the most uneducated person. Dee earned his early reputation as a conjurer from this seemingly innocent bit of stagecraft, and smarted about it for the rest of his life, protesting until his dying day the "vain reports spread abroad" about how he had managed to pull off the special effect.[7]

Those who witnessed or heard of Dee's beetle regarded the ensuing fracas as a prime example of the dangerous intersection of mathematics, mechanics, and magic.[8] Many considered mathematics too powerful and dangerous for the undereducated, because some biblical passages indicated that God molded everything according to principles of number, weight, and measure at the moment of creation (Wisdom 11:21). And as the sad tale of Dr. Faustus made clear, not even Latin and a university education made you completely safe from the dark side of mathematics. Dee could grumble all he wanted about the unfairness of his audience's reaction to his beetle, but until a way was found to demonstrate that mechanical objects were simply concrete manifestations of abstract mathematical principles, he would grumble in vain. When his fellow Cambridge alumnus Henry Billingsley approached him to write the preface to his translation of the

Elements, Dee finally had a proper forum to lay out his beliefs about the practical import, as well as the theoretical intricacies, of mathematics.

Billingsley and Dee were not the first mathematical authors to try to make mathematics appeal to a broad audience. A steady stream of books relating to mathematics began entering the bookstalls soon after the advent of print in the fifteenth century. These included elementary textbooks on arithmetic and geometry; instruction manuals for mathematical instruments like the astrolabe and cross-staff; works on navigation, cosmography, and cosmology; and treatises on astrology and astronomy, especially the popular almanacs and calendars of celestial events known as ephemerides. More than 250 books related to mathematics were printed during the reign of Elizabeth, and the slow trickle of accounting manuals and arithmetic texts at the beginning of the reign turned into a flood of books on all sorts of mathematical topics by the end of the sixteenth century.

As historians of the book know, it is difficult to be precise about the exact number of mathematical books published during the period because so many mathematical works have been lost, especially the ephemerides. Gauging the significance of a discernable increase in the number of published mathematical works is equally problematic given the relatively small number of titles under consideration. Nevertheless, it is possible to acknowledge some trends. After the initial five years of Elizabeth's reign, for example, when mathematics publishing was slight, the number of mathematical titles printed remained relatively steady at around twenty-five per five-year period until 1588 — the year of the Spanish Armada crisis. From 1588 to 1598 those numbers rose for a decade before returning to pre-Armada levels. The higher number of titles published in the period from 1588–98 includes a marked spike in the year 1596, when London printers issued record numbers of mathematics books. While the number of books on arithmetic, geometry, cosmology, astrology, and astronomy remained essentially consistent throughout the period from 1563 to 1588, a growing demand for instrument manuals and navigational treatises played a role in the gradual increase in the mathematical titles available at the bookstores from 1588 to 1597, as well as the spike in the market in 1596, when ten of the sixteen titles available to mathematical readers were concerned primarily with the use of instruments, including their function in navigation and surveying.[9]

As these trends suggest, mathematics books were never as popular, or as plentiful, as the religious treatises, medical handbooks, sermons, and plays that rolled off the London presses into the hands of consumers. The question of who purchased and read mathematics books has been much debated by scholars seeking to understand this small segment of the publishing market in early modern England. Elizabethan mathematical authors, who liked to enthusiastically reel off all the types of people who should be willing, able, and eager to buy their book, may have been more optimistic than realistic when they forecasted their potential audience. Robert Record's happy declaration in his geometry book *The pathway to knowledg* (1551) that "carpenters, carvers, joiners, and masons, do willingly acknowledge that they can work nothing without . . . geometry" was certainly not reflected in an exuberant sales and republication record.[10]

Elizabethans had a more avid taste for some mathematical subjects than for others. Judging by the number of times specific titles were reprinted, almanacs and other interpretations of celestial events were the most popular mathematical publications during the Elizabethan period, followed by books that taught the design and use of instruments, and elementary arithmetic books. Record's arithmetic textbook, *The grounde of artes*, went through seven editions between 1558 and 1603, while his more advanced *Pathway to knowledg* languished on the shelves. Together, these topics were covered in the majority of the mathematical books published in England between 1558 and 1603. Books on astronomy, navigation, cosmology, geometry, and surveying occupied the second tier of mathematical interests, with works on ballistics and military science, magnetism, maps, and mensuration (the geometry of measurement) taking a decidedly back seat.[11] While it has become a commonplace for historians to comment on the popularity of practical mathematics in the period, some highly practical applications of mathematics, like defense and land surveying, were far less prevalent offerings in Elizabethan bookshops than might have been predicted. Instead, it was calendars, technology manuals, and the most basic mathematics books that were issued, regularly and steadily, year in and year out. But this pattern is hardly surprising if we compare these topics to books related to mathematics the average consumer might regularly purchase today: a calendar for the New Year, updated computer manuals, and a book on how to keep track of finances.

It is no wonder that ephemerides, which included complicated astrological calculations, were so popular, or that almanacs were purchased regularly. Almanacs, calendars, and interpretations of celestial events were indispensable living aids for Elizabethans. They were necessary for the proper prescription and administration of medicine, for planting crops at the most propitious time, and for gauging tides and weather. Whether a cheap one-page broadside or a more lavish volume with a multiyear format and diagrams of the human body, an almanac could be found to fit nearly everyone's needs and budget. Even London churches bought almanacs, and St. Benet Gracechurch kept one in the choir for easier access by church officials and parishioners.[12] The minister and almanac maker Thomas Buckminster (1531/32–99) was the most prolific Elizabethan

mathematical author, and his star charts, meteorological predictions, and astrological prognostications went through some twenty editions during the reign.
Other popular almanac writers of the period included Leonard Digges (c. 1515–
c. 1559), whose almanacs continued to be updated and sold long after his death,
and the versatile mathematical writer William Bourne (c. 1535–82), who wrote
works on instruments and navigation in addition to almanacs.[13]

Introductory arithmetic texts were also perennial favorites in the bookshops,
and two frequently reprinted arithmetic authors crowded Buckminster for pride
of place among Elizabethan mathematical authors. Robert Record's *Ground of
artes*, first published in 1543, was reprinted six times during Elizabeth's reign,
while Humfrey Baker's textbook, *The well spryng of sciences* (1562), went through
six editions as well. Couched in Socratic dialogues between a master and a student, these two introductory arithmetic texts took students from basic definitions — what is a number? — through complicated word problems involving foreign currency transactions, fractions, and accounting. The keeping of financial
records, so vital to merchants who engaged in foreign trade and to those with
complicated business partnerships, was a focus of the arithmetic books.[14] More
complicated topics, such as the technique of alligation used to determine the
quantities of ingredients in compound medicines, was touched on in some arithmetic books as well.

The authors of basic arithmetic texts were often teachers who emphasized
that true mathematical literacy depended upon a sequential mastery of skills
and techniques. Record and Baker deplored the habits of readers who skipped
around in their math books looking for a quick answer to a particular problem.
"You may not learn the last as soon as the first," the teacher warned his pupil in
Record's *Ground of artes*, "but you must learn them in that order as I did rehearse
them, if you will learn them speedily and well." Some arithmetic books were
classroom texts — thick, large, and ideally suited to sit on a desk and take students
through a course of instruction. Others, like the anonymous *Introduction for to
learne to recken with the pen, or with the counters* (1566), were so small they could
be slipped into a pocket and taken to work or to the market. These slim ready-ref-
erence books proudly took "pains . . . in the better and more clear declaration
and expressing of the said rules [of arithmetic], and also in the . . . cutting of
divers superfluous and empty things, rather [a] hindrance to the diligent reader,
than a furtherance."[15] So much for the careful explanations and lengthy examples of Record and Baker!

While authors and publishers of almanacs and arithmetic texts could simply
reprint popular titles after some revision, the market for books on instruments —
the third popular genre of mathematical publication — was different. Here Lon-

doners wanted novelty and the most up-to-date designs and easy techniques for using instruments in navigation, surveying, or military affairs. Very few books about instruments or instrumentation were published before the Billingsley-Dee edition of Euclid's *Elements* in 1570. The physician William Cuningham touched on instruments in his book on cosmography, *The cosmographical glasse* (1559), but instrument books were not yet the popular items they would become in the late sixteenth century. Two factors contributed to the relative lack of interest in instruments before 1570. First, the development and use of instruments depended more on a mastery of geometry than arithmetic, and until the Billingsley-Dee Euclid, Londoners had no vernacular geometry book available. Second, instruments were scarce and expensive, which placed them beyond the reach of most citizens. Few instrument makers worked in the City before the 1570s, when waves of religious refugees brought highly skilled instrument makers into London from France and the Netherlands. The instrument makers active in the early Elizabethan period were in high demand from the royal court and London parishes, and their wares, as we shall see, were expensive.

Leonard Digges was the author who first focused serious attention on instruments and the geometrical knowledge that shaped them. The son of a country landowner, Digges was educated at Oxford and had a long-standing interest in the practical application of mathematics to problems of defense and surveying. Even in his 1555 almanac, *A prognostication of right good effect*, Digges explained how to use instruments to observe the stars and to navigate the seas. In *A boke named Tectonicon* (1556), which was reprinted twenty times in the next 150 years, Digges focused entirely on the practical application of geometry to the problems of surveying. On the title page, two surveyors used an instrument called the cross-staff to take measurements. These surveyors represented Digges's audience — he specifically mentions the surveyor, carpenter, and mason — who were unable to read the "many excellent" geometry books "locked up in strange tongues." In addition to covering the basics of geometry, Digges provided detailed instructions on the manufacture and use of specific instruments such as the Carpenter's Rule, which would enable the reader to put newly acquired geometrical knowledge into action. Digges advocated careful reading and reflection when studying mathematics, and instructed his readers to go through his books once, "then [again] with more judgment, and at the third reading wittily to practice" the methods and techniques he described. Frequent "diligent reading," Digges wrote, "joined with ingenious practice, causes profitable labor."[16]

A boke named Tectonicon's title page also promoted the instrument-making skills of Digges's publisher, a Flemish engraver named Thomas Geminus (fl.

1540–62). Digges and Geminus are emblematic of the partnerships beginning to emerge between mathematicians and skilled artisans in the Elizabethan city of London. Their partnerships extended across an ancient division, established in classical times, between the philosophical life of the mind and the vulgar work of the hands.[17] An early modern jack-of-all-trades who dabbled in medicine, publishing, and engraving, Geminus was one of the most technically adept and highly regarded instrument makers working in late Henrician and early Elizabethan London. He kept a shop in the old Blackfriars monastery, a crumbling building in one of the City's first theater districts that was converted into a combination shopping arcade and residential complex during the Reformation. Geminus promised readers he would be "ready exactly to make all the instruments" featured in Digges's book. Seven of Geminus's instruments survive and prove that he could keep his word, including an ornate brass planispheric astrolabe he made for Queen Elizabeth in 1559.[18]

Works by these early Elizabethan mathematical authors — Buckminster, Bourne, Humfrey Baker, Record, and Leonard Digges — were printed and reprinted in the early modern period. Their publication record suggests that copies of an almanac by Buckminster or Digges, as well as an arithmetic book by Record or Baker, were common items in many households. What is striking about the popularity and frequent reprinting of these almanacs and astrological predictions, arithmetic books, and instrument manuals is that they represent a level of mathematical literacy similar to our modern grasp of mathematics. Most of us have a calendar or date book — but few of us could draw up a calendar from scratch or explain the mathematical and astronomical theories underlying its design.

To achieve higher levels of mathematical and instrumental literacy, however, London readers needed to reach beyond the basics of arithmetic to the more abstract principles of geometry. Only then could students attempt for themselves the complicated calculations required to construct a calendar based on astronomical observation, or truly understand the principles that made a quadrant or astrolabe work to calculate the distances between the stars or the height of a tower. Robert Record attempted to provide readers with just such a progression of instruction.[19] While *The grounde of artes* got students started with arithmetic, they could read on through the rudiments of geometry in his *The whetstone of witte*, and then proceed to theories of astronomy, cosmography, and a cursory discussion of several mathematical instruments in his *Pathway to knowledg*. If students, having mastered Record's basic texts, wanted to further their mathematical expertise, they could turn to William Cuningham's (b. 1531) *Cosmographical glasse* (1558), which contained a more academic and advanced discussion of cosmology, emphasizing the classical, geocentric theories of Aristotle and Ptolemy. Copernicus's sun-centered cosmology, published in the *De revolutionibus* of 1547, had yet to make an impression in England's market for mathematical books.[20] Cuningham's *Cosmographical glasse* unfolded in a dialogue between Spoudaeus, a man who attempted without success to learn the science of cosmography solely through reading books, and his more knowledgeable friend, Philo. Philo applauded Spoudaeus for reading the works of Robert Record because this intermediary level of mathematical literacy would make it possible for him to learn the more complicated mathematical sciences, but he urged his friend to read other works as well. These included Oronce Fine's work on arithmetic, Johann Scheubelius on algebra, Theodosius's ancient work on spheric demonstrations, and Euclid of Megara on geometry.

Philo gave Spoudaeus excellent advice, but it was advice that few of Cuningham's readers could follow, since none of these works was available in English at the time. For the vast majority of Londoners, who were able to read and write in English but had little Latin and less Greek, the Billingsley-Dee Euclid represented a turning point in mathematical literacy because it gave vernacular writers a classic text that they could further popularize in cheaper and more affordable editions. The popularization of ancient texts like Euclid's *Elements* in Elizabethan England was far from unique, but was instead part of a European movement to translate important mathematical works into vernacular languages. There are striking similarities between the Billingsley-Dee edition of Euclid's *Elements*, for example, and such Italian translations of the work as those by Niccolò Tartaglia and Federico Commandino.[21] What Tartaglia, Commandino, Billingsley, and Dee recognized was that until Euclid and other classic mathematical works were available to a wider range of students in the vernacular, higher levels of mathematical literacy would be confined to the universities and to students at the very best grammar schools who were taught Latin and Greek.

The Billingsley-Dee Euclid, while neither the first nor the most popular Elizabethan mathematics book, is the best known today. When it was published, however, the book's size and expense put it beyond the reach of many urban consumers.[22] The publisher made every effort to increase potential sales through a clever marketing plan that hinged on the eye-catching title page (Figure 3.2). John Day recycled the intricate title page borders from Cuningham's *Cosmographical glasse* but replaced the old-fashioned Gothic black-letter type of the earlier work with the chic new typefaces popular on the Continent, enabling the Elizabethan buyer to purchase something both familiar and stylish. The great antiquity and divinity of mathematics was brought home to the potential purchaser in the vivid image of the lounging gods Sol and Luna—the sun and moon — while Time and Death peered curiously at the work being performed by a host of real and mythical characters associated with mathematics, including Ptolemy, Strabo, Hipparchus, and female figures representing the mathematical disciplines of geometry, astronomy, arithmetic, and music. Seven mathematical instruments appear on the title page, including an astrolabe, an armillary sphere, and a map. An elfin Mercury, the winged messenger, sits on a cloud that looks like piped icing, promising to communicate arcane knowledge to the reader. If this host of mythical creatures was not inducement enough for Londoners to achieve higher levels of mathematical literacy, the names of two of the City's prominent citizens were splashed across the title page. Henry Billingsley, "Citizen of London," was placed in the center of the page as the first, faithful translator of the work into English. In a panel below, the publisher advertised John Dee's "very fruitful preface" which "disclosed certain new secrets, mathematical and mechanical." Open and secret, classical and modern, familiar yet new — this was a book that promised something for everyone.

To usher readers gently into the subject, Billingsley wrote a brief foreword in which he declared that true mathematical literacy depended upon instruction in "the principles, grounds, and elements of geometry." The problem for students, Billingsley admitted, was that this necessitated "diligent study and reading of old, ancient authors," especially Euclid. Without some knowledge of Euclid, students would be forever trapped in basic arithmetic, unable to see their way to higher and more abstract levels of mathematical reasoning and analysis. For their sake, Billingsley asserted, he had "with some charge and great travail, faithfully translated into our vulgar tongue, and set abroad in print, this book of Euclid." But it was not *only* Euclid, for Billingsley added edifying "examples by figures," glosses, annotations, and inventions he gathered from other mathematical texts.[23]

Billingsley was assisted in his effort to increase English mathematical literacy by John Dee, beetle maker, who by 1570 was England's most famous natural philosopher. Impeccably educated by his Mercer father, Dee had been in and out of royal favor for three successive reigns but still clung to high hopes that Elizabeth would see the value of having a court philosopher on staff to advise her on all things natural and supernatural. Elizabeth, pragmatic to the core, showed little interest in having Dee haunting the corridors of her palaces, although she did ask him to examine a suspicious effigy of her stuck full of pins that had been found in a field. Elizabeth's reluctance may have been due in part to Dee's popular reputation as a conjurer—a reputation he may have enhanced by making repeated references to the mysterious power of numbers in his "Mathematical

Figure 3.2. The title page of Henry Billingsley's translation of Euclid's *Elements* (1570), with a preface by the mathematician John Dee. The publisher, John Day, recycled the borders from an earlier imprint, William Cuningham's *Cosmographical glasse*. Reproduced with the permission of the Henry E. Huntington Library.

Preface" to Euclid. For if Billingsley's goal was to make the *Elements* as legible as possible, Dee's goal was to make the mathematical sciences as alluring as possible. Dee achieved his desired aim by writing a textually dense and labyrinthine exposition of the virtues and dangers of mathematics, liberally sprinkled with classical, occult, and biblical references, and by bestowing bizarre monikers on each mathematical science such as mecometrie and embadometrie. Happily for his readers — then and now — Dee constructed an early modern flowchart of the mathematical disciplines and their applications. Based on the pedagogical graphics used by the great logician Peter Ramus, the diagram conveys the meat of Dee's argument far more readably than does his prose. Even though the chart was remarkably concise for the prolix Dee, later authors found that they could reduce the chart's contents further, and William Bourne managed to hone it down to a simple list accompanied by straightforward definitions.[24]

In his preface, Dee first situated mathematics both in the world of ideas and in the world of man. He placed "things mathematical" securely between the natural and supernatural, and imbued them with "marvelous neutrality" and a "strange participation between things supernatural, immortal, intellectual, simple and indivisible, and things natural, mortal, sensible, compounded, and divisible." Rather than downplaying the magical and occult aspects of mathematics, Dee acknowledged them and tried to show how, in the hands of a knowledgeable practitioner, these features of "things mathematical" were not to be feared or despised but embraced. Just as merchants found arithmetic useful, so, too, the skilled astrologer would benefit from knowing arithmetic, geometry, and the mathematical disciplines of astronomy, cosmography, natural philosophy, and music. Arithmetic and geometry laid the theoretical foundations for architecture, navigation, statics (the mechanics of equilibrium), and geography. When Dee surveyed the world and the work that was done in it, he saw mathematics everywhere. "Trochilike," defined by Dee as "that art mathematical, which demonstrates the properties of all circular motions," was used in mechanical devices like mills, saws, and other machines. Pumps and bellows operated through the mathematics of "pneumatithmie" or the vacuum. The strength of cranes and gibbets was calculated through "menadrie." And the mathematics of hydrography helped ancient Romans build aqueducts and Elizabethans divert rivers and streams.[25]

Mathematics could also help to explain phenomena that might otherwise be dismissed as strange and outside the natural order, like Dee's beetle. Visions seen through optical glasses, Dee pointed out, would not be feared by those who understood the mathematical rules of perspective. He encouraged the skeptical reader to visit the house of Sir William Pickering in London to see "with his own eyes" the strange effects that could be produced using optical glasses. Dee's confidence in Pickering's ability to make a clear demonstration was crucial in his efforts to make Londoners aspire to higher levels of mathematical literacy. Subsequent mathematical students and teachers followed in Dee's footsteps and emphasized the clarifying power of eyewitnessing mathematical demonstrations. Seeing, touching, and manipulating instruments, as Gabriel Harvey explained in marginal notes made in his copy of John Blagrave's *Mathematical jewel* (1585), were essential. "Give me ocular and rooted demonstration of every principle, experiment, geometric instrument," Harvey wrote, "astronomical, cosmographic, horologiographic, geographic, hydrographic, or mathematical in any way." Through demonstration, mathematics could be separated from the merely speculative, or the dangerously magical. In a passage that calls to mind Dee's own disastrous experiences with his beetle, William Bourne pointed out that the "strange works that the world has marveled at, as the brass head that did seem to speak, and the serpent of brass to hiss," were done not by enchantment, as some thought, but "by wheels, as you may see in clocks that do keep time, some going with plummets, and some with springs, as those small clocks that be used in tablets to hang about men's necks."[26] Mathematics, put to proper use and amply demonstrated, could be disassociated from magic.

Dee saved his most powerful message for the end of the preface: mathematics, he argued, was profitable to the commonwealth of England. The skill and experience of England's artisans, when coupled with the "good helps and information" contained in Billingsley's new translation of Euclid, would lead to the discovery of "new works, strange engines, and instruments for sundry purposes in the commonwealth." Dee was one of the first mathematical authors to link mathematical literacy to the well-being of England. Here Dee built on the ideals of the "commonweal," a prominent topic of conversation during the reign of Edward VI, when Dee had been a young man. The commonwealth philosophy was in part a response to the economic, social, and religious controversies of the young king's protectorate, especially unpopular efforts to enclose village commons and transform them into privately held grazing lands.[27] A number of pamphlets and treatises had been published on these ills, but the most well-known work on the subject, *A Discourse of the Commonweal*, circulated only in manuscript until the reign of Elizabeth.

The author of *A Discourse of the Commonweal* was Sir Thomas Smith (1513–77), a sharp-witted and pragmatic bureaucrat who, like Dee, had a fine Cambridge education and ties to London.[28] Despite owning multiple properties throughout England and in the city of Westminster, Smith chose to live primarily in the City, on Philpot Lane with his merchant brother. Smith's City connections did not end there: he married first the daughter of a London printer, and second the daughter of a London businessman. Edward VI appointed him secretary of state, and Smith distinguished himself in that office with his energy and combative cleverness. After alienating a number of important court figures (including the lord protector's wife, the duchess of Somerset), he was instructed to spend the summer of 1549 away from the court at Eton, where he wrote his *Discourse*.

Smith's *Discourse* represents a conversation between a lawyer, a knight, a merchant, a craftsman, and an agricultural worker, all interested in identifying — and resolving — the economic problems that faced England. The lawyer, who served as a mouthpiece for Smith and his commonwealth philosophy, outlined a number of factors that contributed to the country's ills. First, he lamented the low esteem that was accorded to learning, and pointed out that valuable knowledge was held by people from all walks of life, not just gentlemen with university educations. If this vernacular knowledge, learning, and experience were held in higher regard, men would be quicker to step forward to offer their services for the betterment of England. "I would not only have learned men (whose judgment I would wish to be chiefly esteemed herein) but also merchantmen, husbandmen, and artificers (which in their calling are taken wise) freely suffered, yea and provoked, to tell their advice in this matter," the lawyer explained, "for some points . . . they may disclose that the wisest in a realm could not [gainsay]." His companions scoffed at the idea that craft knowledge and book learning could benefit the commonwealth, despite the lawyer's convincing historical arguments that even a military man like the knight could benefit from knowing "arithmetic in disposing and ordering of your men and geometry in devising of engines, to win towns and fortresses, and of bridges to pass over; in which things Caesar excelled others, by reason of the learning he had in those sciences, and did wonderful feats which an unlearned man could never have done."[29]

Not only were all branches of learning undervalued in England, the lawyer complained, but those with good ideas received insufficient encouragement to put them into action. Instead, they were plagued with regulations, laws, and restrictions. These constraints constituted the second factor that contributed to England's problems, and in this regard contrasted unfavorably with "Venice, that most flourishing city at these days of all Europe." In Venice, strangers were welcomed, innovation was encouraged, and skill was recognized. "They reward and cherish every man that brings in any new art or mystery whereby the people might be set to work," the lawyer explained, as well as those who "bring some treasure or other commodity into the country." Venetians recognized that not every clever person was a native, so "if they may hear of any cunning craftsmen in any faculty, they will find the means to allure him to dwell in their city."[30] An ideal commonwealth balanced its need to protect the rights of citizens with its need for invention and innovation.

The Billingsley-Dee Euclid was just such an innovation, one that two Londoners brought forward to benefit the commonwealth of England by presenting ancient theories and fresh, practical ideas to "good and pregnant English wits" for their "furtherance in virtuous knowledge."[31] Dee was sure that "no man . . . will open his mouth against this enterprise," and he was largely right. By "setting forth profitable arts to English men, in the English tongue," Dee had silenced many, though not all, of the critics who saw mathematical literacy as something beyond the capabilities of ordinary Londoners. More important, his preface had transformed mathematics from a wicked and dangerous branch of learning to one that was profitable both for individual citizens and for the state in which they lived. This provided more popular vernacular authors with ample ammunition as they strove to expand mathematical literacy in the City. As the Billingsley-Dee edition of Euclid hit the bookshops and passed into the hands of readers, it prompted London's mathematical educators to think creatively about how mathematics could be taught in the City, and how it could be even more successfully marketed as a valuable tool in their urban world of commerce and craft.

EDUCATING ICARUS: MATHEMATICS AND A "PROMPTITUDE OF WIT"

Cyprian Lucar (1544–1611?), the son of a wealthy London merchant, was just the kind of Londoner likely to splurge on a copy of the Billingsley-Dee Euclid when it hit the bookstalls in 1570. Lucar was raised in a house that valued mathematical literacy and where even the women of the family knew mathematics. When his father, Emmanuel, erected a monument to his first wife, Elizabeth Withypool, in their parish church a few decades after her death, the inscription praised her feminine talents with the needle, her modesty, and her devotion to Holy Scripture. While these sentiments were standard in most funerary monuments for women, Emmanuel Lucar also left a permanent memento of his wife's other, less traditional, abilities: she could "speak of algorism, or accounts, in every fashion."[32] These may not have been conventional wifely talents, but they were highly desirable in a merchant household like the Lucars', where the complications of international trade — different currencies, different measurement systems, and different methods of accounting were used in each transaction — required a firm grasp of basic arithmetic and even algebra.

When Dee sounded the call in his preface for students to take up the study of

mathematics, the first to respond were the schoolmasters and mathematical authors, followed by the wealthy and well educated like Lucar. None of these
groups represented a particularly hard sell, since they already appreciated the
many benefits associated with the discipline. Far more daunting was the task of
convincing craftsmen and lesser merchants that they needed to train their apprentices in mathematics. While it was true that carpenters did *use* geometry in
their work, and merchants depended on arithmetic, few would have been able to
articulate geometrical theorems or define algebra. Yet this situation had not hindered them from constructing buildings, crafting items, or trading with foreign
countries. Artisanal understanding traditionally came from doing and manipulating, and had only lately been supplemented by book learning as literacy rates
rose throughout Europe and print technology made books more available to
consumers.[33] After the Billingsley-Dee edition of the *Elements* was published in
1570, however, Elizabethan teachers, along with mathematical authors and instrument makers, began to clarify the benefits that mathematical literacy could
bring to merchants, masons, and other citizens and to market that literacy at a
price that was within the reach of the average consumer.

Specifically, the mathematical cognoscenti focused on how the study of mathematics sharpened problem-solving skills and enabled students to find accurate
and innovative solutions. Once again, the popular story of Daedalus and Icarus
proved instructive. Daedalus ("that excellent geometrician," as William Cuningham had described him) first constructed his fateful wings in order to escape
from prison. Trapped in a tower after losing the favor of King Minos, Daedalus
"prepared wings (through Science's aid)" and used them to flee custody.[34]
Daedalus's mechanical ingenuity and mathematical knowledge enabled him to
develop a quick and effective solution to a problem with which he had no experience. His son Icarus, more interested in playing at his father's feet than learning about the construction of the wings and their quirks, took the wings beyond
their limits and fell to his death. The lesson was clear and could make even the
most cautious merchant or craftsman sit up and take notice: people without
mathematical knowledge could figure out how things worked and use mechanical items, but might suffer disastrous consequences. Those blessed with real
mathematical literacy, however, would succeed and thrive.

Increasingly, mathematical educators and authors argued that should an artisan, statesman, or merchant have to face the unexpected, the discipline of mathematics was an indispensable tool. And there was a great deal of the unexpected
in the Elizabethan period: new worlds, new trades and industries, and a new political and social order as medieval powers proved vulnerable to the challenges of
the times. The theme of easy problem solving began to crop up frequently in

mathematical treatises and curricula as a way to mitigate the difficulties. One author argued that mathematics was "the best whetstone, or sharpening of the wit of every man that ever was invented." Studying mathematics helped to foster a "promptitude of wit" that facilitated all kinds of business transactions and inventive thinking. Someone with mathematical literacy "shall the sooner conceive any matter," argued William Bourne. London educators promised to produce students prepared to deal with unfamiliar and unexpected situations. Word problems positing business or technical challenges appear frequently in affordable mathematics books, each written to guide pupils through currency-exchange transactions, trade between merchants using different standards of weights and measures, and more complicated computations where accounts had to be settled. One of London's most successful math teachers, Humfrey Baker, asked students in *The well spryng of sciences:*

> Three merchants have formed a company. The first invested I know not how much, the second put in 20 pieces of cloth, and the third has invested £500. So at the end of their business, their gains amounted to £1000, whereof the first man ought to have £350, and the second must have £400. Now I demand: how much did the first man invest, and how much were the 20 pieces of cloth [worth that] were invested?[35]

This was a business transaction that might actually face a merchant or tradesman, and it was one that could not profitably be solved through trial-and-error experience.

While experience could provide a merchant with the knowledge of how many casks of wine could be laden on to a specific size of ship, or enable a carpenter to cut boards for a standard size of table, the tried and true was not always useful when venturing into the unknown. Trial and error might eventually yield a useful result, but it was costly and time-consuming. Mathematics, however, was cheap and quick, once basic principles and applications had been mastered. Humfrey Baker highlighted his "brief rules" of problem solving that would allow questions to be answered "with quicker expedition, than by the rule of three." William Bourne explained that even experienced gunners could not accurately judge the distance to a new target without pacing it out, despite the fact that "there be very true and exact ways to know . . . by geometrical perspective."[36] This was the promise of mathematics: not that carpenters needed mathematical knowledge to eke out a living, but that they could use that knowledge to compete more successfully in a tight labor market. Mathematical knowledge, these authors argued, was a crucial weapon in the arsenal of a skilled craftsman.

A growing community of teachers and mathematical authors after 1570, in order to get more students into their classrooms and more purchasers into the bookstores, turned to the message that mathematical literacy went hand in hand with expedient problem solving. London's teachers were well-known proponents of mathematics, as Humfrey Baker explained. "I find not in any nation . . . more ready knowledge in this science than there is in . . . our youth here in England, and namely in London," he wrote, crediting the "great care" of their teachers, who "have trained them up into the practice thereof."[37] While little direct evidence survives to shed light on how students and teachers interacted in the Elizabethan City, by drawing on illustrations, the popular dialogue format of textbooks, and descriptions of London's schools, we can piece together some sense of what it was like to be a student of mathematics in the period. Here I will be focusing on formal education with private tutors, in public lectures, or in grammar schools. Even though we know from examples like that of Elizabeth Withypool Lucar that important educational work took place in the home, it has not been possible to uncover much information on how girls and other young people who were not put into classroom settings (public or private) came to know mathematics.

Efforts made by London's educators and mathematical authors to promote mathematical literacy fit into an effort across western Europe to promote study of mathematics. Niccolò Tartaglia, Robert Record, Giovanni Francesco Peverone di Cuneo, and Petrus Ramus all trumpeted the benefits that mathematics could bring to young minds and prosperous nations. Florence was a pioneer of civic mathematics instruction, in part because of her reliance on banking and commerce for economic survival, and all along the Italian peninsula abacus schools flourished throughout the medieval and early modern periods. In Antwerp, another early modern European financial center, "reckoning masters" from the Schoolmasters' Guild taught young men arithmetic.[38] In London, teachers were needed who could translate difficult concepts to an audience made up of two constituencies: the sons of gentlemen with good classical educations but little hands-on knowledge of how to apply mathematical theories to use an instrument or to solve a surveying dilemma, and the children and servants of merchants who could keep accounts and construct mechanical items but were not well versed in abstract mathematical theories and concepts.

This divergence among social orders emerged after children left one of the City's "petty schools" — elementary schools for girls and boys that taught reading in the vernacular through a familiarity with key religious texts like the Lord's Prayer and catechism. Paper, ink, and quills were expensive, so many from the lowers orders of society learned to read but not write during their time in school from the ages of five to seven. After leaving petty school, some girls continued their formal education at home (especially in merchant families), while the wealthiest boys, along with some highly talented poor boys on scholarships, progressed into one of the City's grammar schools. Grammar schools trained boys in the language skills necessary for admission into one of the universities; the curriculum emphasized Latin, Greek, and penmanship over subjects like mathematics. Some London grammar schools, most notably St. Paul's and Merchant Taylors', had more innovative curricula that included arithmetic and music. Richard Mulcaster, successively master of Merchant Taylors' and St. Paul's, felt it was necessary to train boys in arithmetic and geometry, for example. And the grammar school established by the parishioners of St. Olave's in Southwark was the first to include accounting in its curriculum in 1561. Few grammar schools went beyond the teaching of classical languages and writing to delve into mathematics, however.[39]

Some Elizabethans did call out for greater mathematical education, especially for the sons of gentlemen. Sir Humphrey Gilbert, Walter Raleigh's half brother, proposed founding an academy for the queen's wards and others from the nobility and gentry to train them in the arts and sciences beneficial to the commonwealth.[40] Traditional subjects like military policy, rhetoric, law, and language would be taught alongside natural philosophy and mathematics. Two mathematicians would teach arithmetic and geometry on alternate days, with the geometry instruction dedicated to practical applications such as "embattlings, fortifications, and matters of war with the practice of artillery." Gilbert advocated a similar approach to the more esoteric mathematical disciplines of cosmography and astronomy, which were to be taught alternately with navigation and shipbuilding. Another instructor would be employed to teach students how to draw maps, sea charts, and plans according to "the rules of proportion and necessary perspective and mensuration."

Elizabeth never endorsed Gilbert's ambitious plans, and it was not until another Mercer, Thomas Gresham, put up the financing for a college to educate Londoners that anything like Gilbert's academy was realized. Gresham, who also financed the construction of the Royal Exchange, set aside funds in his will to establish an institution for the education of Londoners from a variety of backgrounds, including grammar schools and the guild system. Gresham College was years in the making — Gresham made out his will in 1575, but it was not until after his wife's death in 1596 that anything was done about the college — and it was hoped that the institution would serve as a center for the exchange of all kinds of knowledge in the City. The early curriculum called for lectures in medicine, divinity, music, law, and rhetoric as well as mathematics. The first men appointed to oversee the mathematical curriculum of the college were Edward

Brerewood of Oxford, in the position of professor of astronomy, and Henry Briggs of Cambridge, as professor of geometry. But the Gresham lectures were never as popular as their founder had hoped, and in the early seventeenth century the lecture halls were often half empty.[41]

If Elizabethan Londoners were becoming more mathematically literate throughout the period, most were doing so by reading books or seeking the guidance of one of the City's private teachers rather than by sitting through public lectures. There were private mathematical teachers for hire in Elizabethan London — ambitious young men who possessed either practical or university training in one of the mathematical disciplines or their applications: arithmetic, geometry, accounting, and the use of mathematic instruments. Shortly after the reign of Elizabeth, when Sir George Buck surveyed the educational options available in the City he called "the third university of England," he mentioned that instruction was readily available in arithmetic, astronomy, geometry, mathematics, hydrography, geography, navigation, cosmography, and the military sciences.[42] Given this amount of competition, it is not surprising that private mathematics teachers, like medical practitioners, advertised their services.

Baker was not the only well-known mathematics teacher in the City who used the power of print to gain more students, status, and visibility. John Mellis, a schoolmaster of Southwark, increased his London profile by editing Robert Record's venerable arithmetic textbook, *The ground of artes*, which had already been reissued with additions by John Dee. In 1582 Mellis "beautified" the work with "some new rules and necessary additions," specifically a new third section dedicated to "rules of practice, abridged into a briefer method." Mellis closely followed in Dee's footsteps, echoing his references to commonwealth philosophy and proclaiming his desire to give students a command of "the infallible principles and brief practices" of mathematics so that they could become "faithful and serviceable to their masters in good affairs, and . . . good members of a common wealth." Where Mellis differed from Dee, however, was in his emphasis on brevity of instruction and the low cost of his publications. As editor of Record's work, he stopped short of making so many improvements that the book would "rise too thick or grow too dear."[43]

Though both Mellis and Dee were interested in the practical applications of mathematical knowledge, Mellis proved far more pragmatic. Knowing his audience's predilection for arithmetic, Mellis published one of the first English-language books to focus on double-entry bookkeeping, *A briefe instruction and maner how to keepe bookes of accompts* (1588). Mellis revised an English translation by an early sixteenth-century mathematics teacher, Hugh Oldcastle, of Luca Pacioli's Italian work on the subject. No copies of Oldcastle's 1543 work survive, but Mellis described how he kept the translation by his side for thirty years "for his own private knowledge and furtherance." Even in accounting, brevity and speed were crucial marketing concepts for Mellis. Every good merchant, Mellis argued, should "be prompt and ready in his accounts and reckonings" and "have the cunning and feat of arithmetic, with pen or counters." Mellis promised to teach any merchants lacking these vital skills, or their children and servants, at his house near Battle Bridge in St. Olave's parish in Southwark, where "they shall find me ready to accomplish their desire in as short a time as may be."[44]

Mellis and Baker worked out of their homes and advertised their services through printed books and other media, but the push for mathematical literacy in London took a different turn in 1588 when the Londoner Thomas Hood (d. 1620) became the City's first public lecturer in mathematics. Due to the pressing demand of training soldiers for anticipated military struggles with Spain following the defeat of the Armada, and with the financial support of merchant and customs official Thomas Smyth, the City agreed that it was time to give soldiers a better understanding of mathematics so that they could form a more effective fighting force. Hood spoke at his sponsor's house on arithmetic, the use of surveying instruments, and astronomy until attentive crowds forced them to move into the more spacious surrounds of the Staplers' Chapel in Leadenhall Market. Like Dee before him, Hood emphasized the benefits mathematics education could bring to the commonwealth and tried to steer his listeners away from a focus on mechanical toys, to a focus on mathematical applications that would be of "some use, or some commodity fit for a commonwealth." Smyth published Hood's inaugural address for those who could not attend the first lecture, which was less a lesson in mathematics than yet another rousing call for mathematical education. Scholars have debated whether Hood's public lectures were a success, and circumstantial evidence suggests that they were not well attended. Eventually, the City's mathematics lecture ceased entirely, prompting George Buck to lament in 1615 that "in the time of Queen Elizabeth there was a lecture of the chief mathematical sciences . . . read in the chapel of Leaden Hall, but now it is discontinued."[45]

While Mellis and Baker provided a background in basic arithmetic and geometry using familiar texts like Robert Record's *Ground of artes*, and Hood did his part to make the commonwealth message of Dee's preface more widely known, mathematical authors began to write fresh titles to supplement these tried-and-true vernacular mathematical classics. The new vernacular mathematical books continued to play an important role in rising rates of mathematical literacy, since they stood in as teachers for students who lacked the time and financial resources to attend school. Even an educator like John Mellis had to admit that his mathematical knowledge was founded on the careful reading of Robert Record's works before he attended arithmetic lectures at grammar school in Cambridge.[46] Like their predecessors, the authors of many of these late Elizabethan books wrote in dialogue form, which made the act of reading seem more like an act of learning, where a master instructed his pupils by asking questions and correcting their mistakes.

Like educators, mathematical authors who wanted to make their works of interest to urban consumers needed to go beyond simply pointing out that arithmetic and geometry had hundreds of applications. In mathematical works published after the appearance of the Billingsley-Dee Euclid, authors began to shape their prefaces and content to portray mathematics as a time-saving set of skills ideally suited to complex problem solving. The group of authors who emerged to prominence during the period from 1570 to 1588 — including William Bourne, William Borough, Thomas Digges, and Robert Norman — relied heavily on this message and included discussions of instruments, which they presented as tools that would help readers solve problems even more quickly and accurately. Despite their varied backgrounds and professional trajectories — Bourne was a local political official and author, Borough a navigator and naval official, Digges a Cambridge-educated mathematical author and engineer, and Norman an instrument maker — they all included instruction on how to use mathematic instruments.

William Bourne, who had practical experience with applied mathematics thanks to his stint as a gunner, turned his hand to writing on navigational matters after publishing some popular almanacs. With the exception of Thomas Buckminster, Bourne was the most published mathematical author in the Elizabethan period, with seven different titles to his credit and fourteen editions in print. Despite this prolific record of publication, not all of his work was written for a wide audience. Bourne knew the queen's chief minister, William Cecil, and the two men spoke about how one could set about "measuring the mold of a ship." These conversations encouraged Bourne to write a book (now lost) on statics that "has been profitable, and helped the capacities, both of some seamen, and also ship carpenters." Bourne also composed treatises which remained in manuscript on both optical glasses and the buoyancy of water for Cecil, as well as a short piece on naval defense.[47] Among his printed works, the most widely available was *A regiment for the sea* (1574), which was reprinted six times during Elizabeth's reign. Bourne's purpose was to write a more practical guide to navigation than was otherwise available to English mariners, in response to Martín Cortés's treatise for Spanish mariners, *The arte of navigation* (1561). Cortés's work had been translated into the vernacular by Richard Eden, but Bourne still felt that there was a need for a text written by Englishmen for Englishmen. Bourne's other published works on mathematics include a book on geography, *A booke called the treasure for traveilers* (1578); a book of engineering ideas entitled *Inventions or devises* (1578); and a handbook for gunners, *The arte of shooting in great ordnaunce* (1587).

In all of these works, published and unpublished, Bourne combined practical, hands-on experience in instruments, navigation, and ballistics with a theoretical understanding of their underlying mathematical principles. Despite his assertions that he was an "unlearned and simple" man who was "altogether destitute both of knowledge and learning," Bourne clearly had read the works of John Dee and Leonard Digges.[48] His great gift was in blending theory and practical application into works that could be easily read and understood by craftsmen, sailors, and gunners. Though it is doubtful that hundreds of men went to sea with copies of Bourne's works in their pocket, the frequent publication of his treatises suggests that they were in demand on shore and in London schoolrooms.

A closer look at the dedication and preface to the reader in Bourne's *Booke called the treasure for traveilers* shows how the author integrated the practical, theoretical, and instrumental aspects of mathematics for his intended audience. Bourne dedicated the book to the queen's master of ordnance, Sir William Winter, a well-educated Londoner who enjoyed all sorts of technical and popular science pursuits, including metallurgy and mechanics. In the dedication Bourne quickly sketched out the contents and emphasized the work's intellectual soundness and range: it was divided into five books, the first concerned with geometrical perspective, the second with cosmography, the third with general geometry, the fourth with statics, and the fifth with natural philosophy. In the preface to the ordinary reader — who Bourne hoped would be a craftsman, sailor, or gunner — he characterized his book's contents differently. "The first book contains the particular conclusions of the scale quadrant or astrolabe," Bourne wrote, as well as the use of the cross-staff and the horizontal or flat sphere, "whereby to draw or take the plat of any country." In the second book, Bourne promised to show "that if you do know the longitude and latitude of any place," you could figure out how far that place was from your own position, and in what cardinal direction. Book three was dedicated to "the measuring of superficials and solid bodies, and how to augment them, or diminish them . . . whether . . . it be the tonnage of any ship, or the bigness of any cask." The fourth book taught readers how to estimate the weight of any ship afloat, while the fifth explained the natural causes of rocks and sandbars that suddenly appeared in the ocean to the detriment of navigators and merchants.[49]

Bourne's dedication and preface were written for two different audiences: the dedication was for highly educated men like Winter who were already conversant with mathematical ideas; the preface was aimed at a wider popular audience intrigued by the problem-solving potential of mathematics and fascinated by instruments and other devices that could shoulder some of the burdens of calculation. Greater theoretical mastery of mathematics was essentially being replaced with a different skill set based on familiarity with and an understanding of instruments. This expertise depended upon "exact trial, and perfect experiments" rather than on a careful reading of Robert Record, thereby switching the emphasis in mathematical books from acts of reading and calculating to acts of doing and manipulating. "Notwithstanding the learned in those [mathematical] sciences . . . in their studies amongst their books," Robert Norman wrote, "there are in this land divers mechanicians, that in their several faculties and professions, have the use of those arts at their fingers ends, and can apply them to their several purposes as effectually and more readily."[50]

As these educators and authors promoted vernacular mathematical literacy in the City, they began to emphasize utility over theoretical mastery. Whereas John Dee had provided a grand overview of the applications for mathematics in his preface to Euclid's *Elements*, he was not as persuasive as his successors when it came to explaining how and why such mathematical skills mattered. Bourne, Borough, Thomas Digges, and Norman, by paying acute attention to how mathematics provided quick and accurate methods to solve unforeseeable or unusual problems, underscored the practical benefits of mathematics for their readers. They also turned Londoners more resolutely toward mathematical instruments and fostered a new appreciation for those "mechanicians" who knew how to use and construct them.

DISPLAYING DAEDALUS: INSTRUMENTATION AND MATHEMATICAL LITERACY

Elizabethans had long been interested in instruments, engines, and machines that might be used to solve — or at least mitigate — the problems of everyday life, as well as mechanical items, like Dee's beetle, that astonished and delighted viewers. In the autumn of 1588, a few years before Baker printed his broadside, Londoners had the opportunity to troop through the Guildhall and see a perpetual motion machine designed by a foreign immigrant, Henrick Johnson. City officials agreed to keep London's main governmental building open to the public for an entire day at the request of the queen's chamberlain, Sir Thomas Heneage, who persuaded the aldermen to let Johnson put his machine in the center of civic affairs so that "such inhabitants of this city [and] others who shall be willing to see the same" could witness the display. Basking in the military triumph of the summer's successful defeat of the Spanish Armada, London's residents were only too willing to go out and view this latest example of the City's instrumental and mechanical prowess.[51]

Even after Johnson's machine was carted out of the City's center, Londoners had ample opportunities to gasp at examples of instrumental and mechanical ingenuity. Elizabethan London's western end — the area inside the walls that included the publishing houses near St. Paul's, the old ecclesiastical liberty of the Blackfriars, and the suburban parishes along Fleet Street and the Strand — was known for its instrument shops. Walking north from the Blackfriars toward the cathedral precincts, then west through Ludgate onto Fleet Street until it turned into the Strand, and continuing on to the parish of St. Martin's near Charing Cross, one would have been able to buy clocks, mechanical devices, and mathematical instruments from French, Flemish, and English makers. Other instrumental neighborhoods were springing up all over town. In the parish of St. Botolph Aldgate you could purchase compasses and clocks from Thomas Hearne (fl. 1592) and Israel Francis (fl. 1597), while John Read (fl. 1582–1610) and Christopher Pane (fl. 1584–1612) sold their mathematical instruments out of an establishment on Hosier Lane in St. Sepulchre.

London was the center of instrumentation in Elizabethan England.[52] The same influx of foreigners that brought James Cole and Valentine Russwurin to the City brought engravers, clockmakers, and other skilled metalworkers as well. In London, immigrant and English workmen manufactured beautiful and costly instruments for wealthy merchants and aristocrats, and fixed London parish clocks and sundials. In their shops Elizabethans saw displays of craftsmanship that would have made Daedalus proud. Buyers could examine the intricacy of the small pocket calendars called *compendia*, inscribed in brass and the electronic organizers of their day. Even important instrument makers like Humfrey Cole made these popular little items. No longer did English buyers need to send abroad for rarer devices, like a musical chamber clock that played a different tune each time the quarter hour was struck. Nicholas Vallin (fl. 1577–1603), an immigrant from Lille in Flanders, could make one of these items in his City workshop.[53] And not all instruments were expensive: astrolabes, quadrants, and

other devices to measure and map out the heavens and the earth could be purchased in fine metals, or in much cheaper wooden versions that were within
reach of the average skilled craftsman.

One problem facing scholars tracing the development of instrumental literacy in London is how little we know about Elizabethan instruments and their
makers. Scholars have pored over annotations and passing references in printed
books, but linking those few names to the often unsigned instruments that have
survived has made the task of fleshing out the lives and work of Elizabethan instrument makers difficult. The best-known instrument makers, such as Thomas
Geminus, Humfrey Cole, Charles Whitwell, Augustine Ryther, and James Kynvin, garner attention because they signed their works and had wealthy patrons
who may have owned instruments as precious curiosities rather than indispensable craft tools. Such items tend to have higher rates of preservation than instruments that were used daily by surveyors. This situation has been complicated
further because there was no single guild to which instrument makers belonged
before the founding of the Clockmakers' Company in 1631. As a result, tracing
workshop genealogies from masters through apprentices has been successful
only in a handful of cases.[54] Despite these difficulties, however, it is possible to
draw a portrait of London's instrument makers that extends beyond Geminus,
Cole, Whitwell, Ryther, and Kynvin and indicates how prominent instruments
and instrument makers were in their own time.

Instrument makers were powerfully associated with certain places in Elizabethan London, specifically the neighborhoods in which they lived and the
parish churches in which some of them worked. While the smaller instruments
that now ornament museum shelves did indeed circulate through the hands of
craftsmen, customers, and courtly patrons in the period, it is clear that the makers of these precious objects were clustered in well-defined and recognizable
neighborhoods scattered throughout the City. Some neighborhoods specialized
in specific kinds of instruments. In the eastern suburban parish of St. Botolph
Aldgate, the specialty was compasses, which would have found ready buyers
among the many area residents who were mariners or others involved in long-
distance shipping out of the eastern ports of the Thames. In St. Margaret Westminster, close to the royal residences of Westminster, Hampton Court, and Richmond, the instrument makers specialized in making and repairing monumental
clocks. In the main instrument-making district—a cluster of neighborhoods
which included the parishes of St. Anne Blackfriars, St. Paul's, St. Clement without Bars, St. Dunstan in the West, St. Martin in the Fields, and St. Martin le
Grand—we find engravers, instrument makers, and clockmakers who produced
all sorts of technical instruments for consumers. This range of workshop produc-

tion was also found in the many smaller neighborhoods where instruments were produced, such as the area adjacent to St. Bartholomew's Hospital and the streets around the Royal Exchange.

Some of these instrumental neighborhoods were dominated by immigrant artisans and craftsmen, and others by the English. The neighborhood surrounding the church of St. Anne Blackfriars south of St. Paul's Cathedral, for example, was home to at least twelve immigrant instrument makers during the reign of Elizabeth: Thomas Geminus, John Mary (fl. 1544–66), Michael Noway (fl. 1568–1616), Francis Roian (fl. 1572–1616), Thomas Tiball (fl. 1573–98), Laurence Dauntenay (fl. 1570), Francis Noway (fl. 1576–93), Nicholas Vallin, Michael Scara (fl. 1582), Anthony Vallin (fl. 1585–93), John Vallin (fl. 1590–1603), and Peter De Hind (fl. 1594–1608). All of these craftsmen came from France or the Netherlands, and many (like the Vallins) came with their extended families and servants. The Blackfriars instrument makers devised ingenious clocks, constructed mathematical instruments, crafted scales to weigh merchandise, and, to supplement their incomes, engraved plates for the printers in nearby St. Paul's. A number of extant examples of items made by the St. Anne Blackfriars workshops indicate the skill and ingenuity of these foreign artisans.[55]

The parish of St. Dunstan in the West, on the other hand, was dominated by English instrument makers. Richard Blunte (fl. 1580–96), John Modye (fl. 1587–1603), Thomas Brome (d. 1598), Robert Grinkin (fl. 1600–1626), and James Ilsberye (fl. 1601) were all described as instrument makers in parish records. Only two instrument makers known in the parish were French, the clockmaker Peter Dellamare (fl. 1523–67) and the instrument maker Adrian Gawnt (fl. 1567–98). And in St. Botolph Aldgate, the five known instrument makers mentioned in the parish records were English, including the instrument makers Richard Stevens (fl. 1569) and William Thomas (fl. 1589–1616), the compass makers Thomas Hearne and John White (fl. 1602–3), and the clockmaker Israel Francis.[56] Among these English instrument makers, it is common to see guild affiliations to the Blacksmiths' Company, the Goldsmiths' Company, and the Grocers' Company. The instrument maker Robert Grinkin, who specialized in making watches, was a member, and later master, of the Blacksmiths' Company, for example. Many of the most prominent instrument makers, who did work for the queen and her court, were Goldsmiths, including Humfrey Cole (1530–91) and Bartholomew Newsam (f. 1568–93). Other important makers of the period, such as Christopher Pane (fl. 1584–1612), Augustine Ryther (1550–1593), his apprentice Charles Whitwell (fl. 1582–1611), and Whitwell's apprentice Elias Allen (1558–1653) were members of the Grocers' Company.[57]

These clusters of instrument makers were sufficiently large to have made an impression on a resident or visitor to the City. Between the shops full of ingenious devices and the intricate craftsmanship that each item required, walking in the Blackfriars or along Fleet Street in St. Dunstan's would have been a feast for the eyes. But instrument makers also displayed their ingenuity through large-scale projects, especially the construction and maintenance of London's church clocks. Historians have long emphasized the role that subsidiary trades, such as engraving, played in the lives of mathematical instrument makers who might have struggled to make ends meet while making expensive instruments for a limited clientele. The Flemish engraver and instrument maker Thomas Geminus, for example, made instruments for Queen Elizabeth, engraved anatomical fugitive sheets for the press, practiced medicine, and worked as a printer. Augustine Ryther, an English instrument worker active in London during the latter part of Elizabeth's reign, engraved maps for navigational treatises along with creating instruments for Robert Dudley. And Charles Whitwell engraved the plates for William Barlow's *Navigator's supply* (1597) as well as making instruments according to the mathematical lecturer Thomas Hood's designs.[58]

For instrument makers who worked on a grander scale, like Henry VIII's clockmaker Nicholas Urseau (fl. 1532–75), London's churches provided additional revenue. Parishes wealthy enough to afford the costly construction and regular upkeep that clocks required were able to make an important contribution to the quality of life in the immediate neighborhood. The hands of the parish clock marked off the hours rain or shine, and if they struck on the hours or quarter hours they could be used to call parishioners to worship much like the more common bells that pealed throughout the City. London's parish clocks were often the economic black holes of parish revenues, seemingly in constant need of repair and maintenance. St. John Walbrook's clerk recorded annual expenses related to the keeping of the clock that rivaled the parish's health-services costs, including a steady stream of expenses related to wire, rope, wheels, staples, and iron pins. Even the healthiest clocks needed a regular supply of oil to grease the wheels, and the springs periodically wore out and had to be replaced. Quarterly visits from a clockmaker could open up Pandora's box, as the vestry of St. Antholin Budge Row discovered in 1596–97 when the clock required a new mechanism, as well as wire, a new padlock for the clock's door, oil, and a carpenter to put the mechanism securely in place. And St. Botolph Bishopsgate, one of the City's poorer parishes, was overrun by workers and shelled out more than twenty-two pounds when it needed a new clock in 1580–81. A new tower had to be constructed to hold the clock, a new bell founded, weights cast, the clock made, new ropes purchased, food and drink supplied to the workmen, and a glass window placed in the tower to cast some light on the mechanism should it break down.[59]

A number of entrepreneurial clockmakers stepped in to meet parish needs for clock maintenance and repair, including Nicholas Urseau, John de Mullin (fl. 1539–76), Bruce Awsten (fl. 1547–72), and John Harvey (fl. 1594–99). The most enterprising of all the large-scale clockmakers was Peter Medcalfe (fl. 1577–87), who often crossed the river from his home in St. Olave Southwark to maintain and construct parish clocks. In the space of a few years he made new clocks for St. Giles Cripplegate, St. Lawrence Jewry, St. Margaret Lothbury, and St. Peter Cornhill.[60] Contracts for these new pieces of craftsmanship were as lengthy and detailed as any artistic commission familiar to students of the Renaissance. The vestry of St. Margaret Lothbury gave Medcalfe only seven weeks to "erect and set up a good and perfect clock, well and strongly wrought with a fair dial towards the street, with all the furniture belonging to it." St. Lawrence Jewry required "painting and gilding" on their new timepiece. Medcalfe also received stipends for maintenance work on the existing clocks in St. Botolph Aldgate, St. Andrew Hubbard, and St. Antholin Budge Row.[61] He sold St. Peter Cornhill a new clock at a discounted price of four pounds, provided he could keep the old one for parts; charged St. Margaret Lothbury the sizeable sum of nine pounds for its new clock; and received an annual retainer of four shillings for keeping his eye on the rapidly disintegrating clock in St. Andrew Hubbard's church, in addition to fees he was paid for specific repairs.[62]

As the contract between the parish of St. Margaret Lothbury and Medcalfe made clear, these costly instruments were intended to be on display and visible to both parishioners and passersby. When St. Botolph Aldersgate found that its clock was in need of repair in 1588–89, the vestry decided to move the clock's face "into the street that it may be better discerned than heretofore it has been." The clockmaker was instructed to set the dial "on the north side of the church into the street five feet at the least or six foot beyond the wall," just like the clock "in the Poultry."[63] Clocks with eye-catching ornamentation were also highly desirable; when the parishioners of St. Mary Woolnoth had to repair their chiming clock, they paid a member of the Painter-Stainers' Company to paint and gild the dial as well as the "two balls and the vane over it."[64]

While London's parish clocks were highly visible and sometimes spectacular examples of instrumentation, it was through smaller instruments that mathematical theory and practice became even more fully linked to efficient problem solving in late Elizabethan London. Greater accuracy and expediency could stem from the use of instruments, but the reliability of conclusions based on them was only as good as the quality of their construction. In 1572 Sir Thomas Smith had to send to France for a case of reliable instruments including compasses, squares, and rulers, but within a decade he could purchase these same items, in like quality, in London. Even with reliable instruments, students needed to be fully versed in their use, and in the many different factors that could influence their accuracy. Only then would Englishmen be able to evaluate the excellence of instruments made by others and also be able to correct their faults and devise new ones. When Thomas Smith received his foreign instruments, for example, he confessed to his friend Francis Walsingham that he had not had the "leisure to understand them all" but hoped soon to understand "their property and use."[65]

As London emerged as an instrumental city during the final decades of the sixteenth century through this combination of large- and small-scale instruments, experienced authors and educators like Humfrey Baker promised students that they would be trained in their use.[66] Knowing how to use an instrument — whether made of gold, brass, wood, or paper — was of great appeal to Londoners. Like a computer today, such instruments could be used by consumers even if their design and workings were not completely understood, and they facilitated the quick problem solving that was such a popular theme among mathematical authors. In the late Elizabethan City, mathematical instruments may have been the original "black boxes," a term coined in the 1940s to explain pieces of equipment that World War II aviators were content to use because they performed an important function, even if they did not fully understand their construction or operation. Think about the gradual switch from sundials to mechanical clocks that was taking place in the period, for example. The practice of watching the shadows cast by the sun on the dial's face in order to track its movements across the heavens was being replaced by watching the hands of a clock move across a dial with no visible connection to the sun's passage.

But the increasing power of instruments in the City made many uneasy, and the jury was decidedly out regarding their relationship to mathematical knowledge and literacy. The craftsman or mechanic who used instruments without a full mastery of mathematical theory became a stock figure in the cautionary tales told by learned mathematical authors and teachers to distinguish themselves from the instrument makers who were competing for the attention of students and consumers.[67] "I do know divers that will have instruments, and yet be utterly void of the uses of them," complained William Bourne, "for . . . if the person does not consider all things with him and against him, he . . . [is] apt to commit error." People who owned instruments, but lacked the understanding of their underlying principles and use, exercised nothing but their own vanity, wrote

William Borough. "I wish all seamen and travelers, that desire to be cunning in their profession, first to seek knowledge in arithmetic and geometry," he urged, "which are the grounds of all science and certain arts, of the which there are written in our English tongue sufficient [works] for an industrious and willing mind to attain to great perfection."[68] While men like Borough and Bourne argued that instrumental literacy should follow on the heels of mathematical literacy, the popularity of instruments both inside the classroom and out made this progression from theory to practice hard to enforce — especially in London, where instruments were readily available.

While some Londoners would have agreed wholeheartedly with Bourne and Borough on the value of traditional mathematical education, others would have found compelling the Elizabethan author and gadfly Gabriel Harvey's observation that "scholars have the books, and practitioners the learning." An assiduous scribbler, Harvey made extensive annotations in many of his books that suggest how intertwined mathematical literacy and instrumental literacy had become. Around 1590 he entered a long list of mathematical practitioners in his copy of Luca Gaurico's astrological guide to the personalities of the rich and famous, the *Tractatus Astrologicus* (1552).[69] Significantly, Harvey's list was headed not by the theoretical giants of mathematics but by more humble, vernacular figures — many of whom could boast a great deal of hands-on experience with instruments. First, Harvey provided the names of five "expert surveyors" who were also "fine geometricians": Richard Benese, Thomas Digges, John Blagrave, Cyprian Lucar, and Valentine Leigh. Harvey singled out Digges, Blagrave, and Lucar for being "the greatest and most commended" among London's "mathematical practitioners and polymechanists." His characterization of the work of these five figures does not invariably conform to their subsequent reputations within the history of science. Digges, for example, is remarkable in Harvey's world not as an early proponent of Copernicanism but as a surveyor and geometer, as is the leading instrumentalist John Blagrave. Only after praising these individuals did Harvey turn his attention to three theoretical mathematicians — John Dee, Thomas Hariot, and Edward Wright — to admire their "cunning points, profound conclusions, and subtle experiments in geometry, astronomy, perspective, geography, navigation, and all finest mathematical operations." Harvey considered Henry Billingsley the best writer on arithmetic and geometry and William Borough the most knowledgeable about mathematical navigation. Harvey even singled out some up-and-coming mathematicians who were just beginning "to carry credit" among those in the know. These included the astrologer Christopher Heydon, the mathematical writer Thomas Blundeville, the teacher and instrumentation expert Thomas Hood, the pioneer of magnetism Robert Norman, and Harvey's fellow Cambridge alumnus John Fletcher. In another volume, Harvey recorded his fondness for the work of an instrument maker recommended to him by John Blagrave: James Kynvin, "a fine workman, and my kind friend."[70]

Harvey's perplexing jumble of teachers and surveyors, geometers and instrument makers, writers and navigators indicates that by 1590 Londoners were finding it difficult to separate a theoretical understanding of mathematics from a practical experience with instruments. This convergence can be attributed in part to the work of four men on Harvey's list: Thomas Hood, John Blagrave (d. 1611), Cyprian Lucar, and Edward Wright (d. 1615). Together they helped to make mathematical literacy synonymous with instrumental literacy in the late Elizabethan period. A teacher who made instruments, an instrument maker who liked writing books, a lawyer who enjoyed land surveying, and a naval man who became a teacher, each came from a very different background but all were devoted to the theoretical, as well as the practical, side of mathematical literacy. Together they participated in a growing community of authors and educators who worked in partnership with publishers and instrument makers to produce new kinds of books, design new instruments, and create new forums for the exchange of mathematical ideas.

While historians have noticed this partnership between mathematical authors and instrument makers in the late Elizabethan period, less attention has been paid to the pedagogical focus of much of their work and their emphasis on the value of hands-on training in how to use instruments. The physical manipulation and use of instruments, some teachers argued, provided a compelling way to make mathematics lessons stick for forgetful students and made unwilling students more eager to learn the difficult subject. John Blagrave promised that his instrument, dubbed the "mathematical jewel" for its ability to replace a host of other devices, would actually lead the reader on a "direct pathway (from the first step to the last) through the whole arts of astronomy, cosmography, geography, topography, [and] navigation" with "great and incredible speed, plainness, facility, and pleasure." Other authors argued that the regular use of instruments helped make difficult concepts easier to master. "But if anything herein shall seem hard to be learned," Cyprian Lucar told his readers, "remember that use makes mastery."[71]

The value of a physical manipulation of mathematical instruments and objects in mathematical education had already crept into the Billingsley-Dee edition of Euclid, which included a number of pop-up features for the reader to construct in order to help transform the book's lessons into three-dimensional paper instruments (Figures 3.3 and 3.4). Many Elizabethan mathematical books had instruments that could be assembled from paper cutouts on their pages. Thomas Hood took these pedagogical examples to heart and in 1597 constructed a vellum instrument from four diagrams that illustrated the theoretical and practical aspects of astrology. Hood found a way, through the manipulation of ingenious revolving gears and overlays mounted on vellum and pinned together, to illustrate in one view the relationship between the planets, the signs of the zodiac, and the parts of the human body they governed (Figure 3.5). Hood's instrument, which appears to have been made as a teaching aid, made it possible for a student to quickly calculate the relationships between the midheaven, the twelve astrological houses, and the fixed stars or constellations.[72] Much like a modern PowerPoint or overhead projector transparency, Hood's instrument was a pedagogical display intended to facilitate efficient and effective education by encouraging his students to actually manipulate an instrument.

The audience for these increasingly technical publications that taught the use of instruments (as well as the mathematics behind them) was clearly outlined in many book prefaces. Cyprian Lucar intended his books "to profit surveyors, land-meters, landlords, tenants, buyers and sellers of land, woods, and trees, travelers, gunners, men of war, builders, and seamen." In *A treatise named Lucar Appendix* (1588) he explained that gunners should be "skillful in arithmetic and geometry" so that they could "measure heights, depths, breadths, and lengths, and . . . draw the plat of any piece of ground." John Blagrave's *Mathematical jewel* (1585) was addressed to "gentlemen and others desirous of speculative knowledge and private practice," as well as to "navigators and travelers." And Thomas Hood's *The use of the two mathematical instrumentes, the crosse staffe, and the Jacobes staffe* (1596) was written to help the "mariner, and all such as are to deal in astronomical matters" and to "surveyors, to take the length, height, depth, or breadth of any thing measurable."[73]

Given their projected audience, it is hardly surprising that mathematics books were fast becoming instrument manuals, with step-by-step instructions on how to manipulate an instrument to solve problems. The emphasis on instrumental literacy gave birth in turn to technical writing, and diagrams and charts began to supplant textual explanations. Cyprian Lucar's instructions to his readers on how to establish whether the ground was level before shooting off a piece of ordnance is entirely dependent on the diagram of a gunner, with a pair of compasses in one hand and quadrant before him (Figure 3.6). The move toward greater technical precision did not always result in the most readable prose, however, and the frequent references by late Elizabethan authors to the plain and easy style of their explanations is often hard to fathom given the intricacy of their books. Thomas Hood's books are indicative of this trend, but they also offer up

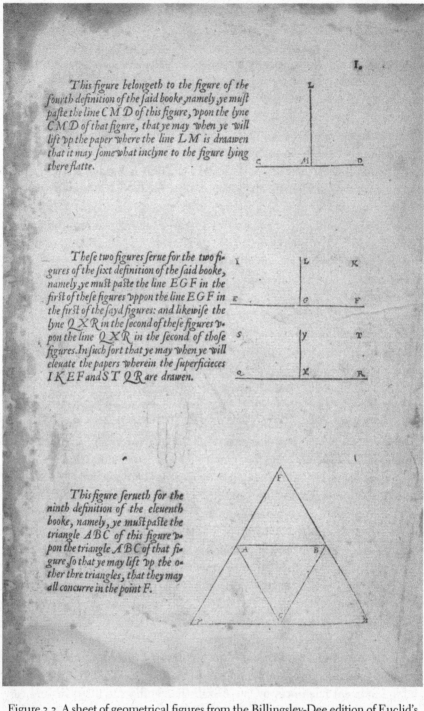

This figure belongeth to the figure of the fourth definition of the said booke, namely, ye must paste the line C M D of this figure, vpon the lyne C M D of that figure, that ye may when ye will lift vp the paper where the line L M is drawen that it may somewhat inclyne to the figure lying there flatte.

These two figures serue for the two figures of the sixt definition of the said booke, namely, ye must paste the line E G F in the first of these figures vppon the line E G F in the first of the sayd figures: and likewise the lyne Q X R in the second of these figures vpon the line Q X R in the second of those figures. In such sort that ye may when ye will eleuate the papers wherein the superficieces I K E F and S T Q R are drawen.

This figure serueth for the ninth definition of the eleuenth booke, namely, ye must paste the triangle A B C of this figure vpon the triangle A B C of that figure, so that ye may lift vp the other thre triangles, that they may all concurre in the point F.

Figure 3.3. A sheet of geometrical figures from the Billingsley-Dee edition of Euclid's *Elements*, which the buyer was instructed to cut out and paste into the appropriate chapters, thereby making a "paper instrument." In this way students of mathematics could begin the process of linking mathematical theory with hands-on applications. Reproduced with the permission of the Henry E. Huntington Library.

ward narower and narower, at length ende their angles (at the heighth or toppe therof) in one point. So all their angles there ioyned together, make a solide angle. And for the better sight thereof, I haue set here a figure wherby ye shall more easily conceiue it, the base of the figure is a triangle, namely, A B C, if on euery side of the triangle A B C, ye rayse vp a triangle, as vpon the side A B, ye raise vp the triangle A F B, and vpon the side A C the triangle A F C, and vpon the side B C, the triangle B F C, and so bowing the triangles raised vp, that their toppes, namely, the pointes F meete and ioyne together in one point, ye shal easily and plainly see how these three superficiall angles A F B B F C, C F A, ioyne and close together, touching the one the other in the point F, and so make a solide angle.

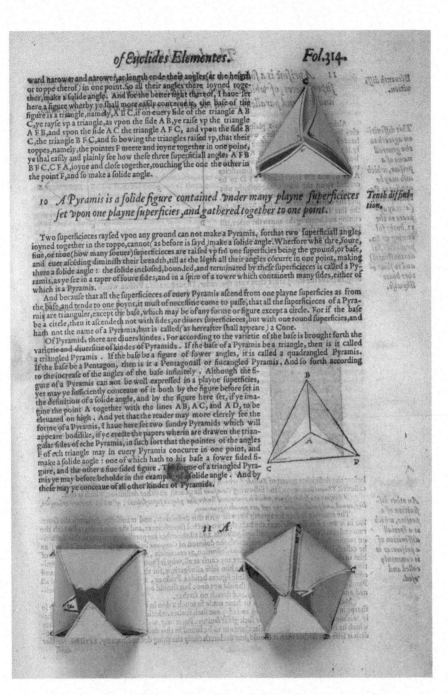

10 *A Pyramis is a solide figure contained vnder many playne superficieces set vpon one playne superficies, and gathered together to one point.*

Two superficieces raysed vpon any ground can not make a Pyramis, forthat two superficiall angles ioyned together in the toppe, cannot (as before is sayd) make a solide angle. Wherfore whē thrée, foure, fiue, or moe (how many soeuer) superficieces are raised vp frō one superficies being the ground, or base, and euer ascéding diminish their breadth, till at the légth all their angles cōcurre in one point, making there a solide angle : the solide inclosed, bounded, and terminated by these superficieces is called a Py-ramis, as ye see in a taper of foure sides, and in a spire of a towre which containeth many sides, either of which is a Pyramis.

And becaufe that all the superficieces of euery Pyramis ascend from one playne superficies as from the base, and tende to one poynt, it must of necessitie come to passe, that all the superficieces of a Pyra-mis are trianguler, except the base, which may be of any forme or figure except a circle. For if the base be a circle, then it ascendeth not with sides, or diuers superficieces, but with one round superficies, and hath not the name of a Pyramis, but is called (as hereafter shall appeare) a Cone.

Of Pyramids, there are diuers kindes. For according to the varietie of the base is brought forth the varietie and diuersitie of kindes of Pyramids. If the base of a Pyramis be a triangle, then is it called a triangled Pyramis. If the base be a figure of fower angles, it is called a quadrangled Pyramis. If the base be a Pentagon, then is it a Pentagonall or fiuecangled Pyramis. And so forth according to the increase of the angles of the base infinitely. Although the fi-gure of a Pyramis can not be well expressed in a playne superficies, yet may ye sufficiently conceaue of it both by the figure before set in the definition of a solide angle, and by the figure here set, if ye ima-gine the point A together with the lines A B, A C, and A D, to be eleuated on high. And yet that the reader may more clerely see the forme of a Pyramis, I haue here set two sundry Pyramids which will appeare bodilike, if ye erecte the papers wherin are drawen the trian-gular sides of eche Pyramis, in such sort that the pointes of the angles F of ech triangle may in euery Pyramis concurre in one point, and make a solide angle : one of which hath to his base a fower sided fi-gure, and the other a fiue sided figure. The forme of a triangled Pyra-mis ye may before beholde in the example of a solide angle. And by these may ye conceaue of all other kindes of Pyramids.

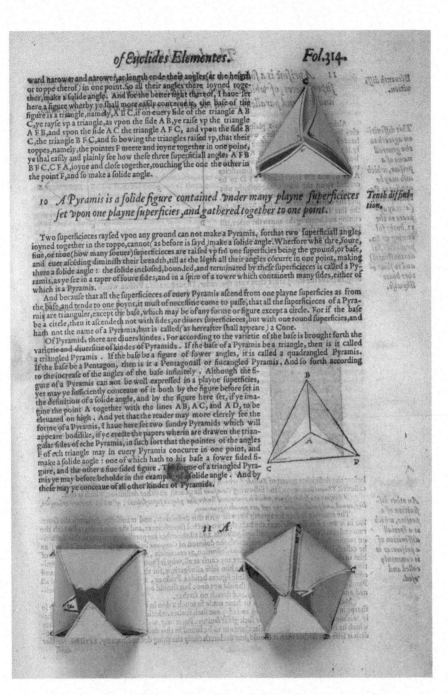

Figure 3.4. Here the sheet of figures opposite has been cut out and pasted onto the relevant figure, making a series of pop-ups that conveyed in three dimensions the nature of the geometrical solids. Reproduced with the permission of the Henry E. Huntington Library.

Figure 3.5. Thomas Hood probably made this paper instrument as a visual aid for use in the classroom. By manipulating the position of the overlays, a student could grasp the relationship between the planets, the signs of the zodiac, and the human body. Additional MS 7495, reproduced with the permission of the British Library.

clues as to why it developed. Hood's books were often printed versions of his lectures. His *Use of the celestial globe in plano* was the published version of a lecture "read last year," for example, and the printed contents would have been more lively and easy to follow if delivered orally with the benefit of such teaching props as the ones Hood himself devised.[74]

In these increasingly technical books, laborious arithmetical calculation and painstaking geometrical mensuration were being effaced by instrumentation. By "black-boxing" arithmetical or geometrical work in favor of the use of instruments, authors seemed to promise even greater expediency in problem solving, as well as greater precision. John Blagrave boasted that his combination walking stick and measuring device would measure heights and distances "exactly and familiarly without any manner of arithmetical calculation." Being amid a "buzzing of dangerous business," Blagrave admitted, was enough to "bring the best learned and skillful man out of his numbers." His new instrument would not leave purchasers "troubled with numbers in such dangerous times."[75] When Thomas Hood updated William Bourne's perennially popular *A regiment for the sea* in 1592, he provided charts of celestial declinations for his readers, which made it unnecessary for "sea-faring men" to do this work themselves. And Edward Wright encouraged the mariners who read his *Certaine errors in navigation* to forgo the common sea chart in favor of the planisphere, which "brings you to more certain truths" in navigation.[76]

The push for greater instrumental literacy in mathematical books led many students into the instrument shops, where they needed to make educated choices from a dazzling display of wares. Mathematical authors took on the challenge of educating their readers about what to look for when they went into such a shop. In some cases the advice was simply about what items should be purchased. A gunner, Cyprian Lucar wrote, should "have a ruler and a pair of compasses to measure the height and length" of the guns in his charge. Readers bewildered by the options could rely on the illustrations of instruments included in most texts, and on the consumer warnings that were included. The gunner's semicircle, Lucar cautioned, should be "made of hard, smooth, and well-seasoned wood"; he especially approved of cypress for its construction "because the cypress wood will not warp with the heat of the sun, nor with any moisture." A few City instrument makers were especially recommended because of their reliability and skill, and Lucar sent his readers to John Reynolds on Tower Street in All Hallow's Barking or John Read and Christopher Pane in Hosier Lane for geometrical tables, frames, rulers, compasses, and squares. John Blagrave, when setting forth instructions on how to use his new "Uranicall astrolabe," explained that the astrolabe should be made of metal according to a design on display at

TO perceiue whether or no a platfourme for great Ordinance, or any other peece of grounde lyeth in a perfeẛt leuell, let vs suppose that L M is the platfoume or peece of grounde vppon which great Ordinance ſhall be planted, & that I am required to tell whether or no the ſaid platfourme is plaine and leuell. For this purpoſe I place my Quadrant or Semicircle vppon a ſtaffe, or ſome other vnmooueable thing, and doe mooue it vp or downe vntill the line and plummet vpon the ſame doth hang preciſely vppon the line of leuell, that is to ſay in the Quadrant vppon the line H L, and in the Semicircle vppon the line R S : and then looking through the ſightes or channell of the ſame Quadrant or Semicircle, I doe ſee N a marke which is leuell with mine eie, and fixed in a ſtaffe or ſuch a like thing perpendicularlie erected. After this I meaſure exaẛtly the heigth of mine eye from the grounde, that is to ſay the length of the line O L, and likewiſe I meaſure the heigth of the ſaid marke N, that is to ſay the length of the line N M, and becauſe I finde by ſo doing that the ſaid line N M is equall to the line O L, and that the ſaid platforme or peece of grounde doth lie vppon the right ſide, and vppon the left ſide according as the line L M doth lie, I conclude that the ſayd grounde L M lyeth in a perfeẛt leuell. For the line L M which lyeth a long vppon that peece of ground (by the 33 propoſition of the firſt booke of *Euclide*) is equidiſtant to the line O N which goeth by the plane of the Horizon, and conſequently the ſaid peece of ground or platfourme vppon which the ſaid line L M goeth is equidiſtant (by the foureteenth propoſition of the eleuenth booke of *Euclide*) to the plane of the Horizon. But if the line N M had been longer than the line O L, I woulde haue concluded that the ſame peece of ground is more lower at M than it is at L. And contrariwiſe if the line M N had been ſhorter than the line O L, I would haue concluded that the ſame ground is more higher at M than it is at L. And after this ſort I will proceede to the right ſide, and to the left ſide, and prooue whether or no the ſaid platfourme or peece of ground doth lie rounde about according as the ſaid line L M doth lie. And ſo by this ſuppoſed worke you may learne to trie whether or no a platforme, or any other peece of ground lieth in a perfeẛt leuell.

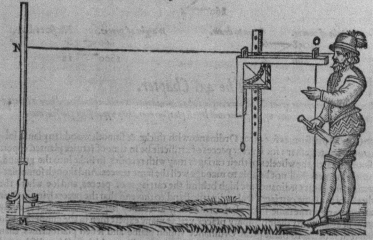

The 48 Chapter.

How Gabbions or Baskets of earth may be made vppon platfourmes in time of militarie ſeruice for the defence of Gunners: and how men vppon a platfourme or vppon the walles of a Cittie, Towne, or Fort, where no Gabbions or Baskets of earth are to ſhadow them in time of militarie ſeruice, may be ſhadowed with cannas, cables, ropes, wet ſtraw or hay, mattreſſes or ſhip ſailes.

Prepare

"Master Matt's the bookseller, dwelling at the Sign of the Plough over against St. Dunstan's church in Fleet Street." Recognizing that his preferred medium might place the astrolabe beyond the reach of the sailors and navigators he was appealing to as potential readers, Blagrave printed up paper templates for each part of the instrument so that they could be "set either on pasteboard or cuttlefish scales" to suit the needs of humbler pocketbooks.[77]

Instruments led to greater precision and accuracy in the hands of skilled users, but catastrophic errors still could be made. Some were caused by human failings, others by faults in design and craftsmanship. In his book on the surveying instrument called the Jacob's staff, for example, Thomas Hood impressed upon his readers the importance of holding the instrument correctly. Hood reported that even surveyors who "thought themselves of no small cunning in the use of the staff" sometimes did not pay attention to this simple, but crucial, detail. Gauging some problematic situations with only one's eyes led to errors. Cyprian Lucar explained how gunners could be deceived into thinking that a far-off army or ship was advancing when it was actually standing still. With the use of an instrument, however, a gunner could truly gauge an enemy's progress or retreat across a visual field. Impediments to instrumental accuracy included not only poor manufacture but also publishing errors. At the end of *The making and use of the geometricall instrument, called a sector,* Thomas Hood apologized for the many faults that had escaped his notice before the book was printed and explained that he had found "by experience that it is a harder matter to print these mathematical works truly" than it was to print other books. Years earlier, when he had published *The use of the celestial globe in plano,* Hood had taken the time to inscribe the stars and draw the constellations himself because "I would be sure to have them rightly placed."[78]

One of the most vigorous assaults against the poor manufacture and unskilled use of instruments was led by Edward Wright. He campaigned against manifold *Certaine errors in navigation* in 1599. Based on his own experiences and observations made in the application of his "mathematical studies to the use of navigation," Wright took on the widely used navigational aids of the compass, sea chart,

Figure 3.6. This illustration from Cyprian Lucar's *Treatise named Lucar Appendix* (1588) demonstrates how books on applied mathematics were in the process of becoming technical manuals. Here the text is difficult to understand without reference to the diagram below, which shows a gunner, compasses in one hand and a quadrant before him, establishing whether a piece of ground is level before shooting off ordnance. Reproduced with the permission of the British Library.

cross-staff, and tables of declinations known as regiments. While some errors made in the charts were inevitable, given "the unsteadiness of the ship, [and] the imperfection of sense and instruments," Wright wanted to do all he could to put reliable and verifiable information in the hands of England's navigators and mariners. His meticulous accounts of observations set a new standard for accuracy and implicitly encouraged replication of results by recounting details about the instruments used and the precise locations where observations were made. "These observations . . . were taken by Mr. William Borough's instrument of variation (published in his book of the variation of the compass)," Wright wrote in a chart of observational data taken in the Azores, "and by a quadrant whose semi-diameter was almost two cubits." Another table contained observations of the meridian altitudes of the sun, "taken by a quadrant of six foot [in] semi-diameter at London."[79] Wright's early attention to the precise location and instruments used to make observations make him a trailblazer of verifiable, reproducible experimental knowledge.

Wright extended the challenge of making reliable observations to other naval men in his preface to his translation of Simon Stevin's *The haven-finding art* (1599), when he described Count Maurice of Nassau's efforts to solve the navigational difficulties that went along with the variation of the compass by requiring all ships to have on board instruments capable of searching out the "declination of the magnetic needle from the true north." Wright argued in his preface, "It is certain that this knowledge cannot be found but by the experiments of divers men compared together, and . . . by divers observations a more easy way may be prepared for science (which from the particular rise up into the universal)." Other authors made similar efforts to transform their readers into makers of knowledge rather than simply consumers. John Blagrave encouraged his readers to abridge his explanation of how to use his unique instrument, the mathematical jewel, and keep it by them for ready reference. "I have always done so with any books I liked," he explained. And Cyprian Lucar encouraged readers to keep a book of calculations on the relative diameters and weights of specific ammunition for their own use and professional advancement.[80]

By the end of Elizabeth's reign, Londoners had fully embraced mathematical instruments, and many of her citizens had achieved levels of mathematical literacy that previous generations would not have dreamed possible. Small libraries of vernacular mathematics books could be obtained, and mathematical instruments—whether made out of costly brass or mounted onto simple pieces of board—could be placed in the hands of urban consumers by London craftsmen and artisans. Mathematics had been transformed from a dangerous private indulgence to a beneficial public concern by Billingsley, Dee, and the scores of mathematical educators and vernacular mathematical authors who followed in their footsteps. The City's interest in all things mathematical and mechanical would have profound implications in the next century, as compasses were perfected and telescope and microscope technology was developed. But it was in Elizabethan London that mathematical authors, educators, and instrument makers had decisively captured the imagination of the urban population by linking mathematical literacy to efficient problem-solving skills and the use of instruments. The links forged in London between theoretical mathematics, the application of those theories to matters of practical concern, and the use of instruments to achieve greater accuracy brought together students and scholars, artisans and apprentices, into fruitful exchange and collaboration.

4

"Big Science"
in Elizabethan London

In the chill winter of 1577 William Cecil, Queen Elizabeth's most trusted minister, held a clandestine meeting somewhere in London with a Venetian merchant and alchemist named Giovan Battista Agnello. The alchemist lived in the eastern end of London, in the prosperous parish of St. Helen Bishopsgate, which was home to many immigrants. There Agnello had established his alchemical reputation as someone with special expertise in the early form of metallurgical chemistry, which put experimental processes within a resolutely allegorical framework concerning the transformation and redemption of matter. Some alchemists toiled away with their potions and equipment to transform base metals like lead into gold; others saw the material transformations as of secondary importance when compared with the transformed substance's ability to perfect nature and bestow immortality on the successful alchemist. Many alchemists, including Agnello, were equally interested in both the material and the spiritual aspects of their pursuits. A devoted experimenter, Agnello had also published a book on the spiritual side of alchemy in 1566, *The Revelation of the Secret Spirit (Apocalypsis spiritus secreti).*[1]

Agnello met with Cecil to discuss a small black rock that the explorer Martin Frobisher had brought back with him from the New World, and which had passed into the hands of one of the voyage's chief investors, the merchant Michael Lok. Margery Lok, Michael's wife, alerted her husband to some alluring properties in the otherwise unpromising specimen when she flung it angrily into the fireplace after he presented it to her (rather like an early modern version of a moon rock) as a gift from an exotic, recently explored locale. In the fireplace, the stone began to burn and, after its removal from the flames and a quick wash with vinegar, it glittered like gold. Lok began to ask expert goldsmiths and other metallurgists in London about the rock's unexpected reaction to heat, and when they all confessed to be mystified, he resorted to Agnello for advice. It is not clear whether Agnello requested an interview with Cecil after examining the rock, or whether the canny minister (whose ear was perpetually tuned to gossip and news from the City) had learned of the strange substance from others in London. Once a housewife, a merchant, several City metallurgists, and a Venetian alchemist had learned of Lok's rock, London gossips must have been actively spreading the news. By the second week of January 1577 the Elizabethan gold rush was off to a strong start as news of "Frobisher's gold" was heard in the streets of the City and new speculators and investors rushed forward to put up the money required for a share in his still unchartered Cathay Company, sure that their investment would be returned tenfold. News about the possibility of enormous riches in the New World blew through the City like smoke from Margery Lok's chimney, and the stakes surrounding Frobisher's first unsuccessful attempt to locate a lucrative trade route to China by sailing northwest into the Atlantic rose so high that the crown was bound to become involved.

William Cecil did what he so often did in cases where significant investments of men, money, and materials coincided with interesting reports of a natural event or technological development: he began, quietly and competently, to gather information, vet competing proposals, and carefully guide the inquiries toward some definitive conclusion that would benefit the crown as well as individual investors. Cecil was an ideal person for this job: he was well educated and had an abiding interest in the natural world. When a new star appeared in the constellation of Cassiopeia in 1572, Cecil asked the mathematician and astronomer Thomas Digges to report on its significance. Cecil's gardens were carefully cultivated and contained rare plants brought in from every corner of the globe. And like many men and women of his class, he took an active interest in the medical care that he and his family received.[2]

Where Cecil made his indelible mark on the history of English science, however, was in his interest in what can be described as Elizabethan "Big Science," an early modern analogue for a term developed in the twentieth century to describe the governmental and industrial patronage of science and technology that led to big budgets, big staffs, big machines, and big laboratories.[3] Elizabeth I and Cecil, like leaders of industrial nations after World War I, were passionately interested in transforming the fortunes of England financially, militarily, and geopolitically by investing in science and technology.[4] Elizabeth had a vested interest in promoting what she described as "all good sciences and wise and learned inventions tending to the benefit of the commonwealth of our said

Realm and Dominions, and serving for the defense thereof." While the tools and techniques that Elizabethans employed to study and exploit the powers of nature were different from the university laboratories, atomic bombs, and particle accelerators familiar to students of twentieth-century Big Science, the crown's interest in developing mining, metallurgy, navigation, and military technologies required similarly large financial outlays, involved issues of secrecy and national defense, and included taking calculated risks with money, manpower, and machines. In the end, nearly £25,000 was spent on the Frobisher voyages and the assays of Lok's black lumps of Newfoundland ore, for example — a vast sum in early modern terms — and hundreds of workmen both in England and the New World were enlisted to lend their technological expertise to the project.[5]

While the analogy between modern Big Science and the Elizabethan interest in mining, metallurgy, navigation, and technologies offers up intriguing parallels, Elizabethan Big Science was a different enterprise in some important respects. It involved large-scale projects — some technical and some military — which employed scores of workmen ranging from the most humble assistants and unskilled laborers right up the social ladder to university-educated gentlemen and aristocratic investors. These workers and investors, not the crown, provided the majority of private funding for the projects.[6] Participants in the projects were committed not to an open-ended pursuit of natural knowledge but to garnering economic profits from their exploitation and mastery of nature. And while there is evidence that Cecil and perhaps even the queen saw each individual project as part of a larger, collective endeavor, the technicians and investors who were more directly engaged in the work did not always share their grand, panoramic view.

London was the central, though not single, corporate institution through which Elizabethan Big Science projects passed from projectors and speculators into the hands of William Cecil, and then on to the queen's desk. Londoners proposed all kinds of schemes to Elizabeth's government, from plans to restore the value of the English currency through better refining and assaying methods, to projects designed to resolve trade imbalances by manufacturing (rather than importing) crucial staples like salt, to ambitious voyages of discovery and exploration. Shipyards, furnaces, foundries, glass factories and tileworks dotted the City's landscape and provided venues where ideas could be debated, new techniques developed, and problems resolved among the workers. London also had wealthy citizens with the ready capital to invest in such projects. And the City's close proximity to popular royal palaces — at Greenwich, Richmond, Westminster, and Hampton Court — gave projectors and speculators easier access to aristocratic patrons and bureaucrats who could smooth ruffled feathers and lend additional support to the ventures. From 1560 to 1580 London played an important role in both the development and the administration of these large-scale projects. As the stakes grew higher and crown interest and investment mounted, however, Elizabethan Big Science came to an end as protests were voiced over unrealized schemes and the extortive monopolies typically connected to the projects.

Here we will take a closer look at London's role in Elizabethan Big Science, paying particular attention to the ways in which the City's technological literacy helped it function as a clearinghouse and a meeting place. Once London's projectors and investors had developed their strategies for exploiting nature, many of them turned to William Cecil to acquire the queen's letters patent, a process that reveals how important Cecil was both as the queen's chief client and as her chief patronage broker. Nevertheless, the acquisition of letters patent depended on more than a slick marketing strategy and the right connections. While many Elizabethans were interested in exploration, new industries, and defense, the projects of most interest to Cecil were high-risk alchemical and metallurgical ventures that are more difficult for us to appreciate and understand. Yet their presence within the broader range of Elizabethan projects casts a bright light on how sponsors and investors coped with new and conflicting results, who had the last say when investors and technicians had different objectives, and how London's technical experts reflected the wider urban culture in which they were embedded.[7]

WILLIAM CECIL AND ELIZABETHAN BIG SCIENCE

William Cecil (1520–98) was Elizabeth I's most valuable, and most indefatigable, public servant. For forty years, first as secretary of state and then as lord treasurer, he planned policy, pushed paper, and prodded the queen and her court into new ways of thinking about England and its place in the world. Cecil was blessed with a probing intellect and an inquisitive mind. Never content to sit back and accept received wisdom, he was an eager consumer of novelties and inventions. Strewn among the intelligence reports and treasury accounts that passed his desk every day were regular requests that required him to turn his sharp mind to technological problems and projects. Would the queen like to fund the development of a new kind of corn mill, asked one eager inventor? How about a guaranteed method for making armor and other weapons impervious to assault? Another, who sent Cecil models of his battering ram, wrote, "I understand your honor to be delighted with such rare inventions."[8] Reports from voyages to explore far reaches of the globe, requests for monopolies on imported drugs, schematic drawings of bizarre and purportedly useful machines, and end-less opportunities to invest in new and improved ways to transmute base metals, refine salt, make gunpowder, and explore the mineral resources of England poured into Cecil's office.

While proposals for a wide range of projects were abundant, Cecil was most inclined to lend his support to maritime voyages of exploration, the exploitation of mineral resources through mining and alchemy, the development of new and more reliable weapons for England's defense, the introduction of such indus-tries as glassmaking and salt manufacturing, and the invention of new machines and engines.[9] These large-scale Elizabethan technical projects — which required teams of practitioners and ample financial support from investors — have in-triguing analogues to modern efforts to explore space, study the properties of matter with particle accelerators, and develop new weapons, materials, and in-dustries.[10] Cecil's interest in such seemingly paradoxical ventures as the al-chemical transmutation of base metals into copper and the mining of alum for use in cloth dyeing and ordnance manufacture is more understandable when we realize that they represent a search for a coherent solution to a set of interrelated economic and political crises. When Elizabeth took the throne in 1558, she faced bankruptcy, significant levels of domestic poverty, an imbalance between imports and exports, and threats from foreign powers that saw an unmarried queen as an easy target for their own geopolitical aspirations. For Elizabeth, the projects offered a means to address these issues by exploiting natural resources, creating jobs, adjusting the country's trade imbalance, and advancing national defense.

Cecil found that Elizabethan Big Science initiatives such as these demanded a new approach to natural knowledge, one that was based on a different scale of work as well as a different quality of work. There was a stark contrast between the ideal of the medieval natural philosopher working in his study reading texts or crouched over his alchemical vessels, and the teeming world of the Thames, where merchants, pilots, instrument makers, and carpenters worked to get a ship under sail to explore the seas. Though natural philosophers entirely isolated in ivory towers were few and far between in early modern England, and most intel-lectual work was done in semiprivate spaces, these projects were still markedly different in both style and scale from medieval and even early-sixteenth-century efforts to manipulate and control nature.[11] The workforce required to open up a copper mine, assay tons of ore extracted from the New World, or build a new wa-terworks for London required large numbers of experts, artisans, and administra-tors. Issues of teamwork and hierarchy made the pursuit of these large-scale Elizabethan projects difficult and tendentious.[12]

Despite these problems, Elizabeth I vigorously supported large-scale projects, their inventors, and their investors during her reign — though she was not as generous with those who expressed a more philosophical interest in the natural world. The queen had declined, for example, to grant the mathematician and astrologer John Dee a pension in exchange for an ill-defined position as her "court philosopher" and mathematician. Instead, she doled out the occasional purse of gold to Dee and his wife and stalled repeatedly on the question of giving him an ecclesiastical living until he fled the country in search of a patron who would look more favorably on his university education and abstract aspirations.[13] Elizabeth's chary treatment of John Dee has inaccurately colored our sense of the crown's interest in science and technology in the second half of the sixteenth century. It turns out that Elizabeth was not uninterested in natural knowledge and its exploitation, nor was she reluctant to give money and positions to those who claimed some special knowledge of the natural world. What she was averse to was granting someone a title like "philosopher" and installing that person at court as a paid servant.

Elizabeth's patronage was important, and the role that such support played in the practice of early modern natural philosophy and technology has been well studied.[14] Galileo Galilei's experiences in the Medici court in Florence have become totems of the ideal of princely patronage at a time when the pursuit of natural knowledge often yielded such intangible and controversial benefits as the discovery of new planets. It was difficult to translate such a discovery into currency or sell it in the open market. But the intricate, ceremonial strategies employed by Galileo and men like him to obtain financial backing and positions from the Medici and princes like Holy Roman Emperor Rudolf II did not work everywhere. Elizabeth preferred an indirect, mediated form of patronage that did not place her in direct contact with hordes of would-be clients. Instead, she surrounded herself with a tight circle of trusted individuals like William Cecil who enjoyed face-to-face interactions with her and whom she employed as patron-brokers. These men and women were expected to handle hundreds of requests for support and influence in exchange for a privileged relationship with the queen. William Cecil was chief among these clients, whose ranks included Robert Dudley, Earl of Leicester, and Blanche Parry, one of Elizabeth's most loyal retainers. The style of patronage Elizabeth adopted was thus very different from her continental contemporaries. She was not willing to give petitioners the face time the Medici family devoted to Galileo or play the direct role Philip II of Spain enjoyed in his correspondence with engineers, medical experts, and other figures interested in the natural sciences.[15]

By surrounding herself with a tight circle of courtiers who received her patronage directly, she put her most trusted clients like Cecil in a position to serve as patrons in their own right, and as brokers for clients seeking royal privileges like letters patent, court positions, and other appointments. Elizabethan England had a "great chain of patronage," with hundreds of mediating links that stretched between the queen and the scores of individuals who did the work in the mines, engraved instruments, sailed on the voyages of exploration, or put coal in assaying furnaces. Within these complicated relationships, as Sharon Kettering has found in studying patronage in early modern France, "one man's patron was another man's client, and brokers bridged the distances between patrons."[16] In Elizabethan England, there was only one pure patron — the queen — and only the most humble artisans and craftsmen could boast of no clients of their own.

As one of Elizabeth's few direct clients, Cecil owed the queen loyalty and service in exchange for titles, administrative positions, and royal protections that were extended to him and members of his family. Cecil found many ways to express that loyalty and service, including playing a key role in the development of technological projects and fostering the study of nature more broadly. Cecil's interests developed from his education, and from his service to Edward VI. A friend to humanists and educators like the royal tutors Cheke and Ascham, Cecil believed that education and knowledge improved the individual and could benefit the crown. These beliefs were strengthened by his commitments to the ideals of the "commonweal" and by his friendship with Thomas Smith, who wrote *A Discourse of the Commonweal*. Smith argued in his dialogue, as we saw in the last chapter, that England's negative attitude toward learning and inventiveness played a role in the country's economic woes. But later in the dialogue Smith dismissed these attitudes as mere symptoms of the most pressing illness to affect the commonwealth: the inflation resulting from a debased currency. Both Henry VIII and Edward VI had issued coins with increasingly high percentages of alloy in them in an effort to benefit England's bottom line, but this practice had only led to higher prices, higher governmental costs, and further trade imbalances. As Elizabeth's secretary of state, Cecil was in the unenviable position of being expected to find a solution to these problems. It was a complicated matter, for England had to import much of its gold and silver (there was no copper coinage until the eighteenth century), manufacture the coins in the Royal Mint, and try to determine the amount of gold and silver currently in circulation through assays.[17] Cecil's familiarity with, and adherence to, Smith's commonwealth philosophy, coupled with his need to solve these pressing economic problems, fostered his interest in Elizabethan projects.

After Elizabeth took the throne and Cecil learned more about the extent of the currency crisis, the problems appeared insurmountable. The Royal Mint, housed in the Tower of London, was burdened with antiquated manufacturing techniques and a bulky bureaucracy. England's currency was issued under no fixed standard concerning the content of precious metals, and no one seemed to know how much debased currency was in circulation. Mint workers used traditional methods of assaying, melting, and striking coins that were inefficient and inexact. The officer in charge of the work, the master of the mint, was not always someone possessed of both knowledge of metallurgy and a gift for administration. Cecil embarked simultaneously on multiple possible solutions to the crisis. He cleaned house at the mint, putting new workers in place, new officials in charge, and new technologies into action. In 1560 business was in sufficiently good order at the mint for Elizabeth to recall the silver money in circulation in exchange for silver coins of a standard weight. The amount and type of work expected of mint personnel after the queen recalled the currency required subsequent reorganizations and new technologies.[18]

Cecil's experiences reorganizing the mint were typical of those he would face in countless other Elizabethan Big Science projects throughout his career. First and foremost, he faced the problem of technological literacy. Cecil had to be able to identify individuals with technological skills and knowledge before embarking on a costly, high-profile project like reorganizing the mint. Cecil's biggest challenges were how to find skilled workers and how to regulate and control them afterward. Traditional craft organizations and guilds, like the Goldsmiths' and Ironmongers' Companies, could be enormously helpful in locating laborers and overseeing the work. But the blessing of the guilds could also prove a curse, since they monopolized some skilled work and sought to keep foreigners like Agnello from impinging on their privileges and prerogatives. During the reign of Elizabeth, London's guilds might have been in a weaker position than in the Middle Ages, but they still played an instrumental role in policing and administering the capital city. Cecil had to find a way to permit foreigners with European techniques unknown and unused in England, and English inventors offering novelties, to experiment with new methods and undertake projects that threatened to trespass on guild territories.

Nothing pleased Cecil more than finding technologically literate individuals like Agnello who had some valuable skill or vein of knowledge that the crown could exploit. Cecil, unlike many with his background and position, reveled in direct contact with artisans and craftsmen. The embodiment of two of Thomas Smith's maxims in *A Discourse of the Commonweal* — "That from many heads is gathered perfect counsel" and "That every man is to be credited in his own art" — Cecil consulted with a wide variety of skilled craftsmen in his pursuit of the technically literate. Legend has it, for instance, that a leatherworker taught Cecil how to tan hide as part of his efforts to better understand the industry and its problems.[19] Goldsmith and Mint Master William Humphrey advised Cecil on coinage standards and copper assaying.[20] The Janssens, an immigrant family, explained the benefits of their foreign method of making pottery to contain pharmaceutical products. Several engineers outlined their plans for using the power of the water flowing through the arches of London Bridge to drive engines and mills. Cecil kept detailed lists of all known ordnance makers and inquired into the relative merits of brass ordnance over copper. And John Powell, an arrested counterfeiter, explained how his offences could have been prevented by putting a few rudimentary safeguards in place.[21] As far as Cecil was concerned, being found guilty of counterfeiting settled an unmistakable aura of authority on a metallurgist, as Eloy Mistrell (convicted twice) discovered. He was given a position at the Royal Mint, received permission to use his newly invented coin-stamping engine to make coins, and performed critical assays of mineral specimens in the copper mines.[22] Though Elizabeth herself was wary of direct contact with potential clients, Cecil relished this obligation, seeking out experts in prisons, guilds, and the streets of London.

The individuals who supplied Cecil with technical information did not do so out of purely altruistic motives or because they shared his commonwealth philosophy — although they often claimed to do so. Sponsors like Cecil, investors, and technicians all stood to make a profit should their project gain support and prove successful. William Humfrey wanted to use his metallurgical knowledge to set up a monopoly for the making of brass and copper wire and beaten copper vessels that would force competitors out of the market. The Janssens needed to set up a pottery on the banks of the Thames near London to make a living — but as foreign immigrants, they did not possess the freedom of the City and could not exercise any trade without guild permission. Similarly pragmatic and ultimately self-serving considerations influenced most other proposals. Nobody was going to set up a watermill on London Bridge without both City and crown support, the ordnance makers and gun founders sought permission to start their own guild, and John Powell and Eloy Mistrell wanted release from prison. Cecil — perched as he was so near the queen, occupying a seat on the Privy Council, and holding the office of secretary of state — was clearly the person to go to with your project idea if you needed to be made denizen, gain planning permission, obtain a "get out of jail" pass, or win coveted letters patent.

Letters patent were legal instruments granted by the queen that guaranteed both protections and privileges from the crown. The distant ancestor of modern patents, which convey intellectual property rights, in Elizabethan England letters patent could be issued for any number of transactions, including land transfers involving crown properties, denizations of foreign citizens, and the establishment of such corporate bodies as colleges and trade guilds.[23] England adopted and modified the patent system used in many European countries in the early modern period, drawing heavily on Venetian precedents. As Smith's lawyer in *A Discourse of the Commonweal* reported, Venice found an appealing way to reward innovation by granting limited monopolies to citizens and immigrants to encourage project development. Cecil discovered that the continental patent system was particularly useful in incorporating into the labor force immigrant workers — which England was absorbing in high numbers in the late 1560s and early 1570s.[24]

In most proposals the goal of the inventors and investors was to acquire letters patent that conferred a limited monopoly on an invention, process, or product. A loosely defined but consistent set of criteria emerges from successful applications for letters patent. Claims to originality, the enumeration of high expenses associated with developing a new technique or trade route, and the existence of fraudulent wares or practices that put the commonwealth at risk all appear repeatedly. William Cecil sifted through these petitions and their claims like an early National Science Foundation officer, evaluating their strengths and weaknesses.[25] Elizabeth and Cecil remained interested in large-scale projects throughout the reign, but the heyday of projects and petitions can be dated from 1560 to 1580. The queen helped to shape this marked cluster of activity, since she typically rewarded twenty-one-year privileges to inventors and their new technologies. It is not surprising, therefore, that early projectors could exercise a stranglehold for decades on an industry like saltpeter manufacturing or the construction of drainage machines for use in mining. There are other reasons for the sharp interest in big technological projects and the relatively sudden decline, though. First and most notable, many of the projects failed to deliver on the optimistic promises that were made in the early days of Elizabeth's reign, when it seemed like each new invention was just about to make England the undisputed financial and technological leader of Europe. And second, the private investors who funded the projects tended to speculate in waves that ebbed and flowed as their finances were exhausted and then replenished.

Cecil took three chief considerations into account when reviewing petitions for the queen's letters patent: utility, economy, and novelty. Of these three, the most important watchword for the queen and the patron-brokers who surrounded her was utility.[26] The English queen was not interested in ostentatious displays of technical virtuosity, as were some of her counterparts on the Continent; she wanted new industries, better defense, a stronger currency, and more land. While the Medici enjoyed the search for new celestial bodies like Galileo's "Medicean Stars," Cecil and the queen sought more practical natural knowledge. They preferred offers to embark on projects with tangible outcomes, like the one proposed by another convicted counterfeiter, Edmund Jentill. He presented Cecil with several useful devices: an instrument "whereby the distance to any thing, together with its height and breadth," could be measured from a single fixed point; a design for a new mill; and a "Euclidean Compass" capable of accurately describing geometrical figures and spiral lines. Jentill assured Cecil that the "fruits and small ends" of these inventions would not be "common or trivial, but rare and of great use in a state or commonwealth." Jentill was willing to forgo the patents he thought he deserved in favor of a release from prison. He explained that his ventures into counterfeiting had stemmed not from "any vicious or lascivious kind of living" but from his passionate devotion both to his family and to the acquisition of natural and mechanical knowledge. Jentill wrote of the heavy costs associated with "the buying of books, the paying of debts, and the trial of conclusions mathematical and serviceable for my country's good."[27]

In addition to proclamations of utility, assurances of economy and a low initial financial investment also increased a projector's likelihood of receiving letters patent. England was not rich in natural resources, and those she had (in timber, tin, coal, and lead) were being rapidly depleted. When the crown granted Fulke Greville, Lord Brooke, a lease of two iron furnaces, for example, it was on the basis of promises made about the minimal use of timber. While supporting the crown's interest in large-scale projects, Cecil also watched the bottom line, first as secretary of state and later as lord treasurer. Project budgets often rose so precipitously once letters patent were issued that Cecil took an increasingly hard line on expenditures and the use of natural resources. Cecil and the queen were thus inclined to look favorably on inventors like the German soldier Gerard Honrick, who offered to make pumps to drain the water from flooded mines with no financial strings attached to the crown. Honrick promised to "put the same engines and instruments in practice" at "his own charges" if only Elizabeth would grant him a thirty-year monopoly. While these efforts may not have resulted in the "integrated elegance" that budgetary limitations often produce in modern Big Science, they often fostered collaboration and problem solving among Elizabethan technicians and the investors, both of whom were intent on keeping projects viable.[28]

Along with utility and economy, novelty is frequently mentioned in successful petitions for letters patent, but it was not enough to present Cecil with a new or improved gizmo. A projector had to carefully explain how his invention differed from similar items already in use and how its development would impinge on existing monopolies. Proprietary claims over trade and craft processes were prevalent in Europe's medieval guilds, and assurances of novelty given in Elizabethan patent applications show the legacy of this urban ethos in the early modern period.[29] Competition was bound to happen in the race for project support, but controversies had to be handled delicately and with full knowledge of the facts. When John Medley put forward his petition for letters patent for a pump he had invented to drain flooded mines, he was competing with Gerard Honrick's claim for a similar engine. Medley's claim that his instrument was "not heretofore used in this . . . realm" was coupled with assurances that it was also "better than any other yet used in this . . . realm."[30] Even so, Elizabeth granted both men their patents for different, novel engines in the same month, content to let the results speak for themselves and wary of getting involved in a priority dispute among rivals.

Cecil, after weighing the utility, cost, and novelty of a petition, drove shrewd, businesslike bargains with the hopeful inventors who came to him seeking the queen's support. These bargains were far removed from the unspoken, ritualistic, and noncontractual world of noble or royal patronage. The crown's ultimate goal for most projects was the gradual English appropriation of technological inventions and the smooth transfer of their material benefits into common public use, and Elizabeth wanted legally binding arrangements to ensure these goals. Cecil carefully read the language of letters patent before issuing them, and often made changes advantageous to the crown.[31] Cecil found London's sophisticated businessmen especially helpful when it came to lowering a project's overhead or setup costs, as when a committee of merchants headed by Lionel Duckett secured "the best bargain that ever we made . . . these 15 years and more" with the Germans called in to work the Mines Royal. The letters patent Elizabeth granted for large-scale projects typically included clauses that made the privileges dependent on the efficacy of the engine or system and set limits on the length of time that could elapse between proposal and realization, the charges for use of the new technology, and the number of Englishmen who were to be made fully conversant with the invention. These became the template of conditions under which Elizabeth was prepared to bestow her support.[32]

During the Elizabethan period, the number of petitions far exceeded the grants of letters patent, and scores of draft letters patent survive that were never formally issued. The process from start to finish could try the patience of any eager inventor, since the documents had to pass successfully through the hands of several offices and win the approval of the Privy Council. Sir Thomas Smith, who received letters patent to transmute base metals into copper and quicksilver along with Cecil and the Earl of Leicester, secured the queen's signature on the documents only at the end of a day "by candlelight," and quickly sped them through subsequent stages of approval. Greater speed could be achieved by the intervention of one of Elizabeth's direct clients, like Cecil, or by gathering together a company of investors or co-projectors from the City and the court. As a result, Elizabethan projects could be organized in a variety of ways: as individually initiated schemes; through the collaboration of a few individuals, usually working under letters patent which laid out their rights and responsibilities in regulated companies; and in early regulated companies, in which investors purchased shares in exchange for a voice in company decisions and a portion of company profits.[33]

Early letters patent granted to the aristocrat George Cobham illustrate how the ideals of utility, economy, and novelty were put into action to successfully obtain letters patent. In his letters patent Cobham was credited with discovering an instrument to deepen rivers, channels, and harbors — a device of serious interest to the crown and the City as once-navigable inlets and rivers like the Lea silted up due to increased traffic, building projects, and sewage. Any Elizabethan unaware of the catastrophic effects of these natural processes needed only to look across the Channel to the once prosperous city of Bruges, whose harbor had become inaccessible to ships, thereby reducing its status as a leader of international trade and finance. Cobham promised the queen that his invention made the process of "scouring" riverbeds and harbors quicker and more efficient and could be put at her disposal without delays or high cost. His assurances on this point were not sufficient for Elizabeth, however, who explicitly required Cobham to construct the instrument within three months and make it available to her "on reasonable terms." If — and only if — he fulfilled this end of the bargain would the queen grant him a ten-year privilege. During the period the letters patent were in force, Elizabeth would not forbid her citizens from continuing to use traditional methods to widen and deepen canals, rivers, and harbors. But the queen hoped that the issuance of Cobham's letters patent would "give courage to others to study and ask for the knowledge of the like good engines and devices."[34]

A wealthy man like Cobham could benefit enormously from letters patent, but they were absolutely essential for ingenious foreign immigrants because of the trade restrictions placed on them in the City. Elizabeth I's first letters patent for a large-scale technological project were issued to a Stranger, William Berger of Utrecht, who invented a new corn mill. In recognition of the high costs and long period of study required for the mill's development, Elizabeth established Berger's right to use it for a period of seven years. For lucky Strangers, ingenuity could smooth the road to denization, the equivalent of resident-alien status in

Elizabethan England. The German crossbow maker Herman von Bronkers was made denizen after reports reached the queen of his skill in making "machines of war."[35] Federico Genibelli offered up a number of large-scale projects in 1591 to benefit the realm, including plans to increase London's water supply and decrease its risk of fire. The value of the queen's support of foreign projectors was evident when Elizabeth instructed City officials to call off the Dyers and Brewers who were interfering in engineering work under way in London under the guidance of one of her immigrant patent holders, Iacopo Aconcio of Trent. The guilds insisted that he was impinging on their corporate rights and privileges, but the queen's authority was unimpeachable, and Aconcio remained at work.[36]

Only the most intrepid and confident inventors like Aconcio sought individual letters patent. Most projectors recognized that there was safety — both financial and political — in numbers. Petitions for letters patent that requested the establishment of a company were therefore common. Both the petitions and the letters patent that were ultimately issued reveal the corporate pragmatism arising out of the City of London that we have seen at work in earlier chapters. With its insistence on self-government, and its careful maintenance of ancient privileges and liberties, the City was not only an important center for the development of large-scale projects but also the model for how the projects could be governed, financed, and administered. London thrived on self-government, shared risk taking, and the cooperation and competition that existed between native Englishmen and immigrants. These were the hallmarks not only of Elizabethan London but of Elizabethan Big Science.

Self-government was a crucial corporate privilege familiar to Londoners who were members of any of the City's guilds. The letters patent Elizabeth granted to corporate groups engaged in projects emphasized similar rights to self-governance, albeit with crown backup and oversight. Such corporations were expected to hold meetings, elect officers, establish rules of conduct, and settle disputes internally rather than through the English courts. These privileges echo those granted to London's medieval guilds. When Queen Mary bestowed letters patent on a group of merchants and formed the Muscovy Company to explore trade routes to Russia, the group was explicitly instructed to meet in London or elsewhere "as other worshipful corporations in our said City have used to assemble." Letters patent for large-scale projects often referred to members of a common corporate cause as "brothers" and outlined how patentee privileges were to be passed on to their children and "apprentices"; the Merchant Adventurers specifically highlighted their relationship to one of London's ancient livery companies, the Mercers.[37]

Regulated companies developed from some of these early Elizabethan applications for letters patent as a means to spread benefits and risks among share-holders. For fixed sums, individual investors could buy into a company like the Merchant Adventurers in exchange for a voice in government decisions and a slice of the profits. Corporate letters patent for these companies were crafted to protect the organization from the threat of endless litigation stemming from the individual bankruptcies that all too often resulted from the projects.[38] Powerful men like Cecil, who sat on the queen's Privy Council and were her direct clients, often received free shares in regulated companies in exchange for the cachet associated with linking them to a petition for letters patent. Robert Dudley, Earl of Leicester; Henry Hastings, Earl of Huntington; George Cobham, Lord Brooke; and James Blount, Lord Mountjoy, led the ranks of noblemen who were members of corporations set up to support and administer large-scale projects. But they were not the only individuals to receive these benefits: important City figures were accorded similar status, as the financial support of London merchants and civic officials was often as important to the long-term success of a project as a coat of arms and a seat at the queen's council table. When Thomas Thurland and Daniel Hochstetter developed their petition for the incorporation of a company to mine for copper, they offered free shares to three members of the queen's council: the Earl of Leicester, the Earl of Pembroke, and (most crucially) William Cecil. At the same time, however, they offered shares to the City's most active speculator on large-scale projects, Alderman Lionell Duckett.

Because these ambitious projects often depended on a combination of English local knowledge and immigrant ingenuity, corporate letters patent provided a way for citizens and Strangers to work together in a common cause, especially when the invention or project in question might flout guild privileges. Elizabeth granted the Londoner Richard Pratt and Stephen van Herwicke of Utrecht letters patent for their new and more efficient furnace on the understanding that they would practice, use, and teach the "science and knowledge" of furnace making to English subjects so that "much benefit and commodity may grow thereof." On a much larger scale, the Mines Royal Company and the Mineral and Battery Company were collaborative ventures between Englishmen and Germans. The letters patent for the Mines Royal, first established in 1564 by the Englishman Thomas Thurland and the German Sebastian Spydell, passed into the hands of Thurland and another German, Daniel Hochstetter, in 1568. Elizabeth noted that Thurland, under "the skillful direction" of Hochstetter, had "travailed in the search, work, and experiment of the mines and ores . . . whereby great benefit is likely to come to . . . our Realm of England." Not all collaborations that were forged between Strangers and citizens reached the letters patent stage, however. George Cobham's drainage patent began in collaboration with an Italian engineer, Tommaso Chanata, for instance. Together Cobham and Chanata, with "others of their company," approached Elizabeth a few weeks after her succession in December 1558 to ask for permission to construct and use an "instrument to carry away all shelves of sand, banks, ooze, and such like out of all rivers, creeks, and havens."[39] When the letters patent were issued years later in the spring of 1562, they were made out in the name of George Cobham — without mention of Chanata.

Even when corporate letters patent were issued to English and Stranger projectors, a palpable sense of unease often permeated the documents and colored the company's relationship to the crown. In an effort to ensure that foreign knowledge eventually became fully English, like a plant in John Gerard's garden, most Strangers who held patents were required to teach Englishmen how to perform the process or build the engine. Anthony Becku and Jehan Carré discovered this feature of letters patent when they were given a twenty-one-year license to set up glass factories in England provided they taught English citizens the "science of glass making . . . so as the same science . . . may here remain in the Realm and be practiced by Englishmen." Additional safeguards to protect the commonwealth's interests were sometimes put in place. When Gerard Honrick promised to show Elizabeth's subjects his unique method of making saltpeter, she required that he put his method in writing to avoid confusion among his students, concerns among his investors, and contention among his competitors over issues of priority. Despite these efforts, many who administered Elizabethan Big Science projects remained convinced that their Stranger workmen persistently kept secrets. In the copper mines of Keswick relations between English shareholders in the Mines Royal and the German mineral experts they employed disintegrated in the corrosive atmosphere produced by these suspicions.[40]

William Cecil was often able to apply balm to injured egos and allay the fears of both participants and investors. These valuable contributions were made possible through his hands-on approach, and the fact that he took a direct interest in the projects, investigated the projectors and their claims, and did what he could to establish the rights and responsibilities of everyone involved. As the number of petitions grew and he became inundated with requests, however, Cecil had to find more efficient ways to gather information, and more effective means to administer projects. Without guild oversight or a governmental department, Cecil had to set up his own system to track and evaluate complicated projects as they progressed. Anxieties over inventive sleights of hand and the huge cost of many large-scale projects only placed greater burdens on Cecil, who increasingly asked petitioners to make their methods and profit projections crystal clear to both the crown and private investors. Cecil's clandestine interview with Agnello on the subject of the Frobisher ores was an attempt to clarify the strange mysteries of the black rock for the queen, her advisers, and Lok's investors. Others preferred visual proofs, and when the City contracted with a Dutch engineer named Peter Morice to supply water to London by utilizing his patented water pumps and the pressure generated by water as it cascaded through the arches of London Bridge, the engineer prudently demonstrated the soundness of his plan by generating a jet of water so powerful that it arced over the Thames and hit the steeple of the downtown church St. Magnus Martyr.[41] All claims of utility, economy, and novelty made by investors and projectors — whether made behind closed doors or in bravura displays of technical prowess — needed to be verified. The difficulty facing Cecil was the question of who would track down the petitioners and how they would evaluate the proposals in the field.

Cecil developed a network of informants to gather information on large-scale projects. The intelligence gathered by these men — dubbed "roving reporters" by one historian — supplemented with drawings, models, and descriptions included in many letters directed to Cecil, helped him judge the merits and demerits of each proposal.[42] Many of his informants lived in the City, where they had established themselves in networks of technicians and merchants. Cecil's network of investigators was usually the source of the "credible report" of a technician's abilities, or the "credible information" of the hard work required to bring an invention into material being, mentioned in the letters patent.[43] Cecil's two most active informants were Armagil Waad (c. 1510–68) and William Herle (d. 1588/89). Both men cultivated their relationships with spies, technicians, and investigators who haunted the dockyards and Royal Exchange listening for news of would-be Elizabethan Big Science projectors and projects.[44] Waad was active on Cecil's behalf during the initial decade of the reign. A Yorkshire native and Oxford University graduate, he spoke both Spanish and French, which he used to interview native and immigrant projectors. Waad had firsthand experience of the risks and rewards of the projects, since he had been on the 1536 expedition under Captain Hoare to Cape Breton and Penguin Island in North America. His wife, Alice Patten, connected him to the London elite both personally and financially, which increased his usefulness to Cecil, as he could rely on family to help him navigate London's labyrinthine guild politics. After getting his start in the intelligence business with expeditions to Holstein to negotiate commercial contracts, and to the English coast to spy on Huguenots, Waad began to focus on London and the projects undertaken there. While working for Cecil he managed the work of Cornelius de Lannoy, a Polish alchemist, and made recommendations on other patent proposals. Waad himself received letters patent for a limited monopoly on a unique method of making sulfur and for cultivating plants used to make linseed oil in conjunction with another of Cecil's investigators, William Herle. After Waad's death, Cecil diverted Herle from his attempts to infiltrate networks of Catholic sympathizers by giving him the opportunity to stalk would-be projectors in and around London. Herle's first recommendation to Cecil was to support the petition of a Walloon immigrant, Frances Franckard, who had developed profitable new methods for making salt.[45] Herle took early notice of Frobisher for Cecil, reporting on his activities as early as 1572, and also informed Cecil about Valentine Russwurin's skill in surgery.

As the range of the projects grew, however, Cecil's informants became more specialized. The assay master at the mint, William Humfrey, became Cecil's leading source of information on metallurgical projects. Thomas Digges, who supervised the project to rebuild Dover Harbor and was the son of the mathematician Leonard Digges, acted as a general consultant on engineering projects. Inventors increasingly knew that Cecil's commitment to support only those projects that had some demonstrable chance of success meant that they were going to have to produce proof for his knowledgeable investigators. When Emery Molyneux wanted letters patent for his military inventions, he tried to preempt the inevitable sniffing about by Cecil's aides by first showing the designs to his patrons: Sir William Knowles, Sir Henry Knyvett, Sir John Stanhope, and Thomas Knyvett. These men, Molyneux assured Cecil, were able to provide expert assessments of his new cannon and ordnance. Aware that Cecil might not put his faith in their testimonies, Molyneux hastily added that he would let others "your honor shall appoint to survey the same" make a final judgment concerning the utility and quality of his armaments.[46]

Cecil's system for vetting proposals before they were awarded letters patent relied on an extended network of intelligencers, patrons, investors, and inventors. This network switched into action whenever a projector approached Cecil with a plan to drain the fens or construct a new water supply for the City of London. After years of reading petitions and sending out investigators, Cecil knew who was reliable, who was ingenious, who was planning a project that depended on ineffective and inexpert individuals, or (even worse) who had not assembled the necessary administrators, engineers, technicians, artisans, and craftsmen to get the proposed work done. Cecil relied heavily on Londoners to acquire this knowledge. First, he relied on the City's skilled laborers and craftsmen for inside information on a host of technical questions and issues. Second, he depended on its merchants and the urban elite to help him negotiate contracts and agreements that would most benefit the crown, and to invest in projects that demanded huge capital outlays. And finally, he relied on London's own highly developed communities of practitioners for intelligence about inventors and projects.

Both William Cecil and the diverse resources of the City of London were crucial to the development of large-scale projects. Without London, Cecil would have had a difficult time identifying, funding, and administering the projects — but the work itself would not have flourished as it did in the City without Cecil's support. Everyone knew that Elizabeth's chief minister held the key that could unlock the doors of crown financing. But to win his trust, a projector had to have sufficient contacts among England's technologically literate population, her merchants, and other individuals to turn a two-dimensional sketch of a furnace into concrete, three-dimensional reality.[47] And the best place to find literate, skilled, and wealthy individuals was the City of London.

SPECULATING ON ELIZABETHAN BIG SCIENCE: LONDON'S PROJECTS AND PROJECTORS

There were many reasons why London was critical to the development of large-scale projects. London was fast becoming the commercial and economic center of England, and though this process was far from complete, it was not surprising that contemporaries saw in her "a gallimaufry of all sciences, arts and trades," a melting pot of the ingredients necessary for innovation and economic success.[48] It seemed that nearly every prosperous citizen (and many who were not so prosperous) either had a project in hand that promised dramatic yields from strikingly small capital outlays, or was investing in such a project. City officials often reprimanded residents for constructing forges, refining sugar, and making glass in their houses — evidence that the technological bug was biting more than wealthy aldermen.[49] For some, these speculative ventures led to financial disaster, since the optimistic predictions of small investment typically ballooned into staggering sums while the projects consumed ever-increasing resources like manpower and fuel. But the few successes that could be pointed to in the metropolis — the Muscovy Company's lucrative trade with Russia, the Mineral and Battery Works' industrious production of wire and beaten metal for ordnance and household goods, and the new glassmaking plants outside Aldgate and Cripplegate — were enough to keep inventors, investors, and their shared projects alive.[50]

Large-scale technical projects flourished in Elizabethan London because the City boasted a combination of skilled workers and wealthy investors. Like the German, Italian, and Imperial courts, London was an acknowledged node of technical activity in the early modern period.[51] New residents came into the

City from every corner of Europe to introduce novel techniques or launch projects, and once there they met up with wealthy Londoners eager to speculate on such ambitious enterprises as increasing London's water supply through deft engineering or constructing a more efficient brewer's furnace. Thomas Heneage, the queen's vice chamberlain, found himself "much perused" by an inventor named Parret once he left the relatively restricted world of the royal court for his home in entrepreneurial London. And he was probably not alone. Most investors entertained the inexhaustible requests of projectors like Parret because they smelled profit, and nearly everyone in London was trying to make a profit at something. Though the City's streets were not paved with gold, the dissolution of the monasteries under Henry VIII had redistributed significant wealth into the hands of the gentry and merchant classes, and the instability of foreign cloth markets meant that London's elite were actively looking for new investment opportunities.[52] London merchants made significant financial contributions to efforts to explore the oceans, settle colonies in newly discovered lands, help the crown rebuild Dover Harbor, and develop novel technologies.[53]

The City boasted other assets in addition to wealthy and skilled people, namely environmental resources like the Thames and an existing infrastructure that was conducive to large-scale projects. These projects involved sizable teams of investigators, technicians, supervisors, and go-betweens who shuttled between laboratories and furnaces in workshops and backyards, the City's shipyards, the Tower, Elizabeth's mobile court, and civic officials at the Guildhall. Much of this activity was centered on the eastern part of the City, home to the wharves, shipyards, and industrial plants of the time. This infrastructure was vital to the development of new industrial and technological projects which, as Joan Thirsk explained, "clustered in places where facilities already existed to give the enterprise a promising start." The east end of London, sandwiched between the Lea and Thames rivers, had been an industrial center since the Middle Ages. The area was dotted with workshops, each filled with skilled craftsmen toiling to transform London's speculative endeavors into tangible, fungible reality. The east end's geographical location facilitated many Elizabethan Big Science projects. Shipping, navigation, and cloth dyeing all required water. So did the manufacture of brick and tile, which also required enough open space during the winter months for the unfinished products to weather in the open air.[54] Gunmaking, ordnance manufacturing, bell founding, brewing, and glassmaking all depended on engendering a great deal of heat in large furnaces — and a ready supply of water to put out opportunistic fires as well.

Much of the existing infrastructure in the east end supported England's bustling defense industry. This part of the City was home to the Tower of London, the Artillery Garden, and a number of commercial and military docks. The Artillery Garden and the Tower of London provided highly visible focal points for interest in the technical and practical elements of defense. Every month, the master gunner of England received a delivery of shot, powder, and matches "for the exercise of his scholars and gunners shooting . . . at the Artillery Garden."[55] Farther downriver, the Deptford shipyards employed scores of craftsmen and mariners who lived in the parishes that flanked the Tower, St. Botolph Aldgate and St. Katherine's. The defense industry prized novelty—new weapons, techniques, and skills were always in demand—and embraced foreign as well as native contributions. Adventurers, soldiers, and inventors were abundant in London and relished the risks associated with Elizabethan projects. The discovery of new lands and the invention of novel engines of war were not far removed from each other, in either practical or imaginative terms. The list of Elizabethan projectors with a military background included Gerard Honrick (fl. 1555–78), Gawin Smith (fl. 1588–90), Sir Ralph Lane (d. 1603), Rocco Bonetto (fl. 1571–87), and Edward Helwiss (fl. 1545–1612).[56] These men lived among communities of soldiers, mariners, metallurgists, compass makers and gun founders, and had long entrepreneurial careers. The instrument maker James Kynvin, who contributed so much to instrumental literacy in the City, was prepared to offer King James the fruits of his forty years of practical expertise in engines of war even though he had "one foot in the Grave." At the ripe age of ninety-four, Edward Helwiss, who had served under Henry VIII in Boulogne and had been held prisoner in Bayonne under Edward VI, was still hawking his military inventions to Prince Henry, the son of James I, in the early years of the seventeenth century.[57] These military men had the skills and experience on which many large-scale Elizabethan projects depended.

Like Kynvin and Helwiss, the Londoners who dominated the large-scale projects mastered the ability to don many caps and shift roles at lightning speed, from expert informant to humble supplicant, from investor to inventor, from speculator to cautious businessman. Novelty and inventiveness required this versatility, since projectors and investors operated beyond what was known, expected, or proscribed. The opportunity to explore different occupations and acquire different kinds of technical knowledge in the City made it possible for men like Ralph Rabbards to enumerate all the "pleasant, serviceable, and rare inventions" he had "by long study and chargeable practice found out": perfume manufacturing, distilling medicinal waters, making oil and saltpeter, devising fireworks, making new kinds of ordnance, and constructing "naval engines."[58]

Versatility like Rabbards's did not always assure success, however. Ralph Lane's experiences illuminate what was—and what was not—likely to win the support of Elizabeth. Lane, like so many other projectors, was a military man. He tirelessly pestered the queen, her secretaries of state, and any other members of the Privy Council he could think of to support his projects, which included methods for mustering and keeping track of soldiers, designs for new fortifications, and plans to dig silver mines. Some of his projects fell well short of crown objectives and expectations, and received little or no attention despite Lane's best efforts to show how profitable these projects would be to England. Such was the fate of his plan to rigorously prosecute Strangers who impinged on statutes designed to protect native English workers, for example. His proposals to launch a military expedition against the Turks, and to follow up on forfeited bonds of obligation, were met with stony silence or criticism.[59]

It was only after he caught the eye of Walter Raleigh and returned safely from Roanoke in the new colony of Virginia that Lane's proposals received serious consideration. After his return, he presented Cecil with a plan for coastal defenses that could be built to protect England in only thirty-one days. In April 1587, when Lane sent his petition to Cecil, anxiety was mounting over Spain's plan to invade England. Here, finally, was a plan delivered at a timely moment and couched in a way that fulfilled Cecil and Elizabeth's predilections for businesslike agreements. To underscore his main points — speed, economy, and defense — Lane entitled his proposal "An apparent proof by particular demonstration that the offered defense for the coast is universally to be performed along all the coasts of this realm, within the space of 31 days, with small charge to her Majesty." Using impeccable (if optimistic) mathematical calculations, Lane required eighty laborers to be gathered from coastal communities for every mile of defendable coastline. Each day every laborer would be expected to erect a stretch of rampart one yard long and fifteen feet high (though Lane suspected that "foul weather" and hard ground would slow the work), making the coast "impregnable to any army, and not possible by any force to be landed upon." Though the queen never realized this ambitious plan to construct a fifteen-foot high "Great Wall of England," she did employ Lane's considerable energy in the Tilbury defense, assigning him a position as muster master and rewarding him with the guardianship of a royal castle after the Armada was successfully defeated.[60]

In Lane's case what turned the tide with Elizabeth was his demonstrated expertise in fortification. Lane had proven that he could do more than just write endless petitions about building defensive structures — he could actually construct them, even under difficult conditions like those in Roanoke. Despite the increasing emphasis on mathematics and other kinds of theoretical knowledge in London, for Elizabeth and Cecil there was still no substitute for hard-won, empirical knowledge. They clearly agreed with John Wheeler that experience was "the surest Doctor in the school of Man's life." When John Mount started making salt in Suffolk, he sent his apologies to his master, William Cecil, when he could not appear at court to meet with him personally. "I must be here myself to set the [salt] pans, because the workmen are not skillful," Mount explained. Sir Thomas Golding, the inventor of a number of engines and mills he presented to the queen, was deemed to possess "perfect knowledge and means how to drain marshes and low grounds," which he had gleaned from "study and long trial." Some projectors, like the Antwerp native Francis Bertie, were unhappily found not to have the "perfect knowledge by him professed," and were passed over in favor of others.[61]

Cecil increasingly judged the contributions made by London administrators and craftsmen to Elizabethan Big Science projects on the basis of certainty rather than optimistic projections. Empirical knowledge and experience played a growing role in Cecil's calculus of expertise, based as it was on reliable outcomes. The world of the projects was competitive, and London was full of people who made claims that could not be relied upon by investors or the crown. London projectors and administrators realized early on that Cecil expected them to deliver both good and bad news, especially if they were reporting on results or findings not easily observable by investors or other experts. As concerns about certainty increased, the reports from London's administrators and artisans became more detailed and precise. When Henry Pope reported on the saltpeter works in far-flung Fulstone, for instance, he apologized for the delay, writing that he could not report with "certainty thereof until one boiling was finished, which was performed the last week." With the process complete, Pope returned to London and could confidently assert that the "saltpeter which we have made bears no manner of salt, for that saltpeter which comes from beyond the sea and also that which is made here is subject unto much salt, [and] the gunpowder makers are forced oftentimes to refine it . . . to their great hindrance which they shall not need to do with this."[62]

Both versatility and empirical knowledge were vital to the success of London's projectors. But simply knowing how to assay metals or make a better drainage engine did not necessarily ensure that you were capable of running a complicated project. Literacy and numeracy were crucial, and some ability to draw accurate schematics could also be valuable for clarity's sake and for teaching less literate workers how to perform their tasks. One inventor and projector, for instance, sent a drawing of his plan for saltpans into Suffolk "to show my workmen my meaning."[63] But neither mathematical and technical literacy nor the hands-on experiences of the military officer or craftsman provided a projector with the funds to realize his technical dreams. This is where London's elite became invaluable: investing and administering the projects. As guild members, shareholders in corporate bodies like the Merchant Adventurers, and successful businessmen, London elites possessed some familiarity with craft knowledge and a wide network of individuals they could draw on to answer technical questions or to evaluate projectors and their work. Steeped in an atmosphere that prized education, they tended to be highly literate and skilled in mathematics. London's elites also had ample incentives to invest in large-scale projects, for there was enormous commercial potential in exploration, metallurgical experimentation, the introduction of new industries, and defense.

London elites provided the engineers, technicians, skilled craftsmen, and inventors with financial resources and the skills to correspond with, and run interference for, government figures. Since many of them were important City merchants and officials, these urban elites also helped to ensure that the projects could continue without being disturbed by angry citizens or jealous guilds. London's aldermen played an important role in administering and financing projects in the Elizabethan period. Elizabethan Big Science, like modern Big Science, demanded a high degree of coordination and organizational skill. City officials, accustomed to governing and regulating a large and complex metropolitan corporation, were ideally suited to the task of overseeing the projects, even when the work was being done in Cornwall or Virginia. The Society of the Mineral and Battery Works, established to dig for calamine stone used in ordnance manufacture as well as to reduce England's dependence on continental imports of wire and other beaten metalwork, included many Londoners, including Sir William Garrard, Sir Rowland Hayward, Thomas "Customer" Smyth, Anthony Gammage, Sir Richard Martin, and the keeper of the Tower, Sir Francis Jobson. Aldermen Lionel Duckett and William Bonde both invested in the Cathay Company, though it was never officially chartered, and helped fund the Frobisher voyages.[64]

Their wide-ranging experience administering projects from colonization to digging for calamine stone put London elites in a strong position when it came to making claims about the worth of a particular project. When Sebastian Brydgonne, a German furnace maker, set up new seething furnaces near Charing Cross and the Tower of London for two English brewers, word quickly spread that these new furnaces consumed only a fraction of the fuel normally necessary to make beer for thirsty Londoners. The lord mayor and two aldermen were notified of the invention's potential profit by a Stranger involved in the Mines Royal, Cornelius de Vos, and Merchant Taylor and the inventor Richard Pratt. Quickly, the aldermen called the two English brewers before them to give depositions about Brydgonne's furnaces. "Experience hereof we have none," the lord mayor wrote cautiously in his statement to crown officials regarding Brydgonne's furnaces, "but only by their reports."[65] When such reports came from sophisticated and experienced London officials, however, Cecil was inclined to take them at face value.

Lionel Duckett (d. 1587) was one of the aldermen involved in the Brydgonne inquiry, and he was London's foremost investor in large-scale technical projects. A member of the Mercers' Company (like many other Londoners who speculated on the projects), Duckett also belonged to the Company of Merchant Adventurers, the Spanish Company, and the Russia Company, and he owned shares in the Mines Royal. He invested in early trading ventures to Africa, the Frobisher voyages, and Adrian Gilbert's subsequent attempts to find a northwest passage. Duckett helped Thomas Gresham build the Royal Exchange, promoted the plans for a whale fishery in the Arctic, and joined in a Keswick copper mining project. On top of this daunting list of speculative enterprises, he was master of the Mercers' Company four times, an auditor of the City for four years, sheriff of London, an alderman, and president of three London hospitals (St. Thomas's, Bethlehem, and Bridewell). Duckett was knighted by the queen, who was heavily in debt to him at the conclusion of his successful, though controversial, stint as lord mayor of London in 1573 — a time marked in the popular imagination by his largely futile efforts to limit the feasting, drinking, and general troublemaking of City residents of the poorer and younger sort.

One did not have to be part of London's ruling elite like Duckett to invest in large-scale projects, as many wealthy merchants and tradesmen discovered. Between 1570 and 1636 nearly four thousand English merchants invested in regulated companies and their related projects, and a significant portion of those merchants came from London. Though Michael Lok never held high City office, his investment in the Frobisher voyages was more than two thousand pounds — the equivalent of more than half a million dollars in today's money. Few Elizabethans — noble or not — could absorb that kind of expenditure. John Wheeler, in his *Treatise of Commerce* (1601), despaired of the contemporary mania for speculating on risky ventures. "We have seen by experience," he wrote, "that many men in our time leaping from their shops and retailing, wherein they were brought up and gathered great wealth, and taking upon them to be merchants and dealers beyond the seas, have in [a] few years grown poor." Wheeler could easily have been writing about Lok, whose inability to pay all of his creditors after the heavy investments he made — and lost — in the Frobisher voyages meant that he was still being sued, as late as 1615, by the merchant Clement Draper for two hundred pounds worth of supplies provided for the third voyage.

In total, nearly one-half of the money invested in the Frobisher voyages came from Lok's friends, relatives, and associates in London, including Sir Thomas Gresham, William Burde, and William Bonde.[66]

Londoners—whether elite administrators and investors or skilled craftsmen and artisans—were motivated to participate in large-scale technical projects for a variety of reasons that included personal gain, devotion to England, and a desire to resolve political, social, and economic problems. Walter Raleigh's half brother, Sir Humphrey Gilbert, succinctly expressed the reasons why so many men were eager to take part:

> Who seeks, by gain and wealth, t'advance his house and blood:
> Whose care is great, whose toil no less, whose hope is for all good;
> If anyone there be, that covets such a trade:
> Lo, here the plot for common wealth, and private gain is made.[67]

Project proposals that survive from the period tend to emphasize equally the profits the scheme will bring to the English commonwealth and those the individual investor can expect. These motivations were consistent among investors, projectors, and practitioners and harked back to the commonwealth philosophy.

But the interests of the commonwealth proved to be a mixed blessing for the Londoners who participated in large-scale projects. Queen and country proved difficult business partners. The queen held regalian rights over precious metals or jewels mined in the realm, for example, and no mining project could be undertaken without her support. Foreigners, who were not protected by the City or the guilds since they lacked the status of citizens, needed royal approval before undertaking any project that rested on skills like metalworking or woodworking that were protected by civic or craft regulations. Both the development of weapons and the construction of military defenses were too politically sensitive to be undertaken without a nod from the queen. And voyages of exploration were simply too costly to succeed without crown intervention and sponsorship, including the donation of royal ships and ammunition stores. While large-scale projects heavily relied on London for financing and empirical expertise, most projects could not be contemplated without the queen's knowledge and approval. In these cases, London had to turn to William Cecil to acquire letters patent or royal authorization before embarking on any plan.

And Elizabeth's patent process, as supervised by Cecil, had its drawbacks and limitations. Like modern physicists, Elizabethan projectors in pursuit of letters patent often "found themselves increasingly spending their time searching for ways to pursue patentable ideas for economic rather than scientific reasons." When the queen was involved, it was difficult to maintain corporate autonomy and self-government in the face of royal interest, or royal outrage. The case of the Cathay Company illustrates the promise and problems associated with the queen's interest in such grand schemes. Elizabeth was not an investor in the first of Frobisher's voyages, but once Margery Lok spotted the glimmer of gold, the queen announced her intention to invest in the Cathay Company. She was entitled through her regalian rights to one-third of all gold found in her dominions — which, thanks to Frobisher, now included portions of the New World. Obtaining Elizabeth's approval was good news for many investors, who were pleased at having a royal on the company rolls. What were company officials to do, however, with a member who was also a monarch? James McDermott persuasively argued that the queen's failure to give the company its longed-for letters of patent stemmed from her desire to ensure that the one thousand pounds she was going to invest in the project was not lost in the mists of corporate mismanagement. Further donations of naval supplies and equipment from the crown's coffers fully transformed what had been a London enterprise into a government project, and when Cecil stepped in to calculate the rate at which a single man could mine ore in the New World, the venture had ceased, for all intents and purposes, to belong to its City investors.[68]

As Elizabeth's interest in the Frobisher voyages grew, the number of City investors declined. Neither William Bonde nor William Burde would support the second voyage after the goal became gold, rather than the discovery of a commercially viable passage to China. In Frobisher's first voyage, 65 percent of the investors were Londoners. In his second voyage, the proportion of City investors had dropped to 34 percent. Concerned with the realignment in priorities and procedures that the queen's involvement would bring to the project, London merchants turned their eyes to other, more commercial, ventures. Court investors rushed in to fill the void in the Cathay Company, their share of investment rising from 35 percent in the first voyage to 66 percent in the second, including first-time investors like the Earl of Lincoln, Sir Francis Knollys, and the countesses of Pembroke and Warwick. When the queen learned of London's defection from the Frobisher project, she tried without success to strong-arm the merchants into higher levels of participation.[69]

The second Frobisher voyage and subsequent assays of the ore brought back to the ports of London from the New World failed to reward court investors as they had hoped. A third voyage undertaken in 1578 made a bad situation worse. Investors defaulted in large numbers after the disappointing results — but those Londoners who remained loyal to the company were not among them. While more than £2,500 was still outstanding in 1580, when the queen's auditors turned the books inside out to ensure that she, at least, was properly repaid, they revealed that the most serious debtors were such court figures as the Earl of Oxford (£540), Martin Frobisher (£280), the Earl of Sussex and his wife (£205), and the Earl of Pembroke (more than £180). Michael Lok, who had invested so much in the enterprise, was in arrears only £27 and 10 shillings — a figure far less than what William Cecil himself owed (£65). In 1581, 80 percent of the unpaid monies due on the voyages could be traced back to court investors who had defaulted on their promises.[70]

Elizabethan Big Science was increasingly caught between the public interests of the crown and the London-based companies established by letters patent. When London corporations were eager to boast some noble or aristocratic members, they only complicated the situation further, since the leaders of organizations like the Cathay Company and the Mines Royal wanted to retain their rights to self-government despite their court connections. Cecil, as in so many aspects of his professional life, was put in the unenviable position of go-between, seeking to bridge the gap between crown and corporations while keeping the crown's priorities at the center of his ongoing deliberations.

NEW WORLDS, ANCIENT HEADACHES: MINERAL, METALLURGICAL, AND ALCHEMICAL PROJECTS

Nowhere is Cecil's role as go-between more evident than in the mineral, metallurgical, and alchemical projects that consumed significant amounts of his time as a government official. Cecil was not spared the metallurgical madness that gripped all levels of English society between the 1560s (when the Mines Royal and Mineral and Battery Works Companies were organized) and 1580 (when the Frobisher voyages finally exhausted the hopes and enthusiasms of its investors). Nor were the English the only victims of the collective hysteria that most Europeans felt in connection with mining, metallurgy, and alchemy in the sixteenth century.[71] The problems associated with locating mineral resources in the earth's depths and then processing the raw ores into recognizable silver, gold, tin, and other metals had troubled administrators and government officials like Cecil since antiquity. Corruption and deception were frequent partners in metallurgical business thanks to the specialized knowledge of the workers and the fine lines separating metallurgical transformation from alchemical transmutation. Mining treatises and mineral handbooks by Georg Agricola, Vannoccio Biringuccio, and others helped investors and government officials sort out the muddle, but old headaches surrounding metallurgical work were made more acute when the exploration of the New World opened up additional possibilities and problems. Spanish control of American silver mines made other European monarchs eager to locate new sources of ore and better exploit the minerals already in their control.

Because of England's pressing currency crisis, Cecil's interest in minerals and metals was acute. He began to scrutinize how England was exploiting — or failing to exploit — its own mineral resources, and facilitated the granting of letters patent to nearly all of the mineral speculators who came forward in the 1560s and 1570s. But Cecil knew that England's mineral resources were no match for the silver mines of Saxony or Potosi. To close the gap between England and her neighbors, Cecil pursued the possibility of alchemical transmutation on a grand scale while continuing to invest in mining operations and expeditions, like Frobisher's, that could locate new sources of ore. His reasoning was that if England did not possess sufficient gold, silver, and copper to meet its needs, it might nevertheless be possible to alchemically produce the metals using the lead and tin that England had in greater abundance.

Cecil could make this apparent leap from mining to alchemy because both were based on organic theories of metallic growth and transformation. The earth's warm interior was believed to provide the ideal environment for precious metals to grow and develop from base metals, but the alchemist's heated chemical vessels provided an alternative location for mineral transformation. Cecil's idea that alchemical transmutation might be able to solve England's economic woes was not far-fetched in the context of the prevailing theories of the time.[72] Though alchemy and mining crisscross the boundaries we use to circumscribe technology, science, and even the occult sciences, they nevertheless shared a theoretical framework as well as a set of common experimental practices that would have made an Elizabethan shake his head at the oddities of our modern categories of knowledge.

Cecil's conversations with Giovan Battista Agnello about the richness of Frobisher's gold represented not the first occasion, nor the last, in which he delved into mineral matters. In 1568 Cecil supported the founding of the Mines Royal Company. At precisely the same time, Cecil used his network of informants to monitor the alchemical work of Cornelius de Lannoy, a murky figure whose treatises on the philosopher's stone appear in many Elizabethan alchemical manuscripts under the name "Alnetanus." Ensconced in the Tower of London, Lannoy promised but failed to produce the stone he assured Cecil would revive the flagging fortunes of the English treasury. Undeterred by the protracted difficulties of copper mining, or by Lannoy's failures, Cecil became a founding member of the Society for the New Art in 1571. The society was a small corporation of City and court figures who promoted efforts to transmute base metals into copper — a task apparently easier than the one Lannoy attempted, and which promised quicker results than digging in the mines. The Society for the New Art also failed to meet its objectives, but this setback did not keep Cecil from again engaging in costly metallurgical speculation, this time in the New World exploration and mining proposed by Lok and Frobisher.[73]

One of the greatest challenges facing Cecil as he undertook these metallurgical projects was that Englishmen did not always possess the skills required to extract and transform mineral resources into materials that could directly benefit the crown. It was one thing to have lumps of unprocessed raw copper or silver dug out of shallow pits in the earth, and quite another to excavate deeper mineral specimens or have vendible copper pots, brass ordnance, or coins. The immigrants flooding into London were of some help, especially Dutch and Flemish engineers who were experts at extracting water from low-lying land, and who could turn their hands to the challenge of extracting water from deep mine shafts or using water power to drive battering rams into copper. Still, Cecil needed to bring in experts from faraway Saxony to help, and they took up key positions in the mines, the mint, and other industrial areas. The presence of these foreign experts in the historical record can overshadow the Englishmen who, by working at their side and combining foreign expertise with their own engineering and metallurgical skills, began to emerge as leaders in Elizabethan Big Science. Running through Cecil's projects like a fine vein of silver are the names of William Humfrey, Robert Denham, Humfrey Cole, George Needham, Thomas "Customer" Smyth, William Burde, William Bonde, Lionel Duckett, William Medley, and Sir William Winter. From the copper mines of Keswick to the Frobisher assays in London and Deptford, these men crop up again and again, as Cecil authorized more metallurgical projects. Some, like Humfrey, Denham, Cole, Needham, and Medley, played a role in assaying and testing the specimens; others, such as Smyth, Burde, Bonde, Duckett, and Winter, were investors and project administrators.

What all these Englishmen shared was London. While the City's central role in these enterprises may seem geographically questionable, given the fact that the mines were located in Devon, Cornwall, Dorset, and other regions, as long as the pyx chests (the coffers that held sample coins to be publicly assayed and proved to contain proper amounts of gold and silver) and the mint were located there, and the Goldsmiths' Company continued to play a central role in establishing currency standards, London would continue to serve as a hub of metallurgical and mineralogical projects. And the ongoing presence of London officials and merchants — especially Smyth, Burde, Bonde, Duckett, and Winter — assured that the bureaucratic and administrative center of these projects remained resolutely in the capital.

The Londoners involved in the metallurgical projects were comfortable serving as middlemen between the workers and Cecil, and they helped to ensure that miscommunications did not erupt into serious misunderstandings. Still, Londoners involved in metallurgical projects often had to soothe Cecil, Herle, and Waad when they received conflicting, muddled, or alarming news from the workshops, laboratories, and mines. In the case of metallurgists, alchemists, and miners, Cecil and his informants could exhibit strong prejudices against the worker who (even if English) belonged to a different social class, spoke a specialist's patois, and followed different cultural norms.[74] Despite the best efforts of Londoners like Winter and Bonde, however, some miscommunication was inevitable. This was especially true of Cecil's beloved alchemical projects. Alchemists put themselves in a difficult position by signing detailed contracts to perform work that was typically explained only in allegorical and symbolic terms. That the material transformations they promised to complete could not be achieved over a charcoal fire only added to the stress, and most of Cecil's alchemical projects led to the utter undoing of their central participants.

The schemes of Cornelius de Lannoy provide a case in point. When he proposed to manufacture pure gold worth the staggering equivalent of one million pounds in the winter of 1565, he was walking a tightrope strung between alchemical optimism and fiscal fraud. In exchange for access to the raw metals and technical apparatus in the mint, Lannoy promised to do his work honestly and to communicate his results to Cecil.[75] Elizabeth accepted his proposals, and Cecil swung into action supervising and overseeing the details of the process through his servant, Armagil Waad, who began to gather information about Lannoy's work and work habits.[76] Waad's first surviving reports were written in March 1565, and by the following spring, Lannoy's failure to produce the promised gold was wearing on Cecil's nerves. Lengthy delays resulted from Lannoy's dissatisfaction with the quality of English laboratory supplies, such as chemical glasses and pots which, the alchemist sniffed, were of insufficient strength "to sustain the force of his great fires." When Lannoy received an encouraging visit from Princess Cecilia of Sweden, who was in England to escape financial difficulties of her own, he was prompted to jump ship — a plan communicated to Cecil by Waad.[77] Cecil managed to detain Lannoy in England by confining him to the Tower, where he continued to have unimpeded access to raw metals, assaying tools, and furnaces. In May 1567 Cecil's patience was finally worn out. Waad was dispatched to the Tower to fetch Lannoy to court to face his irate investor the queen.[78]

While Cecil was tangling with Lannoy, he was just beginning his relationship with two new metallurgical collectives — the Mines Royal and the Mineral and

Battery Works. The two began to establish their separate identities in 1565, and both were chartered in 1568. With these projects, Cecil added the equally speculative realm of mining and assaying to his interest in alchemical transmutation. The Mines Royal had its roots in the Mercers' Company, the same London guild that was so instrumental in forming the Merchant Adventurers. Both had governing structures that were closely modeled on that of the London guilds, with governors, deputy governors, and assistants seeing to the administration of work and finances.[79] The two companies pursued distinct objectives, however. The goal of the Mines Royal was to extract ores — especially copper ores — from English soil. The goal of the Mineral and Battery Works was to process the ores "into divers forms . . . in such abundant manner as the whole state of the Realm might feel thereof and some resting place for artificers [be] made famous." A Goldsmith and assay master of the mint, William Humfrey, reported regularly to Cecil about the activities of the Mineral and Battery Works' assayers and laborers.[80] Humfrey was comfortable working in partnership with a German technician, Christopher Schutz, the wealthy London merchant and customs official Thomas Smyth, and another enterprising member of the mint's staff, the Goldsmith and instrument maker Humfrey Cole.

Though Cecil's support of Lannoy had led to nothing of value, the Mines Royal and Mineral and Battery Works did repay their investors and successfully extracted and processed English ores. Cecil remained convinced of the potential in metallurgical and alchemical projects and entered into an unlikely partnership with the Earl of Leicester to found the Society of the New Art in 1571/72. Led by the author of A *Discourse of the Commonweal*, Sir Thomas Smith, and the metallurgical speculator William Medley, and joined by Walter Raleigh's half brother Sir Humphrey Gilbert, Cecil and Leicester embarked on a project to alchemically transmute lead into copper and antimony into quicksilver. The queen heartily approved and exclaimed that this "notable invention will be very profitable to Us, our heirs, and successors for the making of ordnance and other munitions for the wars and for many other like uses." Cecil and Leicester were too busy to play central roles in the society, so Smith found himself organizing the work and reporting back to his friends about their project's many failures. The society's letters patent remark on his experience with alchemy and metallurgy, including his "long search in books of divers arts, divers trials many times in vain essayed, and manifold expenses of his time and money . . . lost." Smith knew the many problems that could accompany working with metals and minerals, whether that work was based in mining, assaying, or alchemy, and he had a healthy skepticism about any enthusiast's claims. From a distance, he watched as Lord Mountjoy's mineral works in Poole ended in financial disaster, and wrote to

Cecil that they had taught him "to trust little to words and promises, or experience . . . or accounts of men of that faculty."[81]

Smith's skepticism was well founded, and the work of the society foundered once he was unable to keep his shrewd eyes fixed on the project after being assigned to the post of ambassador to France in 1572. These duties forced him to hand the management of Medley's work over to Gilbert, who lacked Smith's experience with foreign experts and believed the metallurgical speculator's lengthy and increasingly far-fetched promises. Smith pointed out to Gilbert that Medley was leading him "by the nose" with his "dreams and fanfares."[82] As Gilbert was unwilling or unable to force Medley into a more regular work schedule and to take some accountability for his failures, Smith begged Cecil to supervise the project instead.[83] The work regained some equilibrium, but nothing was ever achieved despite the small fortunes invested in the project by Cecil and the other members of the society. By 1575 the Society for the New Art was, for all intents and purposes, defunct.

Still, Cecil remained optimistic about the potential value of alchemical and metallurgical projects, and when he was alerted to the potential value of Martin Frobisher's strange black rocks, he once again swung into action.[84] Cecil had learned a few things from men like Lannoy and Medley, however. When it came to Frobisher's gold, he wanted to make sure that the work done to transform the ore was speedy, accurate, and reliable. After turning to Giovan Battista Agnello for advice, he encouraged Michael Lok to employ the Italian in a series of assays and tests of the substance. In his laboratory, Agnello was able to produce "a very little powder of gold" from the stone in early 1577.[85] Despite these promising results, a controversy erupted that focused on three related issues: the ores' value, the personnel who possessed the skill to assess it, and the methods that should be used to assay and refine it.

The first element of the controversy was made public once London's rumor mills began to speculate on the value of Frobisher's gold. The queen's master of ordnance, Sir William Winter, became so interested in the ore's value that he sponsored his own trials of the substance at his house near the Tower. Winter's trials were conducted by his personally selected metallurgist, a Saxon named Jonas Shutz, and assisted by his alchemically minded friends Sir John Barkley and Sir William Morgan. After the tests were performed, Winter and his friends were convinced that the ore was going to make them rich. The queen's dour and notoriously suspicious adviser Francis Walsingham, fearful of chicanery and alarmed at the growing involvement of powerful court figures like Cecil and wealthy London investors like Winter, urged caution. When his reservations did nothing to diminish speculation, Walsingham decided that if he could not beat the rumors, he would contribute to the debates, sending samples of the ore to "certain very excellent men," all of whom found "nothing therein, but . . . a little silver."[86] Walsingham's experts included the courtier and poet Sir Edward Dyer and a French alchemist named Geoffrey Le Brum (whose Protestantism made him seem marginally more trustworthy than the Italian Agnello, at least to Walsingham).

The crowded streets of the eastern London neighborhoods of Bishopsgate, Cripplegate, and Tower Hill were home to Winter, Lok, Agnello, Le Brum, and most of the other figures directly involved in debates over the value of the ore. Information and misinformation spread quickly in this cramped atmosphere, and rather than dying out, the controversy only intensified as partnerships developed among what had been rival camps. One of the first collaborations formed at Admiral Winter's house on Tower Hill, where the Italian alchemist Agnello and the Saxon metallurgist Shutz began to work together with impressive results. While we might consider their partnership odd and problematic, resting as it did on a union of alchemical transmutation techniques and more conventional assaying and smelting processes, Elizabethans clearly saw the potential in combining these two approaches. Because alchemy had always been considered a "mixed" philosophy that dabbled as much in practical operations as in theoretical contemplations, the notion of bolstering its reliability with an infusion of Saxon metallurgical techniques seemed worthwhile. Combining metallurgical and alchemical techniques, Agnello, Shutz, and Winter convinced Elizabeth and her Privy Council of the richness of the ore, drew up plans for its "charges and melting," received financing for the "buildings and workmen" needed to process it, and elicited promises that "cunning men" would be sent on the next voyage to look for additional veins of the material.[87]

While the Agnello-Shutz collaboration soothed the anxieties of Elizabeth and her council with the heightened promise of profitable results, it also expanded the controversy surrounding Frobisher's gold. Because both an alchemist and a metallurgist were involved, questions arose about which set of practices could accurately determine the value of the ore. Alchemical methods were of particular concern to Walsingham, who still doubted whether the encouraging findings of Agnello and Shutz were anything more than "the devices of alchemists." After several plaintive memos circulated about the rightful place of alchemy in the trials of the ore, another clever compromise between alchemy and metallurgy was reached: Agnello would handle the ore before it was put in the furnaces and supervise the chemical additives that would make the melting process easier; Shutz would then complete the process of melting and refining the gold in furnaces he had invented himself.[88] Traditionally, minerals were processed by subjecting them to a process of crucible fusion, wherein gold and silver were combined with lead and other materials and melted in clay pots or crucibles. This was the stage that Agnello — the chemical expert — was expected to supervise. The molten metals were then cooled until they solidified into a button of gold, silver, and other trace metals (most often copper and zinc), which could be separated from the slag with a hammer.[89] The button of precious metals then underwent a second process, cupellation, to remove any residual copper, zinc, and lead. This is where Jonas took over, maintaining the button at a red-hot temperature in an open cupel or cup made of porous bone ash. Jonas exposed the cupel to the air until all remaining lead, copper, and zinc were oxidized and absorbed into the vessel's porous walls, leaving a silver and gold bead in need of further refining.[90]

With Agnello tagged as the "chemical" man and Shutz tagged as the "furnace" man, the two men should have been able to work together in Winter's house. But the clever compromise failed when a second alchemist, an Englishman named George Woolfe, was brought in to assist Agnello. The three men encountered irreconcilable differences in method and approach, and in late November 1577 Lok wrote that "the three work-masters cannot yet agree together, each is jealous of [the] other" and fears "to be put out of the work." What had been two-way collaboration had turned into a three-way competition, and Lok reported that the men were now "loathe to show their cunning, or to use effectual conference" with each other. Dissention among the workers at Winter's house soon spread to the queen's commission established to oversee the work. The trials of the ore screeched to a halt, and Lok worried about the "schism [that] had grown among [the commissioners] . . . through unbelief, or I cannot tell what worse in some of us."[91]

While the disagreements raged on, other voices asked to be heard. The queen's German physician, Burchard Kranich, "assayed and proved" that Frobisher's gold was not as rich as Agnello and Shutz had promised investors. While Agnello and Shutz tried to "flatter" Mother Nature into giving up her riches via alchemy, Kranich advocated a more aggressive and controlling approach in which the "rough, wild, and foreign" ore would "be well husbanded by a skillful and expert man." Though the physician had been a professional miner and metallurgist, he had been unsuccessful in his previous Elizabethan Big Science project: mining for silver in Cornwall. Frobisher, smelling the first whiffs of disaster at the gap between Agnello's alchemical courtship and Kranich's metallurgical marriage, quickly threw in his lot with the queen's physician and tried to sabotage the rival proceedings at Winter's house by spying on Shutz and Agnello. At court, the Privy Council sought even more input from "the goldsmiths and gold refiners of London and many other named cunning men," all of whom "had made many proofs of the ore and could find no whit of gold therein."[92]

What brought the dispute to a crisis were questions about the proper methods to be used in the final stages of refining the ore, questions which ultimately drove a wedge between the Agnello-Shutz collaboration and brought Kranich back into the center of the controversy. Shutz had been using a three-step process: the crucible fusion of metal additives along with the ore in a "melting furnace"; the cupellation or separation of valuable metals like copper and zinc from the alloy in a "refining furnace," leaving a gold and silver bead behind; and finally, the parting of the gold from the silver.[93] Two stages in this three-step process proved controversial. First, what should be added to the ore in the first stage of the melting to facilitate its breakdown? This, of course, was Agnello's decision, since it had to do with chemicals and not furnaces. Second, what was the best method of parting the gold and silver in the final stage of the refining process? Agnello, Shutz, and their supporters had three options: parting the gold and silver in a furnace through a process of cementation; parting them with the assistance of chemical agents like nitric acid; and finally, employing a new and controversial chemical method known as sulphide parting.

Jonas Shutz subjected his ore to the traditional cementation method. In cementation, the materials were packed in salt and powdered brick and baked until the silver reacted and formed silver chloride. When the materials were taken out of the crucible and then crushed and sieved, gold particles were left behind. Among the many reasons why Shutz decided to use cementation, the most compelling was that since Agnello was the "chemical" man, he worked to keep the final stages of the process firmly within his furnaces and under his control. This decision resulted in frustration, for Shutz couldn't get the furnaces to a sufficiently high temperature to make cementation entirely effective. Shutz contemplated building a blast furnace with water-powered bellows, and even surveyed potential sites for the enormous piece of equipment, but these plans came to naught. Suffering from smoke inhalation and utter exhaustion, Shutz was in danger of working himself to death in his efforts to increase the temperature within his traditional furnaces. Agnello, working in the center of London, began half-hearted experiments with his own "wind furnace," but as Lok sadly recorded, they "succeeded not well."[94]

In the absence of an efficient blast furnace, the only hope of extracting the gold from the ore rested in chemical agents and additives. Yet Agnello failed to come up with the right chemical recipe, so Frobisher began to push Burchard Kranich to devise a solution. Kranich was old, infirm, and testy — but he was the queen's physician, and Frobisher thought that he had as good a chance as any at succeeding. When Kranich settled on the new sulphide parting method, however, it may have been too cutting-edge for his English sponsors. In sulphide parting, Agnello would have been required to add stibnite (antimony sulphide) to the ore in the first stages of the process. Then, in the third stage, the stibnite would react with the silver, forming a new alloy from which the gold could be easily separated. This alloy of silver and antimony would, a few decades later, cause an alchemical sensation as Basil Valentine's "star regulus of antimony," revolutionizing the practice of seventeenth-century alchemy.[95] In the late 1570s, however, Kranich's use of antimony was problematic because witnesses suspected that he was salting the ore.

To reduce the complaints from investors about Kranich's novel methods, Frobisher urged the trusty Jonas Shutz to abandon Agnello in favor of working with the queen's physician. At this point, the controversies surrounding the ore turned into a tale of espionage and skullduggery. Throughout December 1577 and into the first months of 1578 the two Germans began their çollaboration by exchanging insults: Shutz accused Kranich of "evil manners" and of ignorance in "divers points of the works," while the physician responded that "if Jonas had any cunning," the ore should already have yielded gold.[96] The collaboration ended a few weeks later when the two Germans refused to have anything further to do with each other, leaving only one line of communication between the two camps: the English goldsmith Robert Denham (d. 1605).

After all the controversy and contention, only Denham was able to turn the situation to his advantage. Denham was probably spying on both Shutz and Kranich and reporting directly to Cecil and the other members of Elizabeth's Privy Council about the assays and trials of the ore. Only Denham seemed to realize that Kranich was not salting the ore with silver but adding antimony to it. While this procedural irregularity raised the specter of counterfeiting, it did lead to an easier and more effective parting process. Despite Denham's insights, the commission looking into the rapidly deteriorating situation in the assaying workshops had to report that no one but Kranich "knows what he puts" in his pots. Denham responded by conducting an assay before the queen's representatives, thereby demonstrating that Kranich's additives were a mixture of antimony, silver, copper, and lead.[97] Still, the crown's representatives concluded that the queen's physician could not to be trusted with the ore and put Shutz and Denham in charge of the project. The German metallurgist and the English alchemist managed to achieve a good working relationship, but nothing of value was ever gleaned from Frobisher's gold. The only success story is Denham's: after becoming chief assayer on Frobisher's third voyage in the summer of 1578, he went on to become the director of operations in another Elizabethan Big Science project, the Mines Royal.

At the end of Elizabeth's reign, large-scale projects no longer opened up promising vistas of innovation, economic benefit, and military superiority to London investors or projectors. The list of those bankrupted (or nearly bankrupted) by Elizabethan Big Science was long and daunting. Even Cecil became increasingly reluctant to "move her highness any more to make grants whereof nothing did grow." One of his successors, and a chief officer in the letters patent process in the 1590s, Lord Egerton, scowled at projectors and was reputed to be "a great enemy to all these paltry concealments and monopolies."[98] Parliament was increasingly agitated at the labyrinthine combination of letters patent, company privileges, and guild restrictions that posed a legal nightmare and deterred future enterprise. Not all of the abuses were in the realm of Elizabethan Big Science, of course, and the wine monopoly was probably more troubling to the average citizen than were any of the mineral patents. Still, something had to be done to appease popular sentiments against letters patent and to keep the most lucrative large-scale projects viable.

Elizabeth, who knew the value of the preemptive strike, informed angry members of Parliament in 1601 that she was going to see personally to reforms in the patent system and monopolies before they could be demanded of her. Under increasing pressure from outraged citizens who were prevented from manufacturing such common substances as vinegar and salt because of letters patent granted decades earlier that had not produced what they promised, the queen decided to act. In her last triumph of political oratory, the "Golden Speech" to Parliament, she defended her record. "I never was a greedy, scraping grasper, nor a strict, fast-holding Prince, nor yet a waster," Elizabeth proclaimed, cutting off many of the charges that could be laid at the crown's door regarding the failure of Elizabethan Big Science projects and the patent process to pay investors back after such vast sums of money had been spent. Characteristically disingenuous, Elizabeth blamed any abuses on a "want of true information" and assured her subjects that she "did never put my pen to any grant but upon pretext and semblance made [to] me, that it was for the good . . . of my subjects generally." The idea that her "grants shall be made grievances to my people, and oppressions . . . be privileged under color of our patents," she concluded, was something "our kingly dignity" would not suffer.[99]

Elizabeth dissolved the letters patent granted to some individuals and groups on the basis that she had been led astray by "false suggestions" of benefit and profit, while maintaining other patents, primarily those linked to national defense.[100] London's interest in Elizabethan Big Science limped along in the wake of the crisis, breaking into a run of ingenuity and inventiveness at the merest hint of crown support from Elizabeth's Stuart successors. The halcyon days of Elizabeth's reign were thought to be returning when the young Prince of Wales, Henry Stuart, began to take an interest in the natural sciences and technological innovation, but his untimely death in 1612 dashed those hopes. Most turned their attention to smaller projects that could be undertaken in individual laboratories and became committed to such less grandiose, though still useful, aims as treating the plague or improving instrumentation for navigators and mariners. New industries had become synonymous with corruption and abuse, as had the ancient science of alchemy, and neither was seen as particularly desirable within a profitable commonwealth. Only the lingering interest in securing the national defense, and the increasing number of voyages of exploration that were mounted to take colonists into the New World, remained of interest to London investors.

Cecil may have possessed a grand and sweeping perspective on the potential value of Elizabethan Big Science projects, but London technicians and investors remained fundamentally committed to private profits rather than public benefits no matter what they said in their petitions for letters patent. London could boast enormous resources when it came to the projects — including technical literacy, administrative experience, and formidable financial reserves. Yet the urban sensibility that helped to engender these beneficial resources also shaped the tensions that existed in the projects between collaboration and competition, English and Stranger, and market forces and collegiality.

5

CLEMENT DRAPER'S
PRISON NOTEBOOKS

Reading, Writing, and Doing Science

On a Friday night in the spring of 1581/82, Thomas Seafold, who was being held at her majesty's pleasure in the King's Bench prison in Southwark, had a dream. His long-dead tenant Robert Jeckeler appeared to him in his sleep and told him how to prepare a marvelous restorative elixir. The process of manufacture required several stages. After gathering human excrement and putting it into a two-gallon glass vessel, Seafold was supposed to add three or four pounds of quicksilver to the noxious contents and stew them over a gentle fire for a few days before distilling the mess to concentrate its potency and produce a liquid appropriately named "fetid water." Increasing the strength of the fire under the glass, Jeckeler told Seafold, would produce two additional materials from what remained: "fetid sulfur" and a crusty powder he called "magnesia." After chemically recombining these laboriously separated materials, Seafold would at last have his prize in the form of a medicine that could be used in further chemical processes or to treat the sick.[1]

We know of Seafold's strange chemical dream because he recounted it to a fellow prisoner, a middle-aged merchant named Clement Draper (c. 1541–1620) arrested for outstanding debts. Draper was just beginning a period of incarceration that was to last for more than thirteen years, and he whiled away part of his time in prison writing letters to enemies and allies about the complicated and unjust legal wrangling that had led to his imprisonment. During the remaining hours, Draper filled notebooks with information relating to his three great intellectual passions: medicine, mining, and, above all, chemistry.[2] When Seafold told

Draper his vivid dream, the merchant wrote down every detail in a tiny, meticulous hand, recording it for posterity along with the vintner William Hudson's remedies for colds and flu, the deathbed alchemical remarks of a prisoner held in Newgate, and the Dutch immigrant Adrian van Sevencote's accounts of his friend's successful attempts to transmute copper and tin into silver. Using his pen like a magic wand, Draper conjured up long-dead alchemists, the ghostly presences of absent friends, and the nightmares of his fellow inmates, preserving each one of them on paper. Once they were captured on the page, Draper could return to them for inspiration and information, muse over complicated chemical experiments, and write his own responses to the medicines and other cures that he had collected.

When Draper sat down, assembled a stack of paper into a notebook, ruled the pages, sharpened his quill, and dipped it into a pot of homemade ink, he was doing an important type of work for early modern science — he was writing and recording natural knowledge. Reading and writing about nature was an active practice in the early modern period, comparable to constructing a distillation apparatus or molding a pot. One of the stories we tell ourselves about the development of science in early modern Europe is that it depended upon the growth of experimental culture. Clement Draper's notebooks provide us with an opportunity to see how traditional humanist practices of reading and note taking should also be considered among the practical activities oriented toward the acquisition of natural knowledge. Juxtaposing the living and the dead, the textual and the experiential, the world outside the prison and the world within, Draper may not have "known" (in the traditional sense) the fourteenth-century alchemists who taught him so much, may not have shaken hands with the physicians and chemists whose recipes and procedures he eagerly collected from near and far, but he nevertheless felt an intellectual kinship with them and believed that they were engaged in common intellectual pursuits. Though Draper's prison community was in part imagined, it was no less important than an actual face-to-face social community just because it was partially constructed from ideas and ideals. Draper's notebooks provide us with an entry point into his world.[3]

The vibrant intellectual world that we have seen elsewhere in London — and the urban sensibility present on Lime Street or under the arcades of the Royal Exchange — extended into the gloomy corridors of the City's prisons. When one inmate, Geffray Minshull, wrote from the King's Bench prison in 1618, he described the "little world of woe" inside its walls, and painted a portrait of its culture that was bleakly inhumane and based on cutthroat competition and deprivation (Figure 5.1).[4] Given Minshull's grim account, the King's Bench would not seem to be the ideal place for the study of nature. The contrast between life

ESSAYES AND CHARACTERS
OF A *PRISON* AND PRISONERS.

Written by G. M. of *Grayes-Inne* Gent.

Here is a true layler strips the Diuell in ill.

These that keepe mee, I keepe, if can, will kill:

Printed at LONDON for *Mathew Walbancke,* and are to be folde at his fhops at the new and old Gate of Grayes-Inne. 1 6 1 8.

Figure 5.1. A dishonest jailer at the King's Bench prison from Geffray Minshull's *Essayes and characters of a prison and prisoners* (1618) illustrates the dangers and challenges facing a man like Clement Draper, who was incarcerated in the Elizabethan prison system. Minshull knew of what he spoke — he too was a prisoner in the King's Bench. Reproduced with the permission of the Henry E. Huntington Library.

inside the prison walls and the bustling, cosmopolitan, and lively world of intellectual exchange I describe in the previous chapters is acute. By entering
Draper's prison community, however, we confront again the rich variety of experiences and practitioners that characterize science as it was practiced in the Elizabethan City. Despite the apparently daunting obstacles facing him, Draper did
manage to construct a lively intellectual community to sustain him during his
imprisonment, a community that included the authors of books he read and
transcribed, other prisoners who shared experimental nuggets gleaned from
their own experiences, and visitors to the prison who brought in news and information from London and the world beyond.

Draper was not the only prisoner — or even the first King's Bench inmate — to
have an interest in the natural sciences during the sixteenth century. In the
Tower of London, both Edmund Neville (c. 1555–1620) and Walter Raleigh
(1554–1618) inquired into the workings of the natural world from their cells.[5]
Robert Record (c. 1510–58), the Welsh physician and mathematician who introduced the equals sign into common use and wrote popular mathematical texts,
was also arrested for debt and died in the King's Bench. And the physician
William Bullein (d. 1576) wrote *Bullein's bulwarke of defence against all sickness,
sorenes, and woundes* (1562) while in a debtor's prison in the City.[6] We know little of the prison experiences of these men. What sets Clement Draper apart from
his many contemporaries who spent years, even decades, in one of the City's
prisons is that he left us rich accounts of his intellectual activities. From a set of
his surviving notebooks, we are able to piece together how even the King's
Bench prison could serve as an important site for the production, evaluation,
and propagation of natural knowledge in Elizabethan London.

By reading, collecting, and recording his insights into nature, Draper was able
to transcend the walls of the King's Bench and engage in the work of science.
Through his notebooks, Draper managed to situate himself among an array of
experts that he consulted on questions of natural knowledge. He was thus able to
instantiate an intellectual community that provided him with new ideas and
valuable information, which in turn helped him to develop his interests. Some
of the notebook entries suggest that Draper also engaged in London's blossom-
ing world of experimentation while in the King's Bench. The juxtaposition of
these two ways of knowing about nature — one textual, one experiential — illuminates not only their points of tension but also their considerable compatibility,
especially when it came to working through early modern "thought experiments." Writing up experimental processes, whether they happened within the
prison or only within the pages of Draper's notebooks, provided him with a
means to process, query, and develop natural knowledge.

To understand Draper's intellectual activities, we must first understand his world, and this chapter begins with a reconstruction of his prison community and his underlying philosophy about what knowledge was and how it should be used. Draper's approach to the study of nature was predicated on the belief that all information — even from the most reliable sources — needed to be sifted, sorted, and digested before it could be accurately judged and appropriately applied. A method of note taking that was related to the early modern passion for collecting biblical commonplaces and bons mots provided the ideal method for Draper to come to terms with the information that his community provided him about the natural world, as I illustrate in the chapter's second section. Finally, I consider the fine line that for Clement Draper separated reading, writing, and experimenting, and probe the consequences of undertaking the study of nature within this framework.

DRAPER'S "LITTLE COMMONWEALTH": FRIENDS, NEIGHBORS, AND AUTHORS

Clement Draper began his life far from the London prison where he was held in the 1580s and 1590s. Born into a wealthy landholding family in Leicestershire with a tradition of officeholding in the City, Draper was part of that up-and-coming group of middling Elizabethans who were neither poor nor rich by the standards of the time. His father, Thomas, died when he was a toddler, leaving funds in his will for the maintenance of his three children.[7] Little is known about Draper's early years, and we do not know how or where he was educated except that he knew Latin and could both read and write. These were valuable skills, which could be employed in a number of lucrative trades and professions, including the practice of law and the pursuit of a merchant fortune. Draper's education may have been overseen by his paternal grandmother, Agnes Aynsworth, or by his uncles, William and Christopher Draper — all of whom would have had a vested interest in seeing their young charge was well schooled. There is no record of him attending university, however, or any London or Leicestershire school, so it is likely that Draper was tutored at home by private schoolmasters.

As Draper approached maturity, his father's relatives helped him to get started in business so that he could support himself. Clement Draper was admitted to the Ironmongers' Company under the auspices of one of his father's brothers, a membership which gave him the right to exercise a trade in the City, and cemented his ties to the London branches of his sprawling family tree.[8] Chief among Clement's London relatives was Sir Christopher Draper, elected lord mayor in 1566 after years of serving the City in a variety of public offices, including sheriff. Sir Christopher was a deeply pious man and, at the time of his death, he bequeathed substantial sums to London's poorest residents, its prisoners, to its sick, to the Ironmongers' Company of which he was a member, and to the City itself, to repair bridges and roads and clean the filth from the Thames. Despite this widespread largesse, his nephew Clement received as a legacy the paltry sum of forty shillings and a suit of mourning clothes — the same amount Sir Christopher bestowed on his servant John Roberts. The remainder of Sir Christopher's substantial estate was divided among his three well-married daughters (one of whom was wedded to Euclid's first English translator, Henry Billingsley). Other Draper relations who supported Clement during his years in London included the Clitherowes, and he partnered with his paternal first cousin Henry Clitherowe in a long-distance shipping business that transported flax, leather, saltpeter, and gunpowder between London and the northern European cities of Lübeck and Gdansk.[9]

As James Cole had discovered when he wed Louisa de L'Obel, blood relatives could be invaluable in getting an ambitious man started in business, but an advantageous marriage could significantly improve one's future prospects. After establishing himself as a businessman in the City, Draper cast his eye about for a good family to marry into and selected the Gartons. His wife, Elizabeth, provided Draper with ties to prominent noblemen like William West, Lord Delaware, and a wider circle of London notables. The Gartons were prosperous landowners, with estates in Kent and Sussex, who were also associated with London's Ironmongers' Company. They invested heavily in the iron industry, and it may have been the Gartons who introduced Draper to mining and fostered his interest in metallurgical projects. Draper, like so many who speculated on Elizabethan Big Science, found that it only led to his financial ruin.

Clement Draper's investments in metallurgical projects played a major role in the events leading up to his arrest and imprisonment in Southwark. In his first disastrous venture he lent some ships to the Frobisher enterprise — ships that had been profitably employed shuttling between London and the North Sea and which were then diverted to the unprofitable task of shuttling lumps of black rock from what would become Newfoundland to London. Overextended financially after his foray into Elizabethan Big Science, locked in controversy over outstanding debts with men like Michael Lok, and then evicted from his house in the parish of St. Dunstan in the East upon his uncle Christopher's death in 1580, Draper began to invest heavily in yet another project in an attempt to keep his head above water: alum and copperas mining in Dorset. It was unfortunate that these investments put him squarely between two tendentious noblemen, the Earl of Huntingdon and Charles Blount, Lord Mountjoy, who were struggling over rights to the mines. Both Huntingdon and Mountjoy were in dire financial straits, as well, and neither was averse to ruining Draper in his pursuit of what would turn out to be yet another completely profitless enterprise.

Though Draper was arrested late in 1581 or early in 1582, the chain of events leading to his arrest began in 1558, when James Blount, Lord Mountjoy, inherited a Dorset estate at Canford. Financially overextended, Mountjoy mortgaged the properties on the estate but retained the right to explore Canford's mineral riches, specifically alum and copperas. The two crystalline minerals were used in cloth production and a host of other applications, and imported alum was particularly expensive.[10] After beginning to work the mines in 1562, four years later Mountjoy received the queen's letters patent, which granted him a monopoly on alum production for twenty-five years. With the enthusiasm of most speculators, Mountjoy invested money that he did not have in order to import Italian mining experts and equipment. He erected a large workhouse, Okeman's, and established mines at Alum Chine, Darling's Chine, Haven's House, Branksea, and Brownsea Island. Mountjoy's mines produced abundant copperas — but very little of the rarer and more expensive alum. Increasingly in the red, Mountjoy was forced to lease the Canford mines to George Carlton, a relative of his wife, and to John Hastings, a relative of the Earl of Huntingdon.

This lease provided the toehold that Henry Hastings, Earl of Huntingdon, needed to begin his own disastrous speculation in metallurgical projects. In 1570 the earl purchased two-thirds of the Canford estate from Mountjoy — but not the mines, from which Mountjoy could not be parted. While Mountjoy would not surrender the rights to the alum and copperas mines, he was willing to lease them to a series of eager speculators, which is how Draper ended up caught between these two powerful noblemen. In 1579 Draper entered a partnership with Richard Leycolt and John Mansfield to lease the Canford mines from Mountjoy for the sum of eight thousand pounds. Draper's partnership took over Okeman's and the Alum Chine mines and set to work. All told, the partnership invested the awesome sum of eighteen thousand pounds (nearly eight million dollars in today's money) in the operations, most of it siphoned from Draper's business profits or borrowed against his excellent credit in the City.[11]

All might have been manageable had it not been for the death in the autumn of 1581 of Draper's Dorset landlord. Most merchants and noblemen of the period were in debt, after all, and experienced in the complicated business of massaging accounts and assuaging creditors.[12] When James Blount, Lord Mountjoy, died, however, his debts were so enormous that the crown seized his estates to repay creditors, and the enterprising Earl of Huntingdon took the opportunity to press for his rights to the mines at Canford. Charles Blount, the new Lord Mountjoy, insisted that the mines were still legally his.[13] The situation of leaseholders like Draper in such a hostile environment was difficult. Factions developed, robbery and assault were frequent, and ill will was pervasive. In 1581 Draper was unable to make an interest payment on one of his many obligations — and the Earl of Huntingdon held the note. At the earl's suit, writs were issued for Draper's arrest, and though he remained at large until after Christmas, his workhouses in Dorset were repossessed by agents working for the earl.

Draper realized too late that his partner, Richard Leycolt, was double-crossing him. Working for the earl, Leycolt encouraged Draper to enter deeper and more perilous levels of investment in the Canford mines, all the while providing intelligence about dwindling accounts, small profits, and the tense situation on the ground among the workers. Though Draper was sent to prison, he stubbornly refused to give up his leases on the mines, while Leycolt remained free and in the earl's pay. Between 1582 and 1586 Draper kept up a regular, angry correspondence with his adversaries and government officials, each letter explaining in full the evildoings of the earl and the underhanded behavior of Leycolt. Wisely, the earl worried about Leycolt's shifty role in Draper's downfall and tried to keep the man on his side. In a letter to his cousin Francis (who oversaw his increasingly difficult financial affairs), Huntingdon wrote anxiously, "I pray God Leycolt stands fast. . . . I should have paid him £400 . . . this Michaelmas last." Leycolt had the power to bring the whole scheme down around the earl's ears, and Huntingdon's only financial hope was to somehow wrest the rights away from Draper. The earl tightened the screws by enlisting Draper's in-laws, the Gartons, to further pressure him into releasing the mines in exchange for a settlement of his debts. This ploy was successful, at least for Huntingdon. In 1583, approximately two years after he was arrested, Draper bowed to pressure from his wife's family and surrendered the leases to the earl. Draper's understanding was that a portion of the profit from the mines would be used to pay off his creditors, thereby releasing him from prison, and would also provide him with a small annuity to support his wife and children. Once again, the earl proved untrustworthy, and Draper's creditors were not paid. Draper complained bitterly to the queen's treasurer, Francis Walsingham, about the continued injustice of the situation and sent letters to well-placed courtiers like Sir Christopher Hatton and the Earl of Leicester, but to no avail.[14]

And so, despite his best efforts, Draper found himself in the King's Bench from the early 1580s until at least 1593. The prison's records do not survive for the period, so his precise dates of incarceration remain elusive, and details surrounding Draper's life there are equally sketchy. Because so little is known about the King's Bench during the Elizabethan period, it is difficult to make direct comparisons between it and better-known institutions such as the City's notorious criminal jail, Newgate.[15] To better understand Draper's experiences at the King's Bench, and how they related to his interest in the natural world, we must resort to other texts written about prison life, the experiences of other prisoners, and his own notebooks.

Elizabethan London was liberally sprinkled with prisons and jails, each linked to a particular category of crime (from petty complaints to treason) and to a particular court that heard motions and decided on cases. London's Compters, for example, housed simple bankrupts, while the King's Bench held debtors to the crown. The White Lion was full of recusants who refused to adopt the new Protestant religion and clung to their Catholic faith, the Marshalsea held dangerous felons and traitors, and the Tower was set aside for the most important political and religious prisoners.[16] The King's Bench prison was on the south bank of the Thames in the London suburb of Southwark — a neighborhood where playwrights, criminals, brothels, and prisons flourished in the relatively lenient atmosphere of Surrey — behind high brick walls constructed during the reign of Henry VIII. Southwark housed five of Elizabethan London's eighteen prisons, as John Taylor the Water Poet (1578–1653) explained:

> Five Jayles or Prisons are in Southwarke plac'd
> The Counter (once St. Margrets Church defac'd)
> The Marshalsea, the Kings Bench, and White Lion . . .

The fifth Southwark prison, the Clink, dated back to the Middle Ages and was officially under the jurisdiction of the Bishop of Winchester. By the Elizabethan period it housed not only those accused of bawdy crimes like prostitution and fornication but additional malefactors as well. Crammed into a few blocks, the Marshalsea, King's Bench, and White Lion formed a large prison complex that only confirmed the area's bad reputation in the eyes of London's upstanding citizens.[17]

During Draper's imprisonment, the King's Bench facility held mainly debtors, as well as a handful of religious recusants, traitors, and wrongdoers who could not be housed in nearby prisons due to overcrowding. Soon the King's Bench itself was feeling the pinch of overpopulation, and the rising number of inmates forced officials to put some into lodging houses and inns adjacent to the prison in an area of looser control known as "the Rules."[18] Local residents complained, and in the spring of 1580 London officials asked the queen to expand the prison so that the King's Bench would not be forced to "lodge . . . the prisoners abroad . . . for lack of room." Debtors like Draper were by far the least dangerous prisoners in the Southwark prisons, and were often housed in the Rules rather

than the prison proper, in part because their prison sentences were considered
less a form of punishment than a temporary detention that kept a debtor from
fleeing the jurisdiction or being arrested by different creditors.[19]

As in the larger Elizabethan culture which the prison reflected, life in the
King's Bench differed for those possessing titles or money. Prisons were fee-for-
service institutions, and those who administered and provisioned them expected
to turn a profit, just like any Londoner doing an honest day's work. Each prison
was divided into wards according to social and financial status, with better ac-
commodations, more palatable food, and greater freedoms granted to those in
the best wards, while not even basic subsistence was provided to prisoners in the
most impoverished, such as the "Hole" in the Wood Street Counter or the ironi-
cally named "Bartholomew Fair" at the Fleet. Perks in London's prisons were entirely determined by how much money an inmate could spend on bribery, tips
(or "garnish") to prison employees, and payments to outside providers who were
willing to keep the prisoners well stocked with fine wine and supplied with ample reading materials in exchange for exorbitant fees. When the money ran out,
prisoners had to rely on their wits for survival, as Thomas Middleton (1580–1627)
reported in *The roaring Girle* (1611):

> A Counter!
> Why, 'tis an University! . . .
>   With fine honey'd speech
> At's first coming he doth persuade, beseech
> He may be lodg'd with one that is not itchy,
> To lie in a clean chamber, in sheets not lousy;
> But when he has no money, then does he try,
> By subtle logic and quaint sophistry,
> To make the keepers trust him.

When Geffray Minshull wrote about the King's Bench prison earlier in the seventeenth century, he was equally censorious about life inside the walls. From
corrupt prison officers who demanded fees for every service and privilege, large
or small, to dissolute inmates who drank too much and cared too little about
the state of their souls, Minshull painted a dark portrait of prison life. All of the
evils of the world collapsed into the building and flourished in a terrible inver-
sion of life outside the walls. Grasping for metaphors to describe the complexity of his prison experience, he reached for "microcosmos," "little commonwealth" and "city" — terms that contained vestiges of the familiar and orderly
world of friends and neighbors outside, but were twisted and contorted by prison
conditions.[20]

Draper, though in debt up to his ears, was clearly able to gather together the funds to purchase some of the amenities of prison life, and to earn enough trust from fellow inmates and King's Bench officials to smooth out the rough edges that would have resulted from temporary shortages in his cash flow. The very real perils associated with the Elizabethan prison system could thus be mitigated for a man like Clement Draper — so long as he could continue to pay for food, clothing, light, heat, and the freedom to visit the streets of Southwark accompanied by appropriately tipped guards. One of the more expensive privileges that Draper apparently purchased was permission to "go abroad" from the prison during daylight hours (for which inmates were charged approximately four shillings). It was probably while Draper was "abroad" that his wife, Elizabeth, conceived a child, for example.[21] A debtor like Draper could rack up considerable further obligations during his incarceration by paying for such freedoms, and before he was released all debts owing to members of the prison staff had to be cleared. Draper would even have paid a fee to the prison's marshal simply for being set free.

Given the deprivations and rapaciousness of prison life, it is perhaps unsurprising that Draper resorted to the activities common among well-born and well-educated men of his time who found themselves behind prison walls: he began a program of reading and writing, discussed his intellectual interests with members of the prison community, and even conducted experiments. Draper constructed his own London neighborhood from the friends who wrote to him, the visitors who came to the King's Bench, and the books that found their way to Southwark as gifts or loans. The one requirement for admission into Draper's neighborhood was that you were interested in the properties of nature and happy to share your insights with him — it did not matter whether you were living or dead, present or absent. Some transmitted information "by word of mouth," such as Giles Farnaby, who told Draper how to sublime *sal armoniac* (ammonium chloride) in a urinal — a vital step in the manufacturing of the philosopher's stone.[22] Others gained entrée by lending him books on chemistry and medicine, which Draper copied out by hand, adding them to his growing collection of prison reference materials. Paracelsus hovered in the air in Draper's rooms (in translated form), and there is evidence that the inmate had access to numerous printed works, including medical texts and books of experiments. And a few people in Draper's "little commonwealth" shared developments in the world outside the King's Bench prison, outlining experiments performed in France and Prague and sharing medical formulas gleaned from prominent London physicians.

Draper played many roles within his prison community. One was to serve as a willing pupil eager to learn from expert teachers. King's Bench prisoners like

Thomas Seafold were happy to supply him with their hard-won wisdom and to share unusual skills and methods. The King's Bench prisoner William George taught Draper how to heal stubborn wounds with plants and kitchen staples like honey and salad oil. A prisoner named Lovis shared his cure for syphilis, which involved downing for fourteen consecutive days a medicinal drink made of Rhenish wine and the New World miracle drugs sarsaparilla and rhubarb. Draper was able to "learn from a prisoner . . . vehemently vexed with a cough" the formula for a syrup to ease the congestion in his lungs that "he got at the apothecary's in Leicester."[23]

While fellow inmates appear regularly on the pages of Draper's notebooks, his most trusted teachers were John Brooke, Everard Digby, and Humphrey Evans. Brooke taught him a number of useful practices, including how to make white pomanders to keep the strong and unpleasant smells of prison at bay, and how to purge tin of its imperfections so that it would resemble silver.[24] Brooke's method of using a combination of three iron pots to speed up chemical processes was a source of inspiration to Draper's chemical work. Sir Everard Digby, the grandfather of the inveterate seventeenth-century experimenter Sir Kenelm Digby, taught Draper how to heal fresh wounds and knit together broken bones using a poultice made from a rare plant. The recipe depended on locating "*scala celie*," which Draper noted was beneficial "for many infirmities," including kidney complaints and menstrual irregularities. Digby shared visual clues to identify the plant, which grew in Hartbarrow in Cumbria — a long way from Draper's Southwark haunts — and even pointers on how the leaves tasted "sweet in eating," in case his eyes failed him. Closer to the King's Bench was Humphrey Evans, the son-in-law of a joiner who lived in Lame Alley near Cheapside. Evans claimed to be the pupil of Thomas Barker, a man Draper esteemed as a font of wisdom about nature. Evans gave Draper the formula for an expensive purging pill made with saffron and rolled in gold leaf that he acquired from Barker, as well as other recipes gathered from his master's notebooks.[25]

Draper's wife, Elizabeth, was also one of his trusted teachers. Elizabeth Draper was expert in making medicines, both from botanical simples and from chemicals. Draper recorded his wife's method for making a chemical medicine that she used to heal their daughter's sore mouth, for example. Elizabeth brought her husband the recipe for a highly effective diet drink procured from her neighbor Bollande. Elizabethan diet drinks were not for weight loss but for regulating bodily humours and keeping the body in perfect balance. Bollande's diet drink, laced with sarsaparilla, the powerful laxative senna, chamomile flowers, licorice, and raisins, was to be consumed three times a day for nine to twelve days. When Elizabeth learned a new method for making gout medicine from physician and astronomer Richard Forster, and a friend taught her how to transform egg yolks, rose oil, and the powder made from burned cork into a medicine to treat ulcerated chilblains, both of these remedies made their way to her husband in the King's Bench.[26]

Teachers came and went in Draper's notebooks, after making brief appearances to share a medical formula or guide him through a difficult chemical process. Draper's friend William Walden taught him how to make a purging medicine with herbs and Rhenish wine that was guaranteed not to debilitate the body but to cleanse it gently. Instructions on how to preserve fowl were "delivered [to] me by Yaxley from the Lord Delaware." There was even room in Draper's world for advice from adversaries and enemies. Draper recalled that his turncoat business partner Richard Leycolt had been taught how to release the mercurial spirit from lead and copper by a "western man," and recorded the method. Even after Draper was released from prison in the 1590s, he was still acting as a student of nature and seeking out advice from passing acquaintances. Early in 1607 he learned from Charles Sled (one of John Dee's quarrelsome alchemical assistants, who became an agent in Francis Walsingham's network of spies) how to precipitate quicksilver and gold.[27]

As a student of nature, Draper was interested in reports of foreign developments in science and eagerly collected news from the wider world. A Mr. Marshall brought back tales of chemical processes undertaken in France, where a man made gold salts in 1579. William Evans shared a recipe from Paris on how to make a distillation of mercury. Draper was alert to reports of work done by the famous English alchemist Edward Kelly, who had served as John Dee's scryer and stared into his master's crystal stones to see visions of angels, and then established an independent reputation as an alchemist at the court of the Holy Roman Emperor Rudolf II in Prague. From one of Kelly's servants, Thomas Warren, and from Jane Constable, Draper received details of Kelly's efforts to make the philosopher's stone. He also copied a chemical recipe credited to Kelly that combined salt of tartar with quicksilver and could be used to perfect gold and silver. Draper had further contacts in Prague, including a Mr. Norten, who explained how to lime chemical glassware so that it could sustain the heat of an experimenter's fire.[28]

Of course, in Elizabethan London, one did not have to go abroad to find exotic, foreign approaches to nature. Draper knew the Dutch immigrant Adrian van Sevencote, who owned property near the prison in the parish of St. Olave Southwark. Sevencote told Draper that he had "seen another stranger" work a transformation of silver, using copper and quicksilver. Francis Verseline, the son of the Venetian immigrant glassmaker and medical empiric Giacomo Verzilini

(1522–1606), shared medical recipes for a kidney ailment and for treating sexually transmitted diseases. John van Hilton, also a Dutch immigrant, gave Draper recipes for sublimating and rubifying ammonium chloride. Nicholas Chester gave him detailed instructions for drawing oil of sulfur from the experiments of the Dutch physician John Stutsfelde. And an anonymous Dutch man taught him how to make a precious tincture that marked an essential step in the process toward the philosopher's stone.[29]

Among Draper's immigrant teachers, pride of place was reserved for Joachim Gans, a Jewish metallurgist and alchemist from Prague.[30] He entered the country in the early 1580s under the auspices of George Needham and the Mines Royal to consult with the crown on English efforts to explore the country's mineral riches. There he came to the attention of Sir Walter Raleigh, who persuaded him to be part of the 1585 Roanoke expedition. Upon his return to England in 1586, Gans resided among the foreign instrument makers and metallurgists able to ply their trades beyond the reach of the Goldsmiths' and Grocers' Companies in the liberty of the Blackfriars in London. He lived in that hustling and bustling environment of Strangers, players, and craftsmen without incident, but ran into trouble when he traveled to the southern port city of Bristol in the autumn of 1589 and was arrested in a crowded inn after engaging in a heated debate with a local cleric on the divinity of Christ. Shortly thereafter, in 1589 and 1590, Gans gave Draper chemistry lessons in the King's Bench prison.[31]

Insights into nature that Draper could not glean from fellow inmates or gather from teachers and friends outside prison he collected from books, and the authors and owners of those books were his constant companions. As we have seen in previous chapters, the pursuit of natural knowledge in the City was intensely social, but the imprisoned Clement Draper had to be creative to participate in lively exchange and open debates like those taking place among the Barber-Surgeons or the mathematics teachers on London's crowded streets. For Draper the gap between his status as a prisoner and his desire to converse about science with other students of nature was filled by books. Historians of science often turn to early modern books and examine them as material representations of intricate networks of authors, readers, and practitioners who were interested in making and evaluating natural knowledge in early modern Europe. From the vast Renaissance library of John Dee to the bookstalls of London, historians of science and historians of the book have provided us with important insights into how books about nature were read, how they were written, and how they were bought and sold.[32] Whether manuscript or print, vernacular or Latin, full of time-worn ideas or crammed with novelties, books played a crucial role in the formulation of intellectual communities, the exchange of ideas about nature, and the transfer of skills and techniques among London practitioners. Books were bought and sold in commercial transactions, given as gifts, traded with friends, and eagerly sought out by readers hunting down new ideas and hoping for intellectual stimulation. Collecting and publishing books were two important points along what Robert Darnton has described as a "communications circuit" — an orbit of books that looped from author, to publisher, printer, shipper, bookseller, reader, and back to author.[33]

On this grand orbit of books, smaller communication circuits existed between readers that were much like the small epicycles astronomers sketched onto planetary deferents to account for the sometimes confusing movements of the heavenly bodies. One of the most common ways for a book owner to share his bounty and participate in a community of readers was to allow a friend to copy all or part of a given text. Clement Draper, who couldn't roam London's bookstores during the decade he was incarcerated in the King's Bench prison, received a steady stream of new texts to copy from friends, visitors, and fellow inmates. "This book of diverse receipts," Draper wrote in one notebook devoted to alchemy, "I had from Richard Bromhall, along with [an]other translation of George Ripley's work." Bromhall also supplied Draper with the method of practice of "an Italian doctor, faithfully translated by Doctor Delaberis," a book of "diverse receipts," "translations of George Ripley's work," and the *Rosary of the Philosophers*.[34] Draper extracted many items of interest from the works of Paracelsus that had been lent him, including outlines of chemical processes and instructions on how to make a medicinal balsam.[35] Draper also copied a number of "secrets" he gathered from Italian authors Isabella Cortese, Giambattista della Porta, and Leonardo Fioravanti.[36]

Draper, though confined to prison and suffering enormous legal pressures from the Earl of Huntington, nevertheless managed to engage in the study of nature. He did so by stitching together a neighborhood filled with prisoners, friends, and the authors of the books he collected and read. The members of Draper's community kept him supplied with the ideas, anecdotes, and materials necessary to do the work of science. As Draper's experiences show, Elizabethan London's intellectual energy was so powerful it could even seep through the cracks in prison walls. Draper, by tapping into the City's vitality and intellectual industry, was able to sustain his interests in chemistry, metallurgy, and medicine for more than a decade while in the King's Bench.

DIGESTING SCIENCE: DRAPER'S PRISON NOTEBOOKS

While friends, family, and authors provided Draper with the essential raw ingredients to do the work of science — books, recipes, and nuggets of experimental and practical wisdom — he needed to expend further effort to transform his

collection of information into a more coherent understanding of the complicated natural world. Nestled within his community and surrounded by stacks of
books and notebooks, Draper worked away to gain fresher and more reliable insights into nature. He did so by not only collecting experiments but also copying
them into notebooks. As he did so, he pored over them with an eye to clarifica-
tion, comparison, and evaluation in order to gain greater understanding of natural and chemical processes. Draper's notebooks thus delimited his "laboratory,"
the intellectual space where he worked through his ideas and came to terms with
them.

Draper's notebooks functioned like a particular piece of alchemical distillation apparatus called the pelican. The pelican was a vessel used in chemical and
alchemical processes that produced an endless circulation of materials in a
closed container. As the substances were heated in the bottom, evaporation and
condensation occurred in the upper reaches of the vessel, and drops of liquid
would trickle down the sides and into the two hollow arms, falling back into the
lower portion of the pelican. This circulation and recirculation of matter led to
the production of a new substance, but it did so in a gradual fashion. Similarly, in
Draper's notebooks, we see a cycle of collecting, copying, clarifying, and comparing that kept his knowledge in a constant state of circulation through the page
and within his community. Likening Draper's notebooks to a piece of distillation
apparatus is not an empty metaphor, therefore, but a useful way of describing
how his process-based approach to the natural world reflected his interest in
chemistry and his belief that matter could be endlessly combined, separated,
and recombined.

Draper was one of many avid note takers in Elizabethan England. The period
saw an information explosion as book after book rolled off the printing presses
and into the hands of eager consumers. As Elizabethans read and absorbed the
contents of these works, they had to find ways to cope with the overload, and
scores of manuscript miscellanies, medical and chemical recipe books, notebooks, and commonplace books that collected edifying sayings were composed
by women and men trying to organize their responses to this welter of new
knowledge. Note-taking skills were taught in the grammar schools to help pupils
overcome information overload, and to prepare the more advanced and privileged students to enter the universities. Many notebooks were kept like merchant books or intellectual journals, with entries made chronologically whenever a reader found a new tidbit of information or a provocative passage in a
book. Over time, indexing and classification skills developed as notebooks became longer and locating a particular item when it was required grew more difficult. Until recently many of these early modern notebooks have received little

attention from scholars, since their contents seemed too inchoate to be either valuable or completely understood.[37]

If we try to understand the notebooks kept by Draper and his contemporaries as technologies, rather than texts, their value and significance become clearer. Note taking was a personalized technology of collecting, tracking, and sorting. Such a practice was never intended to produce a finite, complete, and orderly single text. Instead, note taking produced open-ended, ongoing, and imbricated documents that stood in relationship to additional volumes both in print and in manuscript. Notebooking and commonplacing, therefore, were not just a means to cope with information overload — they provided opportunities for exploiting it. Draper's seemingly chaotic running commentary on alchemy, medicine, mining, and other topics reveals that he embraced the new information age with enthusiasm. He clearly believed that whoever dies with the most information wins, and he piled up texts, eyewitness testimony, secondhand reports, rumors, and physical demonstrations with glee. Like a chemist hovering over a pelican containing hundreds of ingredients, Draper was never sure what the result of his note taking would be after things stewed and circulated, but he seemed confident that the results would be more interesting — if more hard won — than they would from simply sifting through a handful of sources on a given problem.

Draper's work of textual circulation and gradual distillation reminds us that in late-sixteenth-century England the natural sciences were not yet exclusively, or even primarily, experimental. Instead, they were in the process of becoming more experimental, as an older way of understanding nature based on mastering authoritative classical and medieval texts by authors like Aristotle and Albertus Magnus began to give way to a new philosophy rooted in the combination of theoretical and practical knowledge. Theoretical knowledge had long been considered the dignified output of a mind disciplined through a Latinate study of the seven liberal arts, while practical knowledge was considered to be the vulgar byproduct of handiwork and craft activity. Theoretical knowledge (*scientia* or *episteme*) was certain, systematic, and logically sound; from theoretical knowledge one could arrive at generalized beliefs about the world and everything in it. Practice was based on individual human experiences (*praxis* or *experientia*), or on knowledge gained from labor and manufacturing (*techne*), which could not be gathered into a trustworthy, reliable view of the world.[38]

By the end of the seventeenth century this ancient division between the life of the mind and the life of the hands had been replaced by an interdisciplinary "new philosophy" — experimental science — that linked particular acts of manipulating and experiencing nature with a generalized sense of knowing about nature. The typical question that historians have asked is *how* this new, interdis-

ciplinary approach emerged. Elizabethan London provides us with some answers to that question. By taking to the streets of the City and exploring who was
engaged in the work of science, what constituted human experiences with nature, and how, finally, these people and practices constituted new forms of natural knowledge, we have seen how surgeons, midwives, distillers, apothecaries,
gardeners, technicians, and other Londoners began to juggle theories and practice, thinking and doing. These humble, urban practitioners thus contributed to
the growth of a new interdisciplinary science by sharing their hands-on experience with nature and their hard-earned practical skills with other students of nature, while taking an increasing interest in the theories about how nature worked
and operated that were included in vernacular science publications and discussed in public lectures. It was from the efforts of these practitioners that later
scientific experiments — the replicable, verifiable interaction of theory and practice — were derived.

London's interest in nature thus included acts of making and acts of doing as
well as acts of thinking. Acts of making could include artistic work like sketching
botanical illustrations or making a wax model of the human body as well as concocting medicines or crafting a mechanical object. Defining acts of doing is
trickier, for while some activities clearly qualify — performing a chemical experiment, for example, or dissecting a corpse, or using a mathematical instrument
to measure the distance between two heavenly bodies — other activities need to
be added to this list. These include the work of collecting (whether books or natural objects), sifting and sorting, reading, thinking, and writing. While we might
consider reading, thinking, and writing as forms not of active doing but of passive
reflection, Elizabethan readers were actively engaged with their texts, and the
act of writing required physical activity such as making ink, sharpening pens, assembling paper, and finally sitting down and tracing the letters on the page.[39]

In one notebook entry Draper describes the dynamic relationship that should
ideally link reading, writing, and practice. The entry described an alchemist —
Draper himself? — on a gradual learning curve that began in his late twenties
when he set out to study alchemical theories "by little and little." These textual
efforts were followed by his first efforts to actually practice alchemy by manipulating chemical substances. "Many divers and marvelous things I did see and
know," Draper recorded, but this knowledge gleaned from hands-on experience
only made the alchemist return to his "study and copy of books, whereunto I applied myself to gather the fruits and sayings" of the old alchemists. Then it was
back to the practice of alchemy once again, as a preservative against "fantasy and
imagination," since "many fools . . . may not understand one word and yet they
do read daily our books."[40] In alchemy, at least, the best practice combined sev-

eral kinds of active doing. Good alchemists did not rely exclusively either on hands-on experimentation or on reading to gather knowledge and expertise.

Thus Draper provides an excellent case study of how a variety of activities including reading, writing, and experimenting supported and reinforced one another. When he sat down to begin the process of circulating and digesting ideas about nature by recording an entry on the pages of one of his many notebooks, theory and practice mixed, mingled, and intertwined while the seemingly firm lines dividing a hands-on experiment, a written record of an experiment, and a reading of that record blurred. "I have seen mercury drawn out of lead by these mixtures once myself," Draper wrote at the end of detailed instructions on how to separate a variety of mineral substances.[41] But on another occasion, when he entered the description of a chemical experiment on lead, it was harder to distinguish between what he had witnessed, what he had performed himself experimentally, and what he had been told. Draper once wrote about an alchemical process, "this work I did never prove myself, yet I think it is true, for it seems very likely. And, I saw a fair work which was done [this way] and the party that did it swore to me it was wrought in this manner."[42] Rather than seeing this entry as evidence of a slipshod approach to obvious evidentiary problems, we can more usefully view it as a type of *virtual witnessing* — a term coined to illuminate how descriptive writing vividly captured the hands-on work of experimentation for other readers who could use the account of an experiment to enter vicariously into the experience.[43]

A fine example of virtual witnessing can be found in Draper's notebook copy of John Nettleton's account of the final illness of Lady Elizabeth Jobson, the wife of the lord lieutenant of the Tower. In August 1591 Nettleton described Lady Elizabeth as she had lain dying in the Tower of London in 1569. The queen sent her physicians to see whether they could do any more for Lady Elizabeth, who was an aunt to the royal favorite Robert Dudley. Dudley and his brother John also sent their physicians to treat the sick woman. "Considering her estate and condition," Draper wrote, all of the physicians concluded "that she could not live above four days." Unwilling to accept defeat, the queen sent for the mineral expert and physician Burchard Kranich, who attended Lady Jobson and pronounced that "he would, by God's grace, prolong her life eight days further at least in perfect memory." (Wisely, Kranich abstained from promising to cure her.) Kranich treated Lady Jobson by anointing her with civet oil as she sat by a warm fire. This simple treatment, and God's benevolence, Draper wrote, enabled Kranich to keep Lady Elizabeth alive "nine or ten days beyond the physicians' resolution, and she was in as good reason and memory to read or write or speak her mind in anything as ever she could do in her life." Finally, Draper recorded, "she died, her book in her hands, and was reading."[44] This vivid account of a woman brought back from the edge of certain death to enjoy a few more days of reading and conversation had passed from Lady Elizabeth's son, "who was an eye witness thereof," to Nettleton, and then to Draper, who thus became a virtual witness to the story and provided the means, in his written account, for others to virtually witness the events as well.

The witnessing of experimental work, virtual or otherwise, played an important part in the development of experimental practices from personal experiences and encouraged the circulation and exchange of ideas — even in the relatively close confines of an Elizabethan prison. For Draper that circulation began with his clear desire to gather all the human experiences with nature that he could. He did so by collecting information from living informants, as well as textual informants. Once an account of an experience was in his possession, Draper copied it into one of his notebooks, putting it alongside all the other items he had already collected. Lengthy texts, short recipes, the testimony of his friends, and his own observations, queries, and amendments began to bubble and percolate. But what was the ultimate purpose of this work? Like the philosopher's stone, Draper's goal remains elusive. At times it seems he was intent on verification. At other times, Draper's goal appears to have been the development of the best or most reliable chemical or medical process. These two interests are crucial in the transformation of experiences into experiments. For the most part, however, Draper focused on the process rather than the result: he simply loved the pursuit of information and enjoyed keeping under active consideration as many theories as possible about nature as well as practices that illuminated how nature worked. "I am yet in doubt," he wrote in one entry about an alchemical process, and professed himself content to remain in that state "until I have further knowledge by practice."[45]

Draper's intellectual patterns developed as he collected information from two bodies of sources, one informal and one formal. The informal sources were the members of his community who shared their knowledge by passing on recipes and giving him verbal instructions, as we saw in the previous section. His formal sources were drawn from a wide assortment of medieval and early modern chemical, medical, and metallurgical works. One of Draper's favorite authors was Paracelsus, and he extracted many items from texts by the German alchemist and physician, including instructions on how to perform chemical processes, methods for preparing tinctures of antimony, and recipes for making medicinal balsam. Friends and fellow inmates knew that Draper was always on the lookout for new reading material, and his notebooks were peppered with references to texts given or lent him by others, including a book of "diverse receipts" along with "translations of George Ripley's work" and the *Rosary of the Philosophers*. He had access at the King's Bench to numerous printed and manuscript works, which were lent to him or given as gifts. Jane Constable lent Draper a book (written partially in code) that included instructions on how to create the chemically sensational star regulus of antimony, remove painful corns with the dew gathered at midsummer, cast a bell, and charm an adder.[46]

Motivated by antiquarianism, humanism, or some combination, Draper lovingly described the condition and provenance of the books passing through his hands, especially particularly old volumes. Draper transcribed a collection of fifteenth-century alchemical processes, calling his work "the true copy of an ancient book written in parchment by George Marrow, monk of Nostall Abbey in Yorksire in anno 1437 and now . . . copied the first of July 1600." Draper transcribed a number of fifteenth-century texts during that summer, including the work of Roger Merrifoot and "an old manuscript written by Sir Peter Perry. . . . He lived in anno 1486." He copied a recipe for making false pearls from "an old written book that one did give to his friend for a special thing," and took a recipe for hippocras "out of an old book." An ancient chemical process credited to Richard of Salisbury was taken "from an old book of Mr. Marshall." Draper was curious about how the texts he acquired had been transmitted through long chains of transfer and circulation. At times, the textual genealogies were long and complicated, as is the case with Draper's copy of "The Book of the Secrets of Alchemy, composed by Galid son of Jaziche." This work, Draper noted, was "translated out of Hebrew into Arabic, and out of Arabic into Latin, and out of Latin into English, as here followeth."[47]

Because he was so devoted to his notebooks, Draper was ideally suited to understanding the problems and difficulties associated with transcription. It was common for readers — in prison and out — to "grow" their libraries by allowing a friend to copy all, or part, of a given text. And so Draper transcribed a number of manuscript collections gathered by other experimenters, including "Johnson's Book" of experiments (which included recipes for ink, instructions on how to soften steel, medicines to treat deafness, optical illusions like how to make straw appear to writhe like a serpent if illuminated by a specially made candle, and magical spells) and William Hutton's notebook (which contained medical formulas, husbandry instructions, and tips for finding veins of gold). Sometimes Draper preferred to draw together snippets of related information from a diverse range of sources, rather than transcribing a text in full. But he scrupulously noted that his copy of alchemical information attributed to Ripley and Hermes

Trismegistus was "taken from sundry scattered copies."[48] At times Draper was able to copy a text only imperfectly because of its condition, noting, "here the leaf of the copy was torn" to explain a gap in the text.[49]

Draper displayed a remarkable degree of linguistic facility as he eagerly sought out foreign works to include in his notebooks. The work of translation, like reading and writing about science more broadly, was an integral component of early modern practice, and Draper appreciated the texts made available to him in English by earlier scholars. He copied an English translation of the *Rosary of the Philosophers*, for example, "compiled and made diligently in . . . anno domini 1550." And he was happy to contribute to the tradition by translating European-language texts that passed through his hands. Continental languages, including Latin, posed no barrier to him, with the exception of Italian — a Doctor Delaberis had to translate the work of an Italian physician into English for Draper so that he could accurately copy an account of it into his notebooks. But Draper easily translated excerpts from Johan Isaäc Hollandus's alchemical works "out of a Dutch copy," took a few more medical formulas out of a "French copy," and noted the success of another chemical process that was "tried in Spain." From Hugh Plat's popular book of experiments *The jewell house of art and nature* (1594), Draper translated Latin passages explaining how to make egg whites harden into stone and how to chemically extract the "life" or spirit from vegetables, plants, and herbs. Even Draper's shifty former business partner, Richard Leycolt, was able to provide him with "sundry papers written in the Scottish tongue" that he translated to reveal numerous old charms and recipes.[50]

Draper read the texts he copied carefully, making regular notes about their points of convergence and divergence as he digested the information he was gathering. He collected multiple examples of various medical preparations and chemical processes to discern which were the most reliable and efficacious. After writing a formula for making the philosopher's stone, Draper reminded himself to compare it with "the *Book of the Great Rosary*, together with Ripley's *Twelve Gates* in verse, [and make them all] agree . . . to one sense." Draper checked and rechecked his sources to make sure the texts he was reading and copying were accurate. Next to two different recipes for the final stages of alchemical transmutation, the so-called great red work, Draper noted "although I think this is a true description, I must put some diversity between this and the former written thing because [I] can not find where it is written." Such precautions did not always lead to his perfect understanding of a text, and Draper noted when specific texts left him puzzled. After copying an anonymous treatise on the philosopher's stone, Draper reported that he understood only parts of the work even after extensive reading in the classics of alchemical literature. Because some aspects of the chemical treatise remained murky, he laid out a program for future reading and study: he should compare the text he was copying with his "figured roll of the philosopher's stone pictured out in glasses" (a rare contemporary description of a beautifully illustrated alchemical "Ripley Scroll") and Paracelsus's writings on the operation and effects of simple fire, as well as "the proceeding of the whole work . . . as I have written it in another book."[51]

The high standards Draper applied to textual comparison were also present when he compared experimental processes. In an entry describing the Dutch Dr. Stutsfelde's recipe for drawing oil of sulfur, Draper noted that Joachim Gans had developed a procedural derivation from the more typical method in which melted sulfur and turpentine were combined in a small cooking pot until the oil took on a yellow color. After devoting pages to a lengthy description of alchemical transmutation copied from a friend's book, Draper reminded himself to compare it with other recipes he had collected, especially one involving copper dust "as I saw it done and did it also myself . . . boiling it in water." Once again, the lines between virtual and actual experiments, and virtual- and eyewitnessing, became blurry as Draper ventured deeper into his work. After an account of alchemy done in Spain, Draper noted that a better version of the recipe could be found "in Mr. Goldwell's book," and after copying a recipe for incombustible oil, he noted that he had "seen in another book" that the material "takes away the burning of sulfur" after frequent reapplications.[52]

Comparison was one way to obtain more reliable and efficacious natural knowledge, but it was not the only way. Draper appreciated the value of visual aids to clarify his findings and to assist those who might one day read his notebooks. "Because . . . in books are found certain measures and weights of physic not known . . . to all such as shall read it," Draper wrote, "therefore I have briefly here set them forth showing the value and estimation of them." As Draper pursued more complicated natural knowledge, he frequently decided that a picture was worth a thousand words and sprinkled diagrams and illustrations throughout his notebooks (Figure 5.2). Sketches of the connections between the human microcosm and the celestial macrocosm, illustrations showing how substances became distributed in a glass vessel, detailed plans for alchemical glasses (such as a balneo or water bath, an alembic, a bolt's head, and the hermetic vase or "Philosopher's Egg"), and diagrams of the placement of these vessels in their appropriate alchemical furnaces helped Draper to make his experimental practices as accurate and edifying as possible.[53]

The contents of Draper's notebooks make it clear that he was not interested simply in collecting information. Instead, he was committed to sifting, sorting, assessing, and digesting that information. Each time he wrote in his notebooks

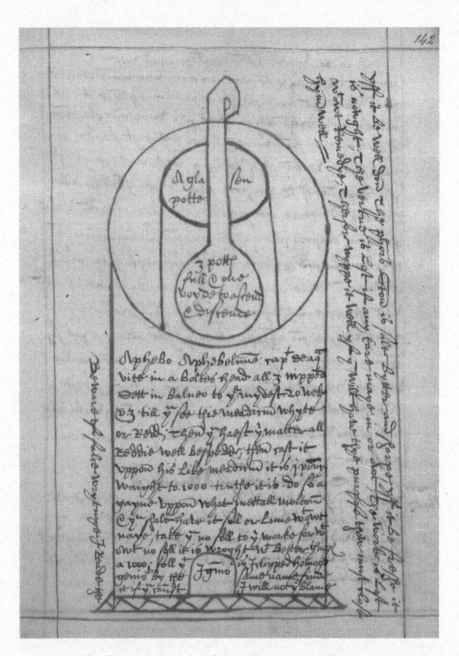

Figure 5.2. This illustration made by Clement Draper in one of his prison notebooks shows a distillation vessel and the chemical furnace used to heat the substances needed to transform base metals into gold. On the left margin of the drawing, Draper noted, "Beware of false writings, I bid thee." Draper often sketched apparatus in the margins of his notebooks to ensure that his record of a chemical process could be repeated with accuracy. Sloane MS 3688, reproduced with the permission of the British Library.

and recorded the experiences of other students of nature, Draper was actively doing the work of science in Elizabethan London. Draper supplemented these important activities with a commitment to testing and verifying this natural knowledge by engaging in the more recognizably modern business of hands-on experimentation.

"BY PRACTICE AND NOT BY FANCY": DRAPER'S EXPERIMENTS

Henry Tyler, a prisoner lodged across town from Draper in the Newgate prison, developed an elaborate method for making the philosopher's stone before he died there in 1590. Somehow, news of his innovation spread to the King's Bench, where Draper eagerly collected the practice and recorded it in his notebooks. Accompanied by three drawings of apparatus and a question-and-answer sheet to clarify certain elements of the procedure, Tyler's work was not something to be undertaken lightly or completed quickly. The plan of the work included twelve steps of transformation that proceeded from calcifying tin through a series of distillations and sublimations until the final stages of fixation and fermentation were reached. Draper admitted that Tyler's twelve-step process bore some resemblance to one outlined by George Ripley in his classic account of making the philosopher's stone. But Draper insisted that Henry Tyler could neither read nor write, implying that the resemblance was coincidental and confirming that the man had been taught the process "by one who had the Stone." Draper concluded his entry by bestowing on the dead Newgate alchemist his highest accolade: Tyler had achieved his knowledge of alchemy "by practice and not by fancy."[54]

While it is clear that Draper saw the active work of science as a spectrum of activities that included reading, writing, thinking, and experimenting, there are signs that he placed a heavier emphasis on a hands-on, practical knowledge of nature. As a reader not only of books but of the Book of Nature, Draper knew that not all useful information was found within texts. At the same time, however, he was acutely aware that practical experiences needed to be recorded in texts if they were to represent more than a one-off personal incident. Both virtually witnessing a practical account by copying someone else's notes into his notebook and eyewitnessing and transcribing a hands-on experience played an important role in his development as a practitioner. But he often followed up these activities by attempting to replicate the work — one of the hallmarks of the experiment. In one entry, for instance, he made note of a "marvelous and delectable distillation . . . the experience of which I both did and saw it done with mine eyes."[55]

The notion that Draper and his fellow prisoners were actually performing chemical assays and distillations within the prison may strain credulity, but his notebooks offer evidence to sustain such a theory. "I have seen mercury drawn out of lead by these mixtures once myself," Draper wrote at the end of detailed instructions for a chemical experiment, "by the help of Nicholas Ryckarde." Following a recipe for making oil of iron with distilled vinegar, Draper noted, "In proving of this oil I obtained Mercury of Iron which stuck in the glass head." Both entries were made while he was in prison, and after his release in the late 1590s Draper continued to pursue experimental work and remained devoted to the practice of alchemy. At his private lodging in early May 1597 he made elaborate preparations to do the great red work of alchemy in order to produce the philosopher's stone, and noted that he "nipped up in a . . . glass and set in lime" the materials "with my own hands by me Clement Draper."[56]

That Draper and his prison friends may have been doing experiments, as well as writing about them, can also be deduced from his regular entries on how to strengthen and repair broken pots and glassware. Heating clay and glass vessels over charcoal fires was a treacherous and expensive business, since the materials were fragile and not always capable of withstanding high temperatures. For prisoners like Draper, finding replacements would have been tricky given their limited mobility and (in some cases) poverty. So Draper gathered Mr. Leigh's cement for broken pots or glasses, which would make the damaged parts "stronger than any other part of the pot or glass." He noted a recipe for a fixative — made from old cheese, roots, pitch, the by-product of boiled horse hooves, and turpentine — that would make cracked vessels perform as well as new. And Mr. Jennings and his father-in-law, Mr. Pockell, gave Draper their recipes for five different cements to mend broken alchemical pots.[57] By assembling multiple techniques, Draper could test their merits and establish the best and most cost-effective means to fix what had been broken.

Despite these entries, which suggest that Draper spent some of his time in the King's Bench performing experiments, it proves difficult to make firm distinctions between his reading, writing, talking about, and doing experiments. Take Draper's ongoing conversations about his most difficult chemical quest — the search for the philosopher's stone — with the Jewish metallurgist, alchemist, and fellow prisoner Joachim Gans. It is not surprising that both Draper (who had invested heavily in mining and metallurgy) and Gans (the "mineral expert" Raleigh selected to accompany him on the Roanoke voyage) would share an interest in the topic. The philosopher's stone had long been the object of serious inquiry for students of nature, and everyone from kings to commoners tried their hands at the lengthy and costly experimental procedures that were meant to take gold, silver, and mercury through a series of purifications and transformations until they reached a state of metallurgical perfection. Once obtained, the stone promised to transform everything it touched — from the human body to a person's pocketbook — into a healthy state of well-being. It was not just Elizabeth who saw alchemy as a potentially lucrative and politically valuable weapon in the arsenal of international politics. In Gans's native Prague, Rudolf II employed scores of alchemists, including the English expatriates John Dee and Edward Kelly, in crown-sponsored efforts to achieve the stone.

Draper's efforts to make the stone were on a much smaller scale than those of Elizabeth or Rudolf II, and involved discussion and the careful reading of texts as much as working over hot fires among chemical vessels. Draper's conversations with Gans focused on detailed descriptions of how best to undertake specific stages of the chemical transformations that led to the so-called white stone (which had the ability to transform base metals into silver) and the philosopher's stone (which had the ability to transform base metals into gold). Draper's notes suggest that the two men focused on the final stages of the alchemical work when, after the alchemical substances were put through many chemical transformations they turned first white, then red. Though many alchemists used symbols — the white queen, the pelican, and the red king — to describe these stages of the chemical processes and the substances that resulted, Draper stuck to less allegorical monikers: "the white work," the "red tincture," and "the Great Red Work."

The detailed information Draper gathered from Gans was largely descriptive and often focused on the appearance of alchemical substances as they underwent transformation. When outlining Gans's method for producing the red tincture, Draper noted that he began by combining the alchemical staple *aqua fortis* or "strong water" with mercury and heating it until the aqua fortis had been "eaten" or consumed. At this point, the material was taken off the heat and left to stand for two days until it resembled common salt sitting in water. Once this visual cue appeared, Draper strained the saltlike substance from the liquid, put it in a crucible, and placed it over coals "until it be red hot." After it cooled, a different visual cue should be evident: its appearance should now resemble a "very yellow" powder. Draper often employed a language of comparison to make Gans's qualitative assessments of transformation clearer and more replicable, explaining how he examined his vessels for a "fair white powder like chalk or snow," and beat a metal until it was "as thin as a groat."[58]

Through conversations like these with Gans, experimentation, and descriptive writing, Draper was training himself to be more discerning when it came to analyzing the developmental stages that his materials passed through during alchemical experiments. Many of the changes, particularly the color differences produced by exposing his substances to heat, could not be measured or quantified. Nevertheless, a skilled practitioner needed to know when to remove his vessels from the fire, when new substances were to be added, and when the elusive goal of the philosopher's stone had been reached. Like a merchant's eye when inspecting a load of gunpowder to gauge its fineness, or a craftsman's instinct in evaluating an apprentice's work, an experimenter's discernment was a crucial element of his skill as a practitioner. Draper found it difficult to master these subjective, qualitative skills, and Gans often had to repeat his instructions before his protégé mastered the procedure.[59] At times, however, Gans's experimental instructions left Draper at a loss for appropriately descriptive language. These moments prompted Draper to resort to illustrations that illuminated specific technical details. This graphic technique was particularly useful for clarifying the elaborate equipment he needed to use for each particular stage of the alchemical work. When Gans shared his design for a furnace that would help Draper separate silver (*Luna*) and gold (*Sol*) from copper (*Venus*), Draper combined verbal instructions with a sketch of how to stack iron grates, ingots of copper, and bricks into a pile a foot high.[60]

As Draper's confidence in his ability to discern these subjective markers of chemical change grew stronger, he began to develop his own experimental modifications and procedures. He was intrigued, for example, by Alan Sutton's lengthy and complicated process for reducing and revivifying a sublimate of mercury by using a combination of chemical vessels that included athanors and retorts. But at the end of the experimental account Draper hypothesized about an improvement to the already elaborate experiment by adding more materials, as well as further stages of sublimation, reiteration, and fixation. He also showed an increased sensitivity to whether an experiment or recipe had proven to be reliable. Draper judged that a "mineral pill given me by a friend" was worthy of being copied into his notebooks because it had undergone testing and was an "excellent approved experiment upon a multitude." The ultimate mark of reliability was self-experimentation, and Draper did not hesitate to put his own body on the line in the pursuit of natural knowledge. In a postscript written after a plague remedy, he noted with great satisfaction, "This medicine was proved upon my self in [the] prison of the King's [Bench], per Clement Draper."[61]

Draper's prison notebooks provide us with an understanding of how practices informed the work of science in Elizabethan London, and illuminate the ways in which individual experiences were transformed into replicable, verifiable experiments. Working with chemical substances and engaging with hands-on practices in a laboratory make up part of the story — but so, too, do the activities of reading, recording, and writing. Draper's notebooks dramatize the challenges associated with juggling information gleaned not only from books, but from people and experiences. In his notebooks Draper was comfortable moving from a fifteenth-century alchemical text to an account of a medical procedure performed on his own knee to insights into the great red work of alchemy obtained from Joachim Gans. But all this information — whether from books or from people — needed to be appropriately credited, carefully examined for points of divergence and convergence, and scrutinized for reliability.

What, finally, are we to make of Clement Draper's boundless enthusiasm for collecting practical experiences and experiments, as he compiled notebooks out of other books, virtual and actual procedures, and prison gossip? The case of Draper's prison notebooks shows that no accurate understanding of early modern science can be achieved by artificially separating reading and writing from additional forms of making and doing. Clement Draper practiced science by capturing and conveying ideas about the natural world from a variety of sources and informants.[62] Because collecting, reading, and writing were important forms of *doing* for Clement Draper, they should not be separated from activities like experimentation or eyewitnessing. While the lines between a hands-on experiment, a written record of an experiment, and reading about an experiment were often blurred in Draper's prison notebooks, this is due less to a lack of focus on his part than to a lack of understanding on ours that reading and writing about science were significant intellectual activities.

There were, however, limitations associated with Draper's approach to the natural world. Draper did not have a clearly articulated end result in mind when he began collecting experiences and experiments. Though he developed a method to digest their contents by keeping the ideas circulating and percolating through further conversation and collection, as well as engaging in the myriad activities that constituted the work of science, this method was not put to a purpose apart from his own edification. Like the alchemical vessel of the pelican, Draper's notebooks were designed not to yield a specific output but to perpetuate an ongoing study of nature. His method of circulating and digesting experimental knowledge was therefore destined to serve only his interests and those of his community. As a result, it is sometimes difficult to appreciate all the hard work that Draper was doing.

Draper's approach might be labeled self-indulgent and of little use, which is precisely what Francis Bacon did when he scoffed at the work of alchemists who toiled away over endless chemical processes. Bacon wanted natural science to be productive, and set out ambitious plans to facilitate such a change. But other

Londoners were also seeking out the productive end results of all the experimental processes that were circulating in the streets and shops of the City. As a prisoner, Draper enjoyed limited access to these resources. It was another Londoner, Hugh Plat, who took the notebooking practices at the center of Clement Draper's work and transformed them from technologies for digesting natural knowledge into a method for producing jewels of reliable, accurate, and replicable experimental practice that could be shared with other Londoners.

6

FROM THE JEWEL HOUSE
TO SALOMON'S HOUSE

Hugh Plat, Francis Bacon, and the Social
Foundations of the Scientific Revolution

We began this exploration of science in Elizabethan London in a busy printer's shop within the City's walls. We end in the leafy western suburb of Holborn in the midst of a handful of collegiate quadrangles known as the Inns of Court. The Inns served as dormitories for wealthy young gentlemen who had not yet succeeded to their lands and titles, and as Elizabethan London's law schools. In the early modern period they were famous for the youthful exuberance of their residents and the sage advice their members provided to the notoriously litigious early modern population. While some residents of the Inns diligently read their law books and worked on their cases, most residents (if contemporary accounts can be believed) spent their time gossiping, reading, writing poetry, seeing plays, and getting drunk.[1]

Some even fit the study of nature into their busy days, and in the 1590s their number included two middle-aged lawyers: Hugh Plat (c. 1552–1608) and Francis Bacon (1561–1626). Plat, the Cambridge-educated son of a London Brewer, had rooms at Lincoln's Inn as well as a house in the City and estates in the suburbs.[2] Given his wealth, Plat needed only to dabble in the practice of law, but he was hardly an Inns of Court layabout. He spent much of his time walking the streets of the City in search of nuggets of practical wisdom about nature, which he copied into small notebooks that he could slip into his pocket before compiling the best and most reliable into published books.[3] Francis Bacon, the Cambridge-educated son of a court bureaucrat, occupied rooms across the street from Plat at Grey's Inn. Bacon lacked Plat's wealth but commanded more status in court circles thanks to his deceased father's legendary and indefatigable service to the crown. Without a large legacy, however, Bacon had to earn his position and fortune, and he did so by practicing the law with skill and vigor while endlessly seeking the queen's patronage and angling for government posts. These efforts left little time for a hands-on study of nature like Plat's, yet Bacon read and thought widely about the subject, driven both by his own intrinsic interest and by the possibilities for advancement provided by the crown's interest in Elizabethan Big Science.

In 1594 each man composed a work related to his study of nature. Like their life experiences, these texts were very different. Plat's *Jewell house of art and nature* (1594) offered readers a smorgasbord of natural knowledge gleaned from hands-on practices. Though he described himself at the time of the work's publication as "a young novice in the schools of philosophy" who had spent most of his life "within the lists and limits of London," Plat was demonstrably well versed in classic and contemporary ideas about nature. In Plat's *Jewell house* the reader learned how to preserve food from spoiling, build a bridge, engrave and smelt metals, distill precious medicinal waters, and even whip up a batch of toothpaste. Plat offered these jewels of experimental knowledge in the vernacular to help "rustic people" better their difficult lives, and to spur "ingenious" citizens to "stir up their sharp wits to a higher contemplation of Nature." Plat collected the information from friends, neighbors, and acquaintances throughout the City, as well as from published books and manuscripts. But every experiment and recipe stored in *The jewell house*, Plat promised, was "drawn from the infallible grounds of practice." The contents of *The jewell house* were therefore more reliable than the contents of other books produced by scholars from their "imagination only, in their private studies."[4] By focusing on practice as the cornerstone of any serious inquiry into nature, Plat articulated what so many of his contemporaries already believed: that a true and useful understanding of nature must be linked to some form of *doing*.

Francis Bacon's *Gesta Grayorum* was a very different work. Every year lawyers and students produced banquets, plays, and masques to celebrate the Christmas season and invited prominent court figures to attend. Bacon wrote an entertainment for these festivities that took the form of a staged debate among six speakers — including a philosopher, a warrior, and a sportsman — each of whom defended his unique way of life. The philosopher, when arguing for the excellence of his life's path, exhorted his listeners to pursue the study of nature by constructing a library, a garden, a cabinet of curiosities, and a laboratory to explore both the wonders of Nature and her potential utility. Exactly what people were to do in these newly constructed buildings was left to the imagination. Bacon also failed to specify who would perform the practical work required: writing up accounts of the activities, tilling the soil, arranging the curios, and mixing substances in chemical vessels. The masque, which sounded a bit preachy, was not enthusiastically received by his audience. After the performance, Bacon put his ideas about institutions for the study of nature on the shelf. But in 1623 and 1624 Bacon dusted them off again for an imaginative utopian adventure story published after his death as *The New Atlantis* (1627).[5]

At the end of *The New Atlantis*, Bacon described a remarkable institution called Salomon's House, which, according to his literary executor, was "a model or description of a college instituted for the interpreting of nature and the producing of great and marvelous works for the benefit of men." At Salomon's House the *Gesta Grayorum's* small clutch of scientific spaces — library, garden, cabinet, and laboratory — were expanded into a complex as sprawling as Bacon's Inns of Court. Specially equipped caves were dug into the earth to simulate natural mines and accommodate refrigeration experiments, and observation towers were erected for studying the heavens. Hospitals and distillation apparatus were erected to foster the study of useful medicinal cures. Orchards and gardens were planted to provide ample opportunities for botanizing, grafting, viticulture, and agriculture. Sound houses, perfume houses, bake houses, brew houses, and engine houses emitted a cacophony of sounds and a sneeze-inducing set of aromas, all in pursuit of natural knowledge.[6] Half a century later, Salomon's House became a model for one of the first, and certainly the most influential, of the early modern scientific societies: England's Royal Society (though the Society's work was less ambitious and exciting than the architectural blueprint sketched out by Bacon). Nevertheless, an early advocate of the Society exclaimed, "You really are what former ages could contrive but in wish and romances; and Salomon's House in the *New Atlantis*, was a prophetic scheme of the Royal Society."[7]

But Salomon's House was not a wishful romance. Instead, it was a dressed-up representation of the real world of science in Elizabethan London. The streets of the City already boasted several libraries, James Garret's fantastic tulip garden, James Cole's curiosity cabinet, and Giovan Battista Agnello's elaborate chemical laboratories and furnaces. St. Bartholomew's Hospital, where Clowes and Baker worked alongside other physicians and nurses, was known throughout Europe for its cutting-edge medicine, and John Hester's shop on St. Paul's Wharf belched out all sorts of aromatic fumes as he made powerful new chemical medicines and herbal concoctions for his urban clientele. The City's workshops produced delicate clocks and mathematical instruments, as well as perpetual-motion machines and large engineering devices. The City of London was already engaged in the study of nature, and men like Hugh Plat did not need Bacon's encouragement to go out into her neighborhoods and consult with her many experts about how to harness the powers of the natural world and put them into the service of England's people.

Somewhere between the late Elizabethan period and the Restoration, however, London's devoted interest in the study of nature was gradually but steadily obscured. Bacon's clear and univocal message in the *Gesta Grayorum*, and later *The New Atlantis*, played a crucial role in reducing this rich, varied chorus of Elizabethan London voices to a barely discernible murmur. And his call for social and political elites to join forces and construct useful scientific institutions spawned the enduring, erroneous impression that he was a scientific visionary articulating a wholly new approach to nature. As I have shown in the previous chapters, Bacon was not calling for something *new*. He was actually calling for something *different* — a science that was located not in the unruly and raucous streets of the City but in the orderly precincts of a college setting, like the Inns of Court. Even more important, Bacon did not want the study of nature to be left entirely in the hands of the Coles, Garrets, Russwurins, Clowses, and Drapers of his world, much less in the hands of gardeners, clockmakers, engineers, alchemists, and women. The message of the *Gesta Grayorum* was aimed at a particular audience of well-educated and well-born gentlemen whom Bacon regarded as the appropriate custodians and arbiters of natural knowledge.

In this chapter we will compare Hugh Plat's belief that any inquiry into nature must be grounded in ongoing practices and experiences with Bacon's belief that any inquiry into nature must be undertaken within a structured, and highly supervised, system of administration. From this comparison Hugh Plat clearly emerges as a better representative of the actual practice of science in Elizabethan London. We will take our final tour of science in London with Plat as our guide, and meet the men and women who helped him to build a jewel house of best practices, experimental wisdom, and useful information. Plat's approach to his many collaborators and the natural knowledge that they shared with him was rigorous, and prefigures in significant ways what we might call the scientific method. His conviction that a true understanding of nature must be based on verifiable, replicable experiments and shared with the larger public is crucial if we are to understand the role that London played in the development of modern science.

While Plat is more representative than Bacon of the inductive approach to natural knowledge some still call "Baconian science," he has been largely forgotten as a contributor to the Scientific Revolution, as have the London streets where he lived, worked, and gathered information. As Baconian science gradu-

Figure 6.1. No portrait of Hugh Plat or the many practitioners that he consulted has survived. Instead, the face put on Elizabethan science is usually that of Francis Bacon. Reproduced with the permission of the Henry E. Huntington Library.

ally supplanted London science, and was adopted as a model by the gentlemen of the late-seventeenth-century Royal Society, it obscured the dense, overlapping social relations that helped to shape science in the City, spotlighting instead the artificial orderliness of Salomon's House. In conclusion, we will consider why Bacon believed that the lively and robust practices of London science required rehabilitation and amendment, and why he set the natural philosopher over the urban practitioner as the appropriate public face of science (Figure 6.1).

HUGH PLAT'S LONDON

In April and May of 1590 a German painter, improbably called Mr. English, met up with Hugh Plat somewhere in London. A font of useful — and not so useful — information, English supplied Plat with potions to remove stubborn spots from clothes and old paintings, and explained how to refresh the appearance of works of art whose varnished surfaces had grown dark and muddy. English also gave Plat his recipes for medicines to treat colic, gout, and the debilitating palsy that affected people who worked with heavy metals in the forge, furnace, or chemical laboratory. He taught Plat a few parlor tricks, such as how to cast frightening apparitions in a urinal, make a man hate the taste of wine, and produce scorpions from putrefying matter. English also shared some eminently practical solutions to common problems, such as how to quench a fire quickly, how to transform salt water into fresh without distillation, and make food crops like beans grow a finger's length every day. Once Plat was satisfied that he had exhausted English's storehouse of information, he moved on, his curiosity unabated about the natural world and how it could be managed and understood, to interview his sister-in-law Mrs. Gore, who was going to give him her tips for the proper maintenance of a dovecote.[8]

The streets, houses, and workshops of London bustled with hundreds of men and women like Mr. English and Mrs. Gore who were investigating the natural world or manipulating some quizzical property of it. When Plat strolled around the City, he felt no compunction in standing and watching their work or interrogating them about what they were doing and why. Plat was a Londoner himself and understood how the City worked and how social relationships were forged, fostered, and negotiated. He was adept at making friends and winning their trust, and was even willing to draw up business arrangements with those who possessed particular kinds of natural knowledge. High or low, rich or poor, London's population responded to Plat's overtures by providing him with an endless supply of information and ingenuity. An Irish saltmaker taught him how to cultivate thin-shelled walnuts that could be easily cracked, and how to stop a chimney from smoking. The queen's surgeon showed him how to make cheap aqua vitae, while her physician demonstrated how to make an ingenious new still. Plat entered into an indenture with Arthur Blackemore to learn his method for preserving the strength of the "logwood" cut from an American tree. Stephen Bateman, the famous preacher, taught him how to make green ink from iris flowers and how to cut glass. Another cleric, the Bishop of Bristol, showed him how to attenuate mercury in water, and Plat's muskmelon vendor explained the secrets of antimony. Immigrant physicians shared their methods of refining May dew and making artificial coral. A husband-and-wife team, Mr. and Mrs. Edgecombe, led him through their techniques to make copper soft enough to take impressions and gave him their recipe for a substance that would preserve metal from rust.[9]

Like a Brueghel painting come to life, Plat's London was a bustling world filled with alchemists and apothecaries, muskmelon men and mathematicians, fruit sellers and physicians — and he was utterly comfortable speaking to all of them. Though he had a Cambridge degree, no amount of time spent in libraries or lectures could erase the fact that an urban sensibility had been bred into his bones and that the crowded streets of London were his home. "I will refuse no conference with the best scholar," Plat wrote; yet he would also converse with soap boilers, tallow chandlers, midwives, and carpenters about their practical experiences. Plat believed that natural knowledge did not belong to the privileged, gentlemanly few but was a divine gift "bound to no age, profession or estate."[10] As a result, Plat had a high opinion of the healthy, thriving state of Elizabethan science that was based on an appreciation for the unique and varied contributions that Londoners of all walks of life could make to the study of nature.

What Londoners like Plat and his informants contributed to Elizabethan science was a practical experience that could serve, like the counterweight on a clock, to move the study of nature along from its largely theoretical medieval foundations toward something more empirical. Experience, Plat wrote, was "the undoubted mother of all true and certain knowledge," and while he knew that most surgeons and gardeners were not grappling with the intricacies of Aristotelian cosmologies or Galenic medical theories, he realized that they were not on that account ignorant or unable to contribute anything useful to students of nature.[11] He often sought out men and women with particular early modern occupations that provided special opportunities to encounter and understand the natural world and its complexities. Physicians, apothecaries, and surgeons were familiar with the human body and medicines, and Plat consulted them for surgical, medical, and chemical knowledge. Goldsmiths and clockmakers knew the properties of metals, and their knowledge could help Plat's investigations into metallurgy and alchemy. Gardeners were adept at grafting, planting, and propa-

gating botanical specimens and knowledgeable about how to enrich the soil —
one of Plat's favorite topics. Wine coopers and brewers had expertise with fermentation, preservation, and storage. Dyers and illuminators understood the
chemical properties of paints and varnishes and knew how to manipulate them
to achieve dazzling levels of verisimilitude when it came to capturing nature on
cloth or wood.

Where Plat gained his appreciation for the practical insights into nature that
could be gleaned from laboring Londoners is open to conjecture, but I believe
that it came from beermaking. Plat's father, Richard, was a member of the Brewers' Company, and though he himself may never have plunged his hands into a
barrow-load of hops or stirred them during fermentation, his considerable
wealth was drawn from the manufacturing of this early modern English dietary
staple. I imagine a young Hugh Plat, accompanying his father on his rounds of
brew houses and listening to discussions of malting, mashing, and fermenting.[12]
The process of taking grain through germination and fermentation to produce
beer does highlight nature's ability to engage in fruitful and dynamic change
when guided by human hands. Throughout his notebooks and published works
Plat showed an abiding interest in the managed alteration of natural substances.
Chemistry, brewing, viticulture, agriculture, husbandry, medicine, and cloth
dyeing — all broadly conceived — were subjects that regularly occupied his time
and his interest.

Plat did not limit his consultants to obscure London artisans and manual laborers; another readily identifiable group more familiar to historians of science
also provided him with information on nature. These included the English
philosophers and practitioners John Dee, William Gilbert, Thomas Hariot, the
Napiers, John Hester, and Thomas Digges. But even among these luminaries,
Plat was drawn more to their practical and experimentally derived wisdom than
to their theoretical sophistication. Plat enthusiastically supported William Gilbert's theories of magnetism, for example, after Gilbert taught him to construct
a magnetic needle using simply a bowl of water, a piece of cork, a wire, and a
chimney piece. By this contraption, Plat noted, "you shall see the sharp end always pointing to the north [and] this proves the earth to be of the nature of a loadstone according to Dr. Gilbert." While the information that Gilbert shared with
Plat focused on what we remember the physician for today, other consultants
provided Plat with more unexpected insights into nature. John Hester, the "alchemist of London" who defended Valentine Russwurin, appears in Plat's notebooks not as a great chemist but as an expert in how to drive fish into a trammel
so that they might be caught more easily.[13]

While obvious names and famous figures do appear in Plat's London, most of
the people he consulted were humble and vernacular practitioners. But the in-

sights Plat gleaned from haberdashers, soap boilers, schoolmasters, glassmakers, and fruit sellers did not always coincide with their occupational expertise. Mr. Basford, an apothecary, explained his method of making fine gold leaf, for example. Plat's muskmelon vendor, in addition to explaining the secrets of antimony, taught him a variety of animal husbandry techniques and agricultural refinements, such as how to geld animals humanely and raise artichokes from seed. Mr. Bromfield, a soap boiler, gave him his recipe for a useful mortar, and a schoolmaster named Mr. Harrison turned out to be an avid alchemist who shared his methods for chemically altering the qualities of metals.[14] Plat's surprising experiences with these individuals remind us of the challenges we face in simply locating Elizabethan men and women with practical wisdom about nature, since we cannot always rely on obvious occupational links or professional training to guide us. Like James Cole, whose English-sounding name and mercantile occupation hid his Flemish origins and obscured his significance as a naturalist, Plat's consultants are not lost so much as they have been misplaced, miscatalogued, and misunderstood.

The fact that natural knowledge could be held by anyone and found everywhere in Elizabethan London is underscored when we consider what Plat's well-known physicians, highly regarded mathematicians, and all the other consultants with occupations that seem obviously linked to particular kinds of expertise actually contributed to his study of nature. The information Plat gleaned from them confirms that there was no necessary connection in Elizabethan London between how one made a living and what one knew. An alchemist named Cotes taught Plat how to steal wild game, not how to manufacture the philosopher's stone. Gawin Smith, a prominent engineer, did not share his considerable mechanical knowledge with Plat but instead provided him with the means to preserve oysters in barrels and transport them on long sea voyages. And when Plat went to visit the mathematician and chemist Thomas Hariot at Sir Walter Raleigh's house in October 1591, the two men discussed ink recipes and the problems they experienced carrying lanterns in high winds, rather than the more obvious topics of calculations and distillations.[15]

Because locating London's most expert practitioners often depended on an insider's knowledge, it is less surprising that Plat's pursuit of natural knowledge took him into every nook and cranny of the City and introduced him to a broad cross-section of her population. As we have seen throughout the stories in this book, London's growing vibrant foreign population was a considerable intellectual and practical resource for students of nature, and Plat frequently approached these immigrants for guidance. Strangers provided him with new and exciting techniques and information, and some of his consultants are already familiar to us from earlier chapters. James Garret, the apothecary who told the

Nortons about the serious problems in Gerard's *Herball,* also brought his considerable understanding of the medicinal properties and side effects of opiates to Plat's attention. A French clockmaker who lived with James Kynvin, the instrument maker esteemed by Gabriel Harvey, taught Plat how to plate surfaces with silver. And Joachim Gans, Clement Draper's prison companion, explained to Plat how to tread water and how to make an especially soft, impromptu bed if one were forced to sleep outdoors unexpectedly.[16] Dutch gardeners and joiners, Venetian glassmakers, Czech alchemists, French clockmakers — all were welcome in Plat's London and in his world of natural knowledge and practice. John Gerard and William Clowes may not have appreciated the input of Strangers, but Plat certainly did, and he integrated their work into his own.

Plat welcomed not only Strangers but women into his world of science. A Mrs. Carlton taught him how she stored apples for the winter so that they would not freeze and spoil. His children's nurses gave him medical advice (how to provoke menstruation and take therapeutic baths, for example), kitchen recipes (how to preserve herbs and fruit and make a popular drink called hippocras), and householding tips (how to draw beer from the barrel). Plat credited his wife with the extremely useful "invention" of a way to have fresh salad greens all year around. Lady Walsingham of Barn Elms and Lady St. John of Battersea were happy to share their tips for planting vines against chimneys and using milk in breadmaking. The advice these women shared with him was tried and true, and Plat noted with satisfaction that Mrs. Bell's medicine for aching limbs "dispersed a humor that did benumb my foot."[17] For Plat gender differences did not play a decisive role in determining who should be included on the City's roster of experts with natural knowledge. Though women would later be barred from the Royal Society (as they were already from the early modern universities), in Elizabethan London they continued to serve as resources in their own neighborhoods and in the wider community of students of nature.

As Draper's notebooks demonstrated, a Londoner's intellectual world need not include only the men and women living in the City but could also include the authors whose books and manuscripts he was able to acquire. Plat's community was similarly diverse, and he collected many books and manuscripts related to his natural inquiries, most notably alchemical, chemical, and natural history titles.[18] From these works he extracted particularly fascinating or rare insights into nature and copied them into his notebooks alongside those gathered from his London consultants. Giambattista della Porta, who so fascinated James Cole and Clement Draper, also earned Plat's attention, for example. The Italian's medical prescriptions, technological tricks like how to make salt, tips on making cheese, and instructions for taming a wild bird are all included in Plat's notebooks. The famous Italian physician and astrologer Girolamo Cardano was another of Plat's favorite authors, whose methods for keeping armor from rusting were among the dozens of tips and treasured recipes that he transcribed from Cardano's published works. Many of these authors wrote books that fit into an early modern publishing genre known as "books of secrets," and Plat was a devoted collector of these titles. In addition to foreign books of secrets like those by della Porta, Cardano, and Isabella Cortese, Plat consulted English books of secrets penned by Thomas Gascoigne, Thomas Hill, and Thomas Lupton. Like his living consultants, his textual consultants helped him to gather jewels of practice and provided additional impetus for Plat to go out in search of still more insights into the natural world.[19]

As this description of Plat's London demonstrates, he is in many ways a perfect representative of the eclectic interest in nature found in the Elizabethan City. Even more important, he embodies the strong social foundations of the Scientific Revolution established there. Born and bred in the City, he trawled through London's streets and neighborhoods gathering insights into nature. What is striking about Plat's encounters with this varied cross-section of London's science practitioners is that he appears to have treated the muskmelon man and James Garret with the same respect as he accorded Sir Francis Drake — as people from whom much could be learned, and with whom much could be shared. We can contrast Plat's interactions with his London consultants — many of whom came from humble origins and possessed simple, vernacular knowledge — with the gentlemanly interest in science in the late seventeenth century. Gentlemen played a crucial role in the development of an empirical approach to nature at that time because only a man of elite status could be trusted to produce reliable, factual knowledge. Robert Boyle, the great seventeenth-century experimenter, employed an extensive staff of expert assistants who were given little credit in his world because their insights were not deemed worthwhile. These technicians have been "almost inaudible" and largely "invisible" — standing mute in the shadow of gentlemen like Boyle — and while their role in the Scientific Revolution has been acknowledged, it seems as though it cannot be fully understood.[20]

As we can see from the case of Hugh Plat, Elizabethan London science adhered to different codes of conduct than those prevalent one century later. Plat, though educated at Cambridge and wealthier than Bacon, was a Londoner born and bred and the son of a member of the Brewers' Company. He illuminates the strikingly distinct way that someone with a corporate, urban sensibility approached the production and evaluation of natural knowledge. First, City residents with guild and craft backgrounds like Plat's fully expected that their work would be shared with, and evaluated by, other Londoners, especially those involved in similar occupations or trades. Trade associations like the Grocers'

Company and the Barber-Surgeons' Company were deeply involved in training apprentices, overseeing the quality of goods and services produced in their members' shops, and policing those individuals outside the company who might impinge on their honor or privileges. And in London's neighborhoods people oversaw the conduct of residents and visitors through a variety of pluralistic authorities and committees from the night watch to the parish vestry.[21]

Second, Plat's London consultants had not yet been relegated to the shadowy world of the "invisible technicians." Instead, they were recognizable individuals with established public reputations. Any city prized collaboration among its residents as a necessary and desirable component of civic order. Whether it involved working together with other parish residents to get an unhealthy ditch cleaned out, joining forces to prosecute a particularly egregious offender of civic ordinances, or sharing the responsibility of policing apprentices with other members of the guild, collaboration was a vital component of getting business done in an early modern city like London — and it was important that you knew whom you could rely on for their wisdom and experience. It is clear that Plat and his many consultants knew exactly whom to turn to for collaborative experiments or investigations. And so Plat could refer with familiarity to "Jacob of the glasshouse" when he talked to the immigrant Venetian glassmaker Giacomo Verzilini, or to "Mr. Digges, that ancient and painful chemist" when he asked the mathematician Thomas Digges for alchemical advice.[22]

Plat clearly appreciated his consultants' vernacular, hard-won, hands-on knowledge about nature and was eager to collect it and convey it to others interested in producing something of profit from their individual and collective labors. Plat's notebooks thus illuminate Elizabethan London's science in all its vital, complex, and messy variety. While his list of consultants may seem bewildering, diverse, and eclectic to modern eyes, his critical approach to the information Londoners and published authors gave him has a far more contemporary ring. In his notebooks Plat's London appears on the page along with his evolving sense of what science should and could be, and it is by comparing the two that we can see even more clearly how the social foundations of the Scientific Revolution extended deeply into Elizabethan London and how those foundations helped to determine not only its social framework but its intellectual content.

TRIALS BY FIRE: HUGH PLAT'S SCIENCE

On 1 November 1582, Plat and one of his chemical teachers, a Lithuanian-born alchemist named Martin Faber, met with John Dee — maker of flying beetles and author of the "Mathematical Preface" — to seek answers to a set of perplexing alchemical questions.[23] Plat devoted several pages in his notebook to the exchange, placing it under the heading "chemical questions and their solutions." First, Plat questioned Dee about the homunculus — an artificially produced "little man" first mentioned by Paracelsus that alchemists claimed to have created over a forty-day period from putrefied semen.[24] Plat was less concerned about how to make the homunculus than he was about how far away from its creator it could be and still be controlled. Dee "answered by speech" that the alchemist and his little man could be twelve miles from each other and still enjoy a connection since the homunculus was "the right familiar" of his creator. In subsequent questions Plat continued to ask about the practice of alchemy. Several queries related to sounds heard while doing the "great work" of alchemy before the creation of the philosopher's stone, while others focused on changes in the appearance of chemical substances in the glass vessels. Dee and Plat's teacher agreed that the philosopher's stone made a sound like "a sensible thunder" when it was made, but in later stages of the process, Dee cautioned that it sounded only like "the frying of grease or tallow."

Despite the inclusion of fascinating details like these, Plat left his meeting with Dee feeling that it had not been a resounding success. Dee's assertion that any metal could be transmuted in only three hours with a speck of mercury no bigger than a "crumb of bread" was met with skepticism by Plat's teacher, who did not think it possible to transmute anything "in so short a time . . . with so small a substance." Faber also observed that some of Dee's answers could be found easily in books and were not "worthy of a philosopher's learning," implying that truly devoted students of nature relied more heavily on information gleaned from their own alchemical practices. Twice Plat was able to engage in a favorite student pastime: could he stump the professor by asking Dee questions he could not, or would not, answer? One of Dee's failures resulted from a trick question Plat posed about an early stage of alchemical transmutation at which the substance was subjected to high heat so that it blackened into a material known as "the Black Crow." Plat wondered when, exactly, the alchemist had to "watch the Crow" and keep it "from sleeping, lest he fall into the Red Sea" of materials that rested in the bottom of the glass alembic. Plat admitted in his notebooks "that there is an error of purpose set down in the question," but felt that it should not trip up a "true Artist" such as Dee. Despite his reported expertise, Dee was taken in, and after a few fumbling attempts declined to say anything at all on the subject.

Plat's critical engagement with one of the period's preeminent alchemists and mathematicians helps to turn our attention from his dizzying array of London consultants to a deeper exploration of how he sifted through and evaluated the many insights into nature that he received from them before making them available to a wider reading public. London's urban environment provided him with all kinds of experts — from ale makers to writing masters — who could lend specialized assistance to his inquiries. The trick was sorting out the good advice from the bad, the precious jewels of real insight from the deceptive and alluring information provided by "experts" like Dee that led nowhere and revealed nothing. While Plat was prepared to accept the help of almost anyone in his studies, he was not prepared to take any reports at face value, no matter how highly regarded his source. In every case, from his muskmelon vendor to John Dee, Plat critically judged and evaluated all claims.

Plat's critical spirit emerged from the densely overlapping obligations, ties, and community affiliations that were part of every Londoner's life during the early modern period. City residents like Plat belonged to complex social networks that included family members, occupational acquaintances, neighbors, members of one's parish and ward, and other friends. Plat was deeply enmeshed in these overlapping communities; and his project of acquiring, selecting, and polishing the experimental practices proved all the more challenging because of the intellectual riches available to him. Just as the urban environment shaped the kinds of knowledge available to Plat and his attitudes toward the men and women who served as his consultants, the City also provided him with a core set of indispensable evaluative skills. As every Londoner knew, any workshop and all the goods or services produced there were susceptible to search, criticism, and evaluation by City officials or members of the guilds. Charges of shoddy workmanship and tagging dangerous goods could result in contention among neighbors, but rituals of fraternity and community (such as drinking together and shaking hands) were always employed to smooth civic relations in the wake of such a dispute. Reputation and honor in the City were firmly linked to the quality of what one produced, and quality was routinely brought into question during the normal course of business.[25] Londoners may not have liked being hauled in front of guild or City officers to answer charges against them, but such interruptions were by no means unexpected, and the honor of the larger collectives of guild and City depended upon each person's ability to answer the charges appropriately and to modify his or her methods if necessary. This urban culture of criticism and accountability, when applied to matters of natural knowledge, is strikingly different from late-seventeenth-century gentlemanly codes of scientific dispute, where to question a gentleman's word was to "give him the lie" and slight the core of his identity.[26]

Plat's manuscript notebooks — those twenty-odd tiny volumes of tiny writing that served as his storage facility for unpolished jewels of natural knowledge —

provide us with glimpses into how he acquired, sifted, sorted, and evaluated the information he received about nature. Like Draper's, Plat's practice of note taking was a way for him to do the collaborative, comparative, and evaluative work of science. Able to walk freely through the City with pen and paper, he jotted down whatever struck his fancy, content to sort out the true jewels from the fakes at a later date. He described how, in a Southwark beer garden, he watched a "pretty nimble chemist" named Dutch Hans stir a pot of molten lead with his finger. Plat was so intrigued that he "wrote the receipt" down "even as I did both see him make it, and use it myself."[27] Plat's sense of urgency and his desire to get the details of this procedure down as quickly and accurately as possible are evident in the hundreds of notebook entries that he made. From these accurate accounts he was able to pose questions about the information he received, explain any adaptations he had discovered that altered the practices for the worse or the better, and suggest possible modifications and further experimental ideas. Plat frankly admitted his preference for the kind of reliable information about nature that could be based, like Dutch Hans's method for stirring lead, only on "the infallible grounds of practice." Insights from other, "speculative kinds of contemplation," Platt asserted, would "when they come to be tried . . . in the glowing forge of Vulcan . . . vanish into smoke." In part, Plat's empirical precision can be related to the pedagogical possibilities of each experiment or demonstration, and he wrote, "There is no truth in philosophy unless I can lead my friend by the hand . . . into the bedchamber of Nature herself."[28] Plat's philosophy of experiencing nature firsthand through experimental practices was a means to understanding nature, and to teaching others how to perceive and judge the truth about the natural world for themselves.

Plat's extensive notebooks allow us to reconstruct his practical, hands-on approach to the study of nature with some assurance. First, he had to select an object of study from among the numerous opportunities nature provided. While Plat was convinced that anything observable in the world could and should be studied, examined, and tested, he was also willing to admit that some aspects of nature would prove merely delightful, others would be useful, and only a rare selection might be able to support broader claims about nature. The natural world was, at least in part, a divine playground where God could dabble in the marvelous and the whimsical and Plat was not averse to collecting reports of Nature at her lighthearted best. Plat culled several delightful natural jokes from Giambattista della Porta, including how to appear to be on fire without posing a danger to life or limb and how to make feline fur spark in the dark. Plat collected a few other ingenious methods for fooling friends, including a French jeweler's design for a ring that could give you some perspective on your opponent's hidden hand of cards when you were gambling, a brewer's technique for getting an egg to follow you like a dog, and the recipe of the founder of the Bodleian Library for manufacturing a candle from boiled toads and hedgehog fat that would give off strange apparitions after it was lit.[29]

Despite his interest in nature's playful side, Plat was concerned with differentiating between pleasant insights into nature and those that were more useful. Daily domestic concerns appear over and over again in Plat's notebooks, such as how to preserve fruit, keep beer from spoiling, or put out chimney fires. Plat haunted John Gerard's garden near the Inns of Court in Holborn, in hot pursuit of ways to grow potatoes and pomegranates in England, force dwarf fruit trees to give up heavy crops in their first year, and graft delicate almond trees onto heartier peach tree stock. This kind of information about the natural world was not only practical but potentially profitable, since it might make food more varied and plentiful, reduce waste, and alleviate common dangers. Still, Plat did have a weakness for more spectacular feats of useful ingenuity, and he took the time to note down a Mr. Atkinson's way of raising a sinking ship to retrieve its cargo. But it was to small yet vital domestic issues that he repeatedly turned for examination and exploration.[30]

Whether large or small, humble or spectacular, of domestic import or of vital use to the commonwealth, when Plat recorded information in his notebooks, he was careful to note whether he had been "credibly informed" about a property of nature, had heard the information straight from a practitioner, had seen an experiment being performed, or had performed it himself.[31] These reminders about how he had come to know each bit of wisdom helped Plat to develop ever more reliable information about the confusing natural world. He was so dedicated to reliability that he, like Draper, was not at all reluctant to experiment on himself, and considered his own body a likely site for developing dependable information about nature.[32] Plat proudly wrote in his notebooks that he had tested one ointment in April 1588 under "the direction of Matthew Ken, surgeon," and it had proven to be "good in my knee." This experience no doubt made Plat even more interested in Ken's recommendation to try "the oil of swallows, rightly made." And he was equally pleased with a concoction of eyebright and calamine water that he developed and used to treat eye complaints. "I cured myself with the said water," Plat wrote, "having a film, knot, blister and inflammation all happening together in my own eye in January 1595." Plat also experimented on his wife and family, and cured his wife's green sickness with a special dietary regimen though she had been afflicted with the disease for seven years and had vainly sought out "the counsel and advice of the best physicians that those times did afford."[33]

Plat's fascination with the use and construction of mechanical and technical devices can also be linked to his interest in reliability and accuracy. His interest in manufacturing more efficient ovens and furnaces, or understanding how a system of weights and pulleys could be used to drive a water pump, fit nicely into London's burgeoning interest in all things instrumental, which was itself part of a move toward greater speed and efficiency. Plat's desire to make better and more effective medicines explains, in part, his request that William Goodrus, the queen's surgeon, make a chalk sketch of the distillation apparatus he preferred to use when making aqua vitae, but it can also be linked to his concern with the instrumental aspects of hands-on testing and evaluation. These concerns led him to collect other designs for similar pieces of chemical equipment, and even to design his own. One of Plat's metallurgical consultants, Mr. Edgecombe, who had become friends with the queen's physician and Frobisher assayer Burchard Kranich, later passed on to Plat the design for his unique distillation apparatus, for example.[34] And in *The jewell house of art and nature* Plat included drawings and descriptions of several engines, machines, and instruments he had invented to extract oils, separate cereal grains from their waste products, and pump water. As always in London, criticism and complaints were sure to follow in the wake of such ingenuity. When members of the Bakers' Company trooped into his house to examine Plat's efficient new bolting hutch, they scoffed at his claims that it would result in finer, cleaner flour. Plat effectively countered their arguments (in front of witnesses), confident that his machine was all that he promised (Figure 6.2).

As the number of consultants, experimental techniques, and recorded bits and pieces of natural knowledge grew to mammoth and unwieldy proportions, Plat needed to devise a way to test and analyze each for its efficacy and merits. Plat's procedure suggests a bare-bones anatomy of the scientific method — although he uses different terminology — including the use of hypotheses, testing, the inclusion of contradictory information or results, the generation of further hypotheses and experimental procedures to test them, and evaluations of their success and failure. Where we might speak of hypotheses, for instance, Plat used "probable conjectures" and "guesses." He made a guess about "the making of indigo or anneal" for use in dying cloth, and advanced his "probable conjecture at the preservation of Malaga raisins," and another on "some profitable use to be made of the maltsters' water" that was a by-product of making beer. Sometimes Plat's observations and experimental practices led not to conclusions but to further hypotheses. One of the most common features of Plat's notebooks is *"qre,"* his three-letter abbreviation of the Latin word for question, *quaere,* and the origin of our word *query.* After observing that the spirit of wine would make rosin

The Iewel-house of

A boulting Hutch.

1 THe price heereof is easie in respect of those good vses for which it serueth.

2 This Engine auoideth al wast of meale and flower, and yet it deuideth the bran sufficiently from the flower.

3 It will bee a meanes to saue boulters, which is a matter of great charge vnto the Baker.

4 But the especiall vse thereof, is to auoid all that grose and vncleanlie manner of boulting which the Bakers forwant of this engine are forced to vse, the particulers whereof appeere more at large in my Apologie.

5 All obiections that were made against this inuention, by the Bakers of the Citty of London, vpon the view thereof at my house, were sufficiently refelled by the Author, in the presence of diuers Cittizens of good worshippe and account, and therefore what

what inconueniences soeuer shall hereafter either by them or by any other be pretended against the same, I would haue them holden for false and malicious.

6 This boulting huch is very durable, neither wil it be chargeable in reparations to the owner.

4 A portable pumpe.

1 IT will be in price one of the cheapest pumps of all that I know or euer heard of, and wil require but small reparations.

2 It is light in cariage and may be transported from place to place, by one single man without any further helpe.

3 With the easie labour of one man, it will deliuer foure, fiue, or sixe tun of water euery hower, according as it is in bignesse, neither can a man possibly be wearie, though he should worke fiue or sixe houres together, without intermission.

4 Being placed in a fit tub, that is bored ful of holes

liquefy into turpentine, for example, Plat's mind brimmed with other possibilities. "Quaere," he wrote, "if spirit of wine will work upon expressed oils" drawn out of aromatic substances, "and to what use the same be applied?" This query led him to consider the use of quicklime to dissolve matter instead of the spirit of wine, and from there to wonder whether one aromatic substance, camphor, could be refined through sublimation, and why surgeons dissolved some aromatic gums in vinegar.[35]

Page after page of the notebooks include this mixture of observation, query, and hypotheses as Plat considered his experimental options. As a result, he often had to refer to items he had recorded at an earlier time, or in a different notebook. In a notebook containing mostly alchemical and metallurgical experiments, Plat recorded a method for extracting "the sweet oil of antimony." At the end of his instructions, he wrote one of his habitual queries. This time, Plat wondered whether you could digest or sublimate antimony as you would lead "according to the manner . . . set down" on a subsequent page. Twenty pages later, in the same notebook, Plat was still intrigued by the sublimation of lead, and now wondered whether you could further distill lead's "volatile fume, by retort or glass."[36] In many cases Plat's observations and experiments generated further questions, but sometimes the question preceded the experiment. Plat began to wonder, for example, "what fructifying virtue rain water has" when combined with powdered lead and then "applied to the roots of plants or trees." Plat suggested modifications to many of the practices he collected, convinced that further manipulation of substances or technique might yield improved results. After pronouncing Sir Ulrick Hutton's recipe for a drink to cure the gout or rheumatism "to be the best of all other that is at this day to be found," he nevertheless boldly added a medicinal bath to the regimen to shorten the time the patient needed to imbibe it. "I think [it] will . . . perform equal if not greater effects," Plat remarked before recording the lengthy process for drawing and concocting the bath water.[37]

Plat gleaned insights into nature from many consultants, but he put their insights to the test experimentally and often came to very different conclusions after repeating one of their procedures or thinking it through after the excitement

---

Figure 6.2. Two of Plat's mechanical inventions, the first to separate grains, the second to set up a portable pump. Note Plat's acknowledgement of his dispute with London's Bakers about the utility and effectiveness of his "bolting hutch" and of its resolution in the presence of "divers Citizens of good worship and account." Reproduced with the permission of the Henry E. Huntington Library.

of witnessing the experiment had passed. In some cases, his musings led him to experiments on alternative substances, suggesting that analogous materials might yield similar, though not identical, results. When an apothecary, Mr. Trowte, shared his cure for gout that depended upon a medicinal plaster made from spruce beer, Plat reported that he thought "that strong ale-wort or beer-wort from the first tap used in this manner would prove much better."[38] Plat clearly differentiated between what he believed based on seeing or doing an experiment, and his untested, hypothetical opinions. After recording several detailed procedures to alter the color of stones, for example, Plat stated "topazes that are naturally white can never be made yellow by art in my opinion," leaving the proof of the matter to further experiments. And when experiments produced no result, Plat was quick to note that the experimental process was "utterly false."[39]

While Plat pursued natural knowledge throughout the City by interviewing consultants and testing their insights into nature, he was also an inventor and ingenious experimenter in his own right. Even with his own novel practices and methods — such as designing new agricultural implements or making his own medicines — Plat employed the same critical procedures. His concern for testing and evaluation is especially evident in his accounts of several proprietary medicines he devised. These included his "characteristic cure" for tertian fevers, his use of embalmed human bodies or *mumia* for quotidian or quartian fevers, and his red powder for burning fevers as well as his famous "plague cakes" — large lozenges made of herbs and chemicals. The last remedy was particularly well known throughout the City, and he claimed that it not only cured the plague but also guarded against contracting the illness. In some cases, Plat's medicines succeeded when the cures of other, more eminent, physicians had failed. Mr. Pennington, a Vintner in Cheapside, was cured of a long tertian fever by taking Plat's red powder, even though "he had taken much physic of Dr. Barrow, [and] Dr. Bredwell," and taken another proprietary medicine, "Anthony's Pill," without result.[40]

Evidence of Plat's sensitivity to issues of testing and proof can be found in a notebook list entitled "Practices," in which he accounted for fifty-one patients cured of their illnesses with his proprietary medicines between 1593 and 1605. The patients included members of his household (Nurse Pane, his son William's nurse; Nurse Price, his daughter Marie's nurse; his maid Joanne; his servants William and Lawrence), his family (his wife; his brother-in-law Robert Albany; his cousin Nicholas's wife; his sister-in-law Mrs. Gore), his neighbors (Goodwife Harsley), some of his most esteemed consultants and their households (Thomas Gascoigne; Thomas Hill's carpenter and servant) and various other people in and around London (Reynold Rowse in Trinity Lane; Richard Thorpe, a Vintner at the Sign of the Miter; John Glover, a painter in Grub Street; Elizabeth Rogers from the Bankside district of the Thames; Mr. Filkins the scrivener; the widow Joanne Gwin; the goldsmith Alexander Prescott; his parish midwife, Susanna Norman, and her son Jeffrey; Roger Alison, a cutler's boy in Holborn; William Cecil's barber "who makes the Gregorian nightcaps"). And, of course, Plat treated himself. According to his accounts, he was entirely successful at curing all his patients. Given how unusual it was for early modern medicines to have their desired effect, if his reports are accurate, it is not surprising that he received many patient referrals. Mr. Susan, a Barbary merchant, recommended that a master of a ship named Mr. Jones go to Plat for a cure, for example. And Mr. Pennington, who had finally been cured of his long illness after taking Plat's red powder, promptly sent his equally sick brother-in-law to him for treatment.[41] Where cures proved successful, gifts or high fees were sure to follow. Since he was not licensed to practice medicine, Plat was barred from charging for his services, but this did not stop him from receiving gifts in exchange for his cures. William Bromley, a coalmaker, was so grateful he sent him "two pullets for a gratuity." Mrs. Brooke was equally overjoyed when Plat cured her husband, the York herald, of his fits of tertian fever simply by appearing at his bedside; she thanked him with a hand-stitched ruff band. Plat grew accustomed to some recompense for his efforts on behalf of his patients, and he refused to treat a Cheapside housewife for a second time after she "did not pay so much as . . . my boat hire to Battsersea."

Plat's confidence in his ability to cure patients only increased after the resounding triumph of his plague cakes during London's vicious 1593 epidemic. The bubonic plague was a frequent, grim visitor to the City, and it was not until after the great fire late in the seventeenth century that the viral disease ceased to be a regular, endemic problem.[42] The plague cakes, which were first developed in Milan before Plat adapted them for English use, contained an awe-inspiring (and expensive) mixture of plants, herbs, chemicals, and ground bezoar stones, the highly sought-after and purportedly medicinal gallstones of Peruvian goats.[43] While a "great number" of cakes were distributed "unto persons whose names are not here recorded," Plat could still account for nearly five hundred dispensed during the summer of 1593 "within London and Middlesex." Sixty of the plague cakes went to the queen's Privy Council; two apothecaries purchased fifty of the cakes to sell in their shops; and Plat gave forty-five to Charles Howard, the lord high admiral of England. The Bishop of Worcester purchased fifty of the lozenges to preserve his diocese from the infection, and a justice of the peace for Middlesex "made one special trial in the parish of St. Marie Abchurch where he himself dwelled." Thirty-three people in the parish were given the drug and all "were preserved from the plague, to the great contentment of the Lords of the Council who sent . . . to be fully informed of the report."[44]

In the wake of his huge success with the plague remedies, Plat planned a major appeal to win the support of leading Elizabethan courtiers for his work in the natural sciences. In the first pages of a notebook, Plat made special mention of three courtiers—Thomas Arundell, Walter Raleigh, and William Cecil—and drew up lists of his own inventions that might appeal to them.[45] He thought Arundell might like one of his perfumes, his medicine for horses, and incendiary ammunition. Raleigh, on the other hand, would be more drawn to his methods for keeping flesh dry, his macaroni recipe (ideal for feeding sailors on long sea voyages), and his list of preserved beverages. All of these experimentally derived items were well suited to a man whose reputation and position were tied completely to the sea and navigation. As for Cecil, Plat drew up a list of offerings that included his own unique method for manufacturing saltpeter, his copious notes on viticulture, and a candle that never extinguished for want of fuel.

Plat was always seeking a wider audience and greater remuneration for his experimental wisdom, and drew up still more lists describing all the ways he could "imagine gaining . . . money, either by artificial ways, secrets, or otherwise."[46] He considered opening his own shop to sell "sweet oils and waters" after finding "out some excellent oils, waters, or spirits." Plat jotted a reminder to ask the apothecary Thomas Basford how many of his looking glasses he could sell, and then to find "some excellent joiner" who could mount his plaster works "in bedsteads, court cupboards, tables, desks, etc." Plat hoped to use the influence of the queen's apothecary, his friend Mr. Huggins, to win him some court revenue for his ambitious plans to victual the navy and keep the nation's ammunitions and weapons in good order. And, like a pharmaceutical sales representative, Plat drew up a dazzling list of "barbers, surgeons, and physicians" to whom he hoped to sell his remedies, including Thomas D'Oylie, Hippocrates D'Otten, William Clowes, Matthew Ken, Edward Lister, and Thomas Mountford. "All my other secrets and wares not mentioned . . . here," Plat noted, were also for sale.

Though Plat seems never to have opened his shop, he did find a ready buyer for his natural knowledge among London's eager publishers, and many of these items eventually went public in the pages of his printed books. Like fellow notebooker Clement Draper, Plat saw his recording and collecting practices as part of the broad spectrum of activities that constituted the work of science in Elizabethan London. To this list of practices Plat added *publishing*, which brought him a more visible role in Elizabethan London as a public man of science. Had Elizabethan England enjoyed television, Plat would no doubt have produced a popular science program aimed at inspiring other students of nature. "I am forced . . . as a scholar first to devise, and then as a politician to bring into public use, that which hitherto has long waited for due deserved favors," Plat explained in the preface to one of his books. Plat produced ten books based on the experimental practices gathered in his notebooks. These included *The jewell house of art and nature,* treatises offering specific technical inventions, and the most popular cookbook printed in early modern England, *Delightes for ladies.*[47] Each of Plat's printed books contained information about nature and methods to usefully employ nature that had been tried, perfected, and polished, and this feature of his writing distinguished his books from many of the other collections of experiments and secrets that were available in the market. Plat had no reservations about publishing his jewels of experimental knowledge, since readers could "either to their great good make use of my labors . . . or else by my example learn to employ both their wits and time in a course more commendable for themselves, and more profitable for their country."[48]

As Plat became more committed to his new role as a public man of science, he resorted to further advertising and publicity campaigns that reached beyond the readers of his books and into the streets of the City. In 1607 he published a broadside advertisement announcing *Certain philosophicall preparations of foode and beverage for sea-men.* Plat divulged that he was willing to part with medicinal waters, preserved lemon and orange juice to fight scurvy, and even his macaroni recipe, which Drake and Hawkins had taken to feed their crews on their last journey across the oceans — but warned that he must gain some personal benefit from the exchange. While Plat assured his audience that his price was within the financial bounds of "seafaring men, who for the most part are destitute both of learned physicians and skillful apothecaries," he threatened to withdraw into more "private and pleasing practices" if no one took him up on his generous offers to aid the commonwealth. "I may happily be encouraged to pry a little further into Nature's cabinet, and so to disperse some of her most secret jewels," Plat wrote, but should he find "no speedy or good acceptance" for his offers, he would "free and enlarge myself from my own fetters." Even though Plat was knighted by King James in 1605, he was prone to bitterness about his thankless devotion to the public good. "Happy men are sometimes rewarded with good words," Plat noted sourly, "but few or none, in these days, with any real recompense."[49]

Plat found his greatest acclaim — though scant financial support, given the rapaciousness of the publishers — in print. Like mathematical authors, Plat identified and filled a niche in the Elizabethan publishing market, one that addressed the public's fascination with nature and provided experimental knowledge couched in vernacular and practical terms. The audience for Plat's published work included other practitioners already bitten by the experimental bug, those curious about nature, and "every gentlewoman that delights in chemical practices."[50] Just as Plat drew on the experiences of women and men when compiling his notebooks, so, too, he imagined an audience of readers that included high and low, male and female. Unlike the authors who claimed a greater readership than their sales records supported, Plat's works were reprinted often enough that he may well have been the only Elizabethan author to actually understate his potential audience.

But not all information was fit for such a mixed readership, and Plat declined to include some experiments in his publications that were "more fit for a philosopher's laboratory than a gentlewoman's closet." Despite these reservations, Plat knew his readers were knowledgeable and critical when it came to experiments and practical, hands-on strategies for coping with nature. Still, he could be taken aback by their demands for greater and more specific proof that the information he provided was accurate and reliable. "I see it is not sufficient to propound the best [experiments] . . . made authentic even by often and manifold Experience herself, the true and undoubted mother of all credit . . . ," Plat sighed, "unless with a manual demonstration . . . the life of all inventions, is associated and joined thereunto."[51]

In spite of the pressure from his readers, or perhaps because of it, Plat remained refreshingly honest in his printed works about his experimental failures and successes. He was also comfortable going to press with a certain amount of hypothesis and contingency still present. His favorite indication of a hypothesis in his notebooks, *quaere*, was sprinkled liberally throughout his published works as well. Sometimes, Plat's great enthusiasm for sharing experimental lore led him to include experiments that he had not yet tested and verified. While he could not resist placing two methods for engraving the shell of an egg in his *Jewell house*, he made it absolutely clear that these were (as yet) unreliable. "I have not proved them," Plat confessed, "but in all likelihood they should seem to be true." His concern for attributing natural knowledge to a particular individual, which was such a feature of his notebooks, could also disappear in his desire to make a piece of information more widely available. "I know not the author of this invention," he noted after he described a wagon rigged out to be pulled by men rather than horses, "but because it came so happily to my hands, and carries some good conceit with it, I think it necessary to be published." Plat even included mystifying stories in his books that left him searching for reasons why a certain technique worked. After a particularly hair-raising story about two English prisoners who survived thirty days without food in an Ottoman prison thanks to the ministrations of a kindly guard, who gave them lumps of alum to roll around in their mouths a few times a day, Plat remained puzzled. Still, he included the information in his book and left the question of how the alum had saved the men to the "reason and probability" of "better magicians." "For though we might suppose that the salt of nature [within man] might receive some strength or vigor from this mineral salt," Plat cautioned in his final remarks on the subject, "yet how the guts should be filled with so small a proportion I cannot guess, much less determine."[52]

Plat's public reputation as one of Elizabethan London's most assiduous experimenters rested on his most wide-ranging and important work, *The jewell house of art and nature*. This volume contained both "rare and profitable inventions" and "new experiments," from which "there may be sundry both pleasing and profitable uses drawn, by them which have either wit, or will, to apply them." Dividing the work in four sections (one on general experiments, one devoted to enriching soil with compost and other additives, one exploring chemical distillations and the manufacturing of medicines, and one on molding and casting), Plat concluded his book with an account of several of his newly invented machines. *The jewell house* is full of helpful hints and practical advice so that readers could accurately replicate the processes he described. Plat's authorial voice was that of a schoolmaster who, thanks to his own experience, has been able to fully anticipate the pitfalls associated with doing the hard work of experimental science. In his discussion of distillation and chemistry, for instance, Plat warned his readers, "If you have cause to draw many oils one after another, having but one alembic for them all, let the oil of anise seeds be one of the last because it will season the alembic so strongly you shall hardly get out the scent but with great labor." Vinegar, he counseled, should be distilled not in lead vessels but in glass, to avoid the "ill touch" that the contact of metal and acid engendered. And when instructing ladies on how to preserve rose petals in a glass jar, he advised them to select red rose buds that were neither fully open nor closed tightly for the best and most pleasing results.[53]

Plat's expertise was hard won, and we should forgive his sometimes pedantic tone, since it resulted from years of compiling, comparing, and collecting experimental processes. He was committed to situating his readers within the body of natural knowledge he painstakingly gathered from books, his own experiences, and the experimental work of others, rather than leading them step-by-step through his own circuitous intellectual journey. We can see how successfully Plat employed this strategy in his discussion of chemical distillation when, after a brief primer on how to extract oils from flowers and herbs, he provided readers with three possible options for a greater understanding of chemistry: they could read the "plentiful discourse" on the subject in George Baker's translation of

Konrad Gesner's *New jewell of health* (1576); continue studying Plat's applications and uses for these oils, "which I have either found out by my own experience, or learned from others"; or "make a trial" of the oils for themselves by visiting the shop of Master Kemish, "that ancient and experienced chemist dwelling
near the new glasshouse, at whose hands they may buy any of the aforesaid oils in
a most reasonable manner."[54] Here Plat sketches out the City's world of natural
knowledge and its routes of exchange, crisscrossing between experimenters,
books, and experts on the streets of London — and saving his readers untold
hours of frustration and mistakes along the way.

Comparing Plat's published works with his manuscript notebooks illuminates
their distinct objectives. Like Draper's notebooks, Plat's manuscript records are
all about the hard processes of sorting and analyzing. His published books, on
the other hand, present a reliable, tested body of experimental knowledge that
he wanted to make public. Putting Plat's published account of a lantern capable
of being carried in high winds without being extinguished next to the corresponding entry in one of his notebooks clarifies this distinction. In the notebook,
Plat included a rough sketch, and plentiful details about the lantern's construction including its size, and the importance of sealing it to prevent air from entering or exiting (Figure 6.3). He explained why the lantern was resistant to wind,
writing that "the reason seems to be because the box is full of air already and so
there is no room for any new air or wind," and added in the top margin an explanation of the parts of the lantern. Plat also included the name of his informant,
Thomas Hariot, and noted that he had seen such a lantern "at Sir Walter
Raleigh's in October last 1591." By contrast, in *The jewell house*'s published account the rough sketch was made more elegant and elaborate, and some of the
lantern's specifications were altered to produce a smaller version (Figure 6.4). To
ensure that readers did not find the lantern faulty due to improper seals, Plat
specified that the lantern should be constructed with "dovetails or cement" to
keep air from entering the chamber. While his explanation for how the lantern
functioned remained in the printed version, Plat replaced Hariot's name with a
vague allusion to "one of the rarest mathematicians of our age." He omitted
mention of Raleigh entirely.

These differences between Plat's notebooks and his published works shed
light on the complicated relationship between his public roles of author and
man of science, and his private roles of editor and compiler. As long as accounts
of experiments were collected, gathered, and circulated through face-to-face
contact and informal networks of exchange, they were always open-ended, collaborative, and contingent. While scholars have conclusively demonstrated that
printed books were also subject to active reading, it is hard to imagine that a

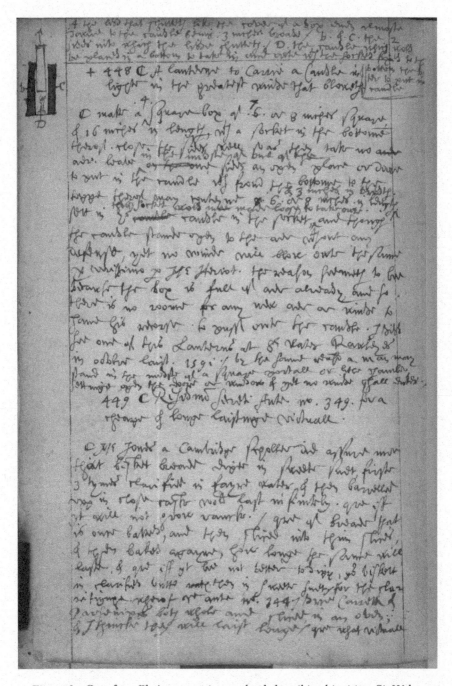

Figure 6.3. Page from Plat's manuscript notebook describing his visit to Sir Walter Raleigh's house in 1591 and the special lantern, devised by Thomas Hariot, which could be carried safely in high winds. Sloane MS 2210, reproduced with the permission of the British Library.

rose-water or some other sweet senting water therin, and therewith perfume your chamber, and by this meanes a small quantitie of sweet water will be a long time in breathing out.

22. How to erect or build ouer any brooke, or small riuer, a cheape and wooden bridge, of 40. or 50. foote in length, without fastening any timber work within the water.

PEece the timber work in such sort, as that it may resemble an arch of stone, make the ioints strong, and binde them fast with crampes or dogs of iron, let this bridge rest vpon two strong pillers of wood at either end, both being well propped with spurres, & at either ende of your bridge make a strong buttresse of bricke, into the which you mustlet your pillers and spurres, that by no meanes they may thrinke or giue backewardes, then planke ouer your bridge and grauell it and it will last a long time. This is already in experience amongst vs.

23. A cheape Lanterne, wherein a burning candle may be carried, in any stormie or windie weather, without any horne, glasse-paper, or other defensatiue, before it.

MAke a foure-square box, of 6. or 7. inches euerie waie, and 17. or 18. inches in length, with a socket in the bottome thereof, close the sides well either with doue tailes or cement, so as they take no aire, leaue in the middest of one of the sides a slit or open dore, to put in the candle, which from the bottome to the toppeth ereof may containe 6, or 7, inches in length,

length, and twoe and a halfe in bredth, place your candle in the socket, and though it stand open and naked to the ayre without any defense, yet the winde will haue no power to extinguish the flame. The reason seemeth to be because the box is already full of ayre, whereby there is no roome or place to conteine any more, neither can the ayre finde any thorough passage, by reason of the closenesse thereof. The socket would be made to screw in and out at the bottom and then you may put in your candle before you fasten the socket, This is borrowed of one of the rarest Mathematicians of our age.

24. How to plom vp a horse, and to make him fatte and lustie, as also how to keepe a Iade from tiring by the way, and to make him to foame at the bit.

TAke enula campana, Comminseed, Turmericke & annis seeds, of each a penniewor th, and seeth them well (with three heades of Garlike amongst them well stamped) in a gallon of Ale, then streine it and expresse as much of the substaunce as you may well wring out, and giue your Horse to drinke thereof of bloud-warme a full quart at once, then ride him til he be hot, then stable him, litter him well, and cuurie him vntill hee bee colde, doe the like two or three morninges together, and so turne him to grasse, and he will thriue wooncerfullie in a short time. Some commend a handfull of grunsell sodden in the afore-said

F 2   said

reader scribbling in the margins of a printed book had the same relationship with her copy of *Delightes for ladies* as she had with her own laboriously collected and transcribed book of medical, chemical, and kitchen recipes. For Elizabethans like Draper and Plat who kept notebooks, the records provided one way to recover the particular contingencies that had brought information to their attention, the records could remind them about the source of an experiment, and they could also serve as a reference point for comparing similar experiments credited to different people, or different experiments credited to a single individual. Once in print, however, it was Plat who was primarily linked to the experiments, and that link bestowed on him a sense of proprietary — if not moral — authorship.[55]

Plat may strike us more as an editor than an author, since so much of his published works were cobbled together from other sources, but the importance of this editorial function should not be underestimated in the development of science and experimental culture. When Henry Oldenburg began the first English science journal in 1665, *The Philosophical Transactions of the Royal Society,* he wanted it to serve as an authoritative source of information about current research into the natural world and its properties. Correspondents and experimenters throughout the world submitted their findings to Oldenburg, who selectively reported their contents through the journal, emphasizing the aspects he felt would be most interesting and useful to his readers.[56] Oldenburg's activities at the *Philosophical Transactions* were editorial — from gathering submissions to selecting and editing those submissions to appeal to a particular, knowledgeable readership — but they helped to shape what kind of experiments came to be performed, as well as how they were recorded and later reported.

Plat's editorial hand is evident throughout his published account of Hariot's lantern — from the changes he made to the lantern's measurements to the insertion of some details and the omission of others. While the changes to measurements may simply be evidence of improvements to the lantern's design, his vague attribution to Hariot and the failure to mention the name of one of the period's most well-known figures, Sir Walter Raleigh, are more puzzling. One tradition of historians might explain this omission as a reflection of negative elite attitudes toward print culture, and the stigma that was associated with putting

Figure 6.4. Experiment 23 from Plat's *Jewell house of art and nature,* where Plat refined his sketch of Hariot's lantern. The two accounts differ concerning measurement and design, and Hariot's name is omitted in the printed account. Reproduced with the permission of the Henry E. Huntington Library.

one's name on a work that would end up circulating through vulgar hands.[57] I argue instead that Plat's published works presented polished jewels of experimental practice that he had tested and found reliable. By the time an experiment made it into *The jewell house*, its authenticity came not from its origins with this skilled craftsman or that credible gentleman but from its proven worth. Plat's rigorous testing process transformed experiments from faithful records of individual experience to authoritative reports of replicable experiments.

Because the pedigree of each experiment was no longer as significant in the printed version as its proven effectiveness, Plat freely added and subtracted details from the careful personal attributions he had recorded in his notebooks and from his painstaking descriptions of how experimental knowledge was produced and exchanged in the City. And so Henrik, a goldsmith whose method for mending broken glasses Plat recorded in his notebooks, became "a Dutch jeweler then living in the Blackfriars but since departed from this world," who cemented "two of her majesty's crystal cups." Plat reported John Hester's strategy for catching fish in *The jewell house*, and attributed it to him, but added that the "ancient chemist" provided him with the technique in exchange for another experimental secret. On occasion, these additions and subtractions could be detrimental to the reputation of the practitioner. Poor Mr. English, the German painter whose experiments Plat recorded without a pejorative word in the notebooks, was reduced to a "Dutch mountebank" in print. And the painter's apparent willingness to share his insights into nature was portrayed in a very different light as well. His experiments "came from him so hardly," Plat confided to his readers, that it was as if "he was extremely costive." Not all of Plat's informants suffered, however, during the transition from manuscript to print. While Plat did not lavish much attention on "Old Croxton" in his notebooks, in *The jewell house* he held him up as a paragon of experimental virtue. Plat informed his readers that Croxton, "that ingenious, though unlearned old man," was a master of "diverse other profitable and ingenious practices" related to fertilizing and improving England's arable land.[58]

Plat's published works reveal an emerging tension between his persona as a public man of science and the unvarnished truth about how reliable natural knowledge was produced in Elizabethan London. While Plat seems so at ease conversing with his informants in the notebooks, print opened up divisions between author and contributor. He sometimes publicly ridiculed the humble, vernacular practitioners from whom he was happy to draw experimental practices in private. Plat could snigger at the "base mechanical workman, whose absurd ear can endure no other music than that which is forced from the anvil with the gross hand, and hammer, of an ordinary vulcanist."[59] Yet on the next page he might praise unlettered and unlearned practitioners like Old Croxton. How can we come to terms with Plat's inconsistent regard for his collaborators and informants? We do so by situating him not only on the streets of London but also within the conventions of print culture and then considering the new opportunities that culture provided for fashioning a public role.

Print culture proved as perilous to the fortunes of Plat's informants as it had been to the residents of Lime Street. Where the failure of Cole and his friends to publish their findings consigned them to the margins of the Scientific Revolution, Plat's decision to gather and publish London's experimental findings drove hundreds of others into the same historical oubliette. Though the practitioners in Elizabethan London laid the social foundations of the Scientific Revolution, published authors like Plat both wittingly and unwittingly erected the walls that would obscure the efforts of people whose work they depended on for technical skills, hands-on practical advice, and experimental information.

While Elizabethan Londoners might have been able to imagine a future story of science constructed on the basis of joint authorship and collaboration, the books published by single authors like Plat have told historians another tale and persuaded us that science was the work of individuals. Writing published books demanded a dramatically new and different social and cultural identity associated with the emergence of the author as an individual, creative genius. Manuscript culture, with its collaborative, composite character, had long resisted such a distinction. As the Latin root of the word reminds us, to be an *author* was to wield a specific kind of public authority and to be accountable to your readers for what appeared in print after your name. While publishers and pirates found ways to co-opt authors' identities to their benefit, Plat's act of publishing—like the acts of reading, writing, and doing—came to be seen as a crucial component of the work of science. As natural knowledge became a commodity that circulated outside the social networks that produced it, and came to be credited primarily to the author who made that knowledge more widely available through print culture, it became inextricably bound to the emerging identity of the public man of science.

STEALING THE JEWEL HOUSE, OR HOW LONDON
SCIENCE BECAME BACONIAN SCIENCE

When Plat established his public identity as a man capable of understanding nature and producing authoritative, experimentally based natural knowledge for the benefit of the English commonwealth, he realized an elusive Elizabethan goal. Poor John Dee spent years trying to achieve this distinction, ultimately failing to win the royal support, or the popular admiration, that he craved. Yet Plat, no matter how much he complained about his lack of both remuneration and appreciation, was extremely successful by the standards of his time. His ability to publish best sellers, catch the attention of Elizabeth's Privy Council, and earn a knighthood from her successor, James I, must have been particularly upsetting to other well-educated men of the late Elizabethan period — men such as Francis Bacon — who still struggled for recognition and respect.[60] Though I've found no definitive evidence that Plat and Bacon knew each other, it seems highly likely given their nearly parallel lives and careers. Both were Cambridge-educated lawyers, and both lived and worked for a time at the Inns of Court outside the City walls. Both were interested first and foremost in natural history and believed that science should improve the lives of ordinary men and women and serve as a tool of state. Both men sought the patronage of Robert Devereux, the Earl of Essex. Despite their similarities, however, there were crucial differences. These differences centered upon the role of the public man of science and what his relationship should be to nature (his object of study) and science (his method of study).

For Francis Bacon, the role and the relationship all boiled down to an issue of power. Bacon attempted to position natural knowledge at the heart of a prince's concerns and thus approached science as a way to achieve political power and might.[61] For Plat, on the other hand, the relationship between a man of science and the study of nature was primarily about process — although he was certainly not averse to receiving fame and monetary gain from his efforts. The comparison is most telling when we see how Bacon and Plat characterized their relationships with a feminized Nature. Plat was happy to accord power to her, and to lead Nature gently into her bedchamber. There, like a romantic suitor, he hoped to persuade her to reveal herself and her secrets through gentle negotiation and conversation. Bacon, on the other hand, expressly did not want Nature to be either a "courtesan for pleasure" or "a bond-woman to acquire and gain to her master's use." The proper relationship between Bacon's man of science and Mother Nature was that of husband to wife. In early modern terms, ideal marital relationships were patriarchal, with the husband in an unassailable position of authority over his wife, who was expected to bear his children, keep his household, and succor him during dark times. In the philosopher's relationship with Nature, Bacon wrote, she should be a "spouse, for generation, fruit, and comfort."[62]

Bacon's endless and largely ineffectual pursuit of unassailable power and position during his own life permeated his work. His dogged attempts to win preferment have even been the subject of popular fiction in *No Bed for Bacon* (1941), a humorous and spot-on send-up by Caryl Brahms and S. J. Simon of his efforts to purchase a secondhand bed that once belonged to the queen as a sign that he was a member of her intimate inner circle. Bacon's difficulties were representative of those experienced by many young men of his social class during the latter decades of Elizabeth's reign, when the road to success resembled all too closely the road to ruin, paved as it was with expectations unmet and promises unfulfilled.[63] Plat made it look easy to garner recognition and respect from the City of London and Elizabeth I's Privy Council — but it was not, as Bacon well knew, toiling away at his unsuccessful dramatic entertainments about the advancement of knowledge. This perceived injustice would have been all the more difficult for Bacon because the two men had so much in common, yet were of distinctly different social classes. Bacon, the son of a gentleman, was the social superior of Plat, who was merely the son of a London sheriff and Brewer. The honors and rewards, by all rights, should have been his.

Bacon was raised to expect to wield the kind of power that eluded him during the reign of Elizabeth and slipped through his fingers amid scandal under James I. Born in the suburban parish of St. Martin-le-Grand between the Inns of Court and the city of Westminster, he spent much of his childhood and young adulthood well beyond the City's walls in Hertfordshire and Cambridge. His superbly educated mother, Anne Cooke, and his administratively gifted father, Nicholas Bacon, provided the patterns he followed in his later life as a scholar and governmental official. When, at the age of twelve, Bacon was sent to Trinity College, Cambridge, he was armed with an excellent grasp of classical literature and languages and a firm sense of his own destiny within English enclaves of power and influence. Three years later, at fifteen, Bacon possessed a Cambridge degree and numerous books related to philosophy, history, and the art of persuasive public speaking. He headed off to the Inns of Court to embark on the next stage of his life, but before he could unpack, his father decided to send him to France in the English ambassador's entourage. There he was able to continue his studies under a tutor's guidance and immerse himself in the tricky world of international politics.

In 1579 Bacon's father died unexpectedly. It was a turning point in the young man's life, and his once-assured prospect of a dazzling career in politics suddenly seemed a long shot. Sir Nicholas died in the midst of arranging his children's legacies, and the estate of Francis, his youngest son, was not yet settled. Francis was left in an uncertain position, and when Sir Nicholas's will was contested, the situation grew even darker. After the legal wrangling drew to a close, Bacon was left without any property or regular source of income. All hope for future advancement lay in his practice of law, and he returned to Gray's Inn. With the help of his powerful maternal uncle William Cecil, Bacon began to inch his way into the good graces of the queen and pave a road to future advancement by securing government positions. This process was complicated by the fact that most of Bacon's contemporaries either loved him or hated him. His advocates thought him intelligent, hardworking, and sincere; his opponents found him arrogant, difficult, and bombastic. In spite of these divergent opinions, he won a seat in Parliament and turned his sharp wits to the most contentious religious and political issues of his time, including the trial and execution of Mary, Queen of Scots. But Bacon's progress up the governmental ladder was slow, and suitably prominent managerial positions were not always forthcoming. In 1592, at the ripe age of thirty-one, he felt that he was washed up, an Elizabethan has-been with no prospects. It was only then that he turned to natural philosophy, announcing to his uncle William Cecil his grandiose intention to take "all knowledge to be my province." If he "could purge it of two sorts of rovers," Bacon optimistically predicted — namely, "frivolous disputations" and "blind experiments" — then he would be able to "bring in industrious observations, grounded conclusions, and profitable inventions and discoveries."[64] It was his first warning shot across the bow for London science.

Given the thorny personal and professional situations he faced, it is difficult not to view Bacon's many philosophical treatises as an extended job application. His letter to his uncle declaring his aspirations with respect to the study of nature was soon followed up in 1594 with the *Gesta Grayorum*. When the queen failed to act on his broad hints that she should get to work reforming the pursuit of natural knowledge, Bacon returned to his legal career, all the while keeping the philosophical fires burning. The overall tone of the works that Bacon produced between 1602 and 1620 suggests why he failed to secure a patron for his visionary aims for the natural sciences: he was a decided pessimist, a man whose glass was perpetually half empty when he assessed the world around him. Much of the work he produced was intensely critical, leavened occasionally with his palpable excitement about the novelties and discoveries that he saw hovering just over the horizon. Bacon felt himself an adventurous explorer in the natural sciences, and drew self-serving parallels between his anticipated discovery of "a world of inventions and sciences unknown" and the discovery of the New World.[65]

Bacon's excitement was already tinged with issues of power and control, and he often remarked upon the sciences' ability to give mankind dominion over nature. In *Valerius Terminus* (1603) Bacon argued that men should seek knowledge not for "the pleasure of curiosity, nor the quiet of resolution, nor the raising of the spirit, nor victory of wit, nor faculty of speech, nor lucre of profession, nor ambition of honor or fame." Instead, they should seek knowledge out of a desire to see a "restitution and reinvesting . . . of man to the sovereignty and power . . .

which he had in his first state of creation." For Bacon, no ordinary London practitioner could possibly guide humanity toward an understanding of nature not
known since the days of Adam. Only the philosopher merited that honor, and
not just any old philosopher. The restoration of man's power over nature would
be achieved only by a new kind of philosopher, one whom historians would later
place between medieval Aristotelians and modern experimental scientists. In
Bacon's writings, the philosopher was a scholar and a gentleman who nevertheless took some interest in adapting and exploiting the work of craftsmen, artisans,
and laborers. While Bacon realized that the philosopher's power over nature
"cannot otherwise be exercised and administered but with labor, as well in in-
venting as in executing," he believed that this hard work would involve the
"sweat of the brows more than of the body."[66]

By 1605, just a few short years after the reign of Elizabeth drew to a close, Bacon was deeply alarmed at the state of knowledge in general and natural knowledge in particular. A gentlemanly outsider to the hurly-burly of activity and industry that had long flourished in the City and the sweat of brows and bodies that
was produced there, he decided to switch tactics. From the prescriptive remarks
he had made about the constructive possibilities of future science, he moved to
a devastating and vitriolic critique of present-day science. Drawing together
years of notes, unfinished treatises, letters, and unpublished manuscripts, he laid
out his scheme for its reform in *The Advancement of Learning* (1605). Much of
the work was concerned with exhaustively cataloging and describing the faulty
logics and errors that riddled early modern knowledge and practice, and many of
Bacon's pointed remarks seem aimed at targets in the City whom we have come
to know and perhaps even admire for their indefatigable efforts to understand
and utilize the natural world.

If we see London — and the Elizabethan practitioners I've discussed in this
book — as the deep, distant landscape against which Bacon foregrounded his reform of natural knowledge and the role that his new philosopher was to play in it,
certain odd features of *The Advancement of Learning* suddenly become more
understandable. In the work, Bacon's criticisms of the current state of natural
knowledge typically took two forms: sometimes he claimed there was scant interest in the natural world; at other times he argued that there was abundant interest in nature but that the wrong people were employing methods of inquiry
that were shoddy, muddled, confused. As Bacon waffled back and forth between
criticizing the Elizabethan practitioners for using the same practices that he suggested would ultimately reform natural knowledge, and deploring the lack of interest in the empirical study of nature, he became wildly inconsistent and terribly confusing to anyone familiar with science in the City. Bacon clearly knew

about London science, its richly diverse practices, and its wide range of practitioners — but he could not believe that any natural knowledge produced without
the supervision of a natural philosopher had worth. And a man like Plat, Cambridge education or no, was emphatically not an appropriate model for Bacon's
new natural philosopher. Because Bacon knew of London science and wanted
to criticize it, but at the same time refused to fully come to terms with the similarities between what was happening in the City and his own reform program, a
reader of *The Advancement of Learning* can experience a bad case of intellectual
whiplash.

Essentially, Bacon wanted a reform of natural knowledge that was based on experimentation, instrumentation, government support, and an attention to utility — all of which he claimed were missing from contemporary science. Yet the
bases of Bacon's "reformed" science were already the bases of London science.
Bacon objected to the fact that "men do not sufficiently understand the excellent use of pure mathematics," which he argued "cured many defects in the wit."
"If the wit be too dull," he wrote, then pure mathematics sharpened it, and if the
wit was "too wandering, they fix it, if too inherent in the sense, they abstract it."
As we have seen, Londoners already appreciated these benefits of mathematical
study, and among urban readers Bacon would have been preaching to the converted. Similarly, Bacon cannot be credited as an original thinker when he argued for the utility of science to the state, and the need for governmental support
for science. Bacon's often-quoted bon mot "knowledge is power" certainly cannot have come as any surprise to his uncle, William Cecil, who had been operating on the principle for decades. When Bacon urged his readers to "descend to
inquiry or meditation upon matters mechanical," he disingenuously overlooked
the instrumental mania that held London in its grip and ignored the instrument
makers who were his neighbors in the parish of St. Dunstan in the West. Bacon
viewed all knowledge as a "rich store-house for the glory of the Creator, and the
relief of man's estate," and scorned those who had treated it as a "couch, where-
upon to rest a searching and restless spirit; or a terrace, for a wandering and variable mind, . . . or a tower of state, for a proud mind to raise himself upon; . . . or
a shop for profit or sale." But the many Londoners who had read Plat's *Jewell
house* a decade earlier, and remembered how he had encouraged readers to
study nature to improve the commonwealth, must have wondered whether Bacon had done his homework. In 1594 Hugh Plat wrote "the true end of all our private labors and studies ought to be the beginning of the public and common
good of our country"; in 1605 Bacon could not invent, but merely repeat, this
message.[67]

Bacon's deliberate refusal to acknowledge the substance of London science was compounded by his dismissal of the contributions made by so many Londoners — the empirics, alchemists, old women, compilers of recipe books, manual laborers, artists, and craftsmen — to the study of nature. Never fully comfortable with the inductive world of experiments and those who performed them, Bacon shied away from too messy an engagement with the natural world. While Bacon claimed that some believed it was a "diminution to the mind of man to be much conversant in experiences," and urged students of nature to delve into empirical study, he showed no inclination to follow in Plat's footsteps and revel in conversations with Londoners about how to build better furnaces and preserve wine and cheese.[68]

Bacon instead promoted what he described as a "middle way between practical experience and unsupported theorizing."[69] Bacon clearly wished to place value on both the practical and theoretical aspects of natural knowledge, but he often found it difficult to explain their proper relationship. He criticized Plat's friend William Gilbert for making "a philosophy out of the observations of a loadstone," for example, though this empirically grounded effort would seem to be a model of the reformed philosophy he was trying to promote.[70] Bacon's later arguments suggested that experiences and experiments should play a decidedly secondary role in the new and improved natural philosophy. Experiences and experiments, Bacon stated, were merely the "light" that after being "ordered and digested" (the mixed metaphor is his, not mine) led to axiomatic statements about the natural world. Bacon's practitioner neither got bogged down in particulars of nature nor flew impatiently to the widest possible generalization. Instead, the ideal science practitioner tried to plot a course midway between these two extremes, crafting what Bacon called "middle axioms" that could always be grounded in particular experiences.[71]

After heaping scorn on the futile, ill-conceived labors of humble practitioners, and charting an ambiguous and inconsistent course between experiment and theory, Bacon announced that the reform of knowledge depended upon fashioning a new style of philosopher who, stripped of all prejudices and predispositions, would be charged with getting the work of science done. The philosopher's role would be managerial and supervisory rather than experimental and messy, thus avoiding all that ungentlemanly sweat of the body, and it would depend on a broad spectrum of literary technologies. The work of the new philosopher included indexing and organizing natural knowledge, collecting important texts, compiling them "painfully and understandingly" in their entirety rather than "dismembered in articles" like Clement Draper or Hugh Plat might do in their notebooks, and drawing up lists of errors and faulty information. While the new philosopher's day would be entirely consumed by these tasks, Bacon remained oddly ambivalent about their significance, which seemed to him to be "second prizes." Through these editorial practices "the patrimony of knowledge comes to be sometimes improved, but seldom augmented," Bacon thought — but he still defined the new philosopher's role in the reform of natural knowledge fundamentally through his performance of these tasks.[72]

Bacon's own note-taking and compiling habits are preserved in a notebook entitled "The Loose Commentary" (*Commentarius solutus*), which he wrote over the space of a few weeks in the summer of 1608.[73] Bacon intended his private notebook to function "like a merchant's waste book, where to enter all manner of *remembrancia* of matter, service, business, study, touching myself, service, others . . . without any manner of restraint." As with so much of his vision for the philosopher's role in the new science, Bacon's notes suggested a procedure for science that would operate through long chains of delegation. He wrote out long lists of men to consult in order to extract their experimental wisdom — but showed little interest in experimenting for himself. He enumerated some books he wanted to read so that he might "note and underline in the books themselves." But, it turned out, Bacon was in the habit of passing his annotated books into the hands of a servant "to be written forth . . . and so collected into this book."[74] Though his new philosopher was supposed to do the work of compiling and editing natural knowledge himself, it was not something that Bacon himself was prepared to do.

Draper, Plat, and hundreds of other London practitioners were already engaged in the kind of textual work he prescribed, and Bacon must have known that the work he was just setting out to do was already being done in the City. Still, Bacon resorted, as was his pattern, to faulting his contemporaries for doing badly what he insisted was not being done at all. He particularly singled out the university-trained physicians for their failure to write out and print up their medical recipes "partly out of their own practice, partly out of the constant probations found in books, and partly out of the tradition of the empirics." Yet he also criticized practitioners who *were* doing this active work of sifting, sorting, and editing. The industrious replication and recopying of textual and practical variants that Draper and Plat used to process and digest their findings in their notebooks was a source of irritation to Bacon, who considered such practices a form of "tedious curiosity" that led to incessant questions but never to reliable answers. Bacon concluded that the compilers of recipe books and notebooks had "no knowledge of the formulary of interpretation, the work whereof is to abridge experience and to make things as certainly found out . . . in short time, as by infinite experiences in ages."[75]

Bacon wanted to base his reform of natural knowledge not only on the role of the new philosopher but on the discipline of natural history which he argued was in need of a significant overhaul. For Bacon the problem with the field's current state was that its practitioners had not exercised sufficient choice and judgment in sorting out appropriate and inappropriate objects of study. Bacon intended to redress this situation and construct a natural history "as may be fundamental to the erecting and building of a true philosophy, for the illumination of the understanding, the extracting of axioms, and the producing of many noble works and effects." He took umbrage at "those natural histories, which are extant . . . gathered for delight and use, [and which] are full of pleasant descriptions and pictures, and affect to seek after admiration, rarities, and secrets."[76] It is hard not to hear the title of Plat's *Delightes for ladies* as we listen to Bacon's remarks, or to recall the community of naturalists on Lime Street poring over their fossils and cabinets of curiosities.

What would Bacon himself have contributed to his project to reform natural history? His collection of recipes called the *Sylva sylvarum, or A naturall historie in ten centuries*, published posthumously in 1627, offers some substantive examples of his own activities in the field of natural history. Compared to Plat's *Jewell house of art and nature* or Cole's letters to his uncle on plants or fossils, Bacon's *Sylva sylvarum* was a pale reflection of the rigors of the discipline as it was actually being practiced around him in the City of London. It was based (as we would suspect, given his attitudes toward doing the work of science and his reliance on the work of others) on reported facts and generalizations drawn from anonymous "daily experience." The *Sylva sylvarum* also contained a few anecdotal reports about Bacon's successful medical treatments but did not try to assess systematically their success or applicability, as Plat did in his account of cures.[77] Bacon's literary executor, William Rawley, admitted in the preface to the *Sylva sylvarum* that his master had in fact been reluctant to publish the work, "For it may seem an undigested heap of particulars, and cannot have that luster, which books cast into methods have."[78] In the *Sylva sylvarum* Bacon had not been able to find his ideal "middle way" between theory and practice, nor had he found it possible to use the "formulary of interpretation" that he assured his readers was the province of the new natural philosopher.

Upon finishing both the *Sylva sylvarum* and *The Advancement of Learning* one cannot help but conclude that Bacon was profoundly ambivalent about both experimental culture and the utility of natural knowledge — two of his claims to fame — at least insofar as they were practiced and understood in the City of London. As far as Bacon was concerned, experimental programs pursued by simple, unlearned people were entirely too disorderly. But by ignoring the considerable contributions these humble Londoners were making to the study of nature, Bacon was skating on perilously thin ice. In 1605, with London science in a state of robust health following the reign of Elizabeth, most of Bacon's readers would have been puzzled why anyone needed to meddle with it all, never mind launch a full-scale reformation. Perhaps this explains why few of Bacon's contemporaries seemed to be interested in what he had to say on the subject.[79] Those who commented at all tended to criticize Bacon for his unwillingness to *do* the work of science, as well as for his lack of appreciation for what was already being done.

Sir Thomas Bodley, founder of the Bodleian Library and a man who supplied Hugh Plat with numerous insights into nature, was one of those who criticized Bacon for his assessment of contemporary science. Writing a few years after the death of Elizabeth in the winter of 1609, Bodley responded to Bacon's *Thoughts and Conclusions* (*Cogita et Visa*), a preliminary version of ideas later featured in another of Bacon's works on the philosophy of science, the *Novum organum* (1620). In the work, which would become influential among early members of the Royal Society, Bacon not only set out a new inductive method for the study of nature that would replace Aristotelian models, he yet again criticized the study of nature being undertaken at the time. In a letter thanking Bacon for the gift, Bodley praised him for prompting every student "to look more narrowly to his business . . . by aspiring to the greatest perfection, of that which is now-a-days divulged in the sciences," and also by "diving yet deeper . . . into the bowels and secrets of nature."[80]

Yet Bodley urged Bacon to adopt a more circumspect tone, and to give credit where credit was due for work already being undertaken. In his letter to Bacon, Bodley wrote, "We profess a greater holdfast of certainty in your science, than you by your discourse will seem to acknowledge," and defended both the contributors and the contributions already being made by students of nature such as Draper, Plat, and the many Londoners they consulted. According to Bodley, Bacon spoke longingly of acquiring "a knowledge [of nature] more excellent than now is among us, which experience might produce, if we would but essay to extract it out of nature by particular probations." But, Bodley pointed out, this only admonished readers to do that which "without instigation, by a natural instinct men will practice of themselves." There already were "infinite [numbers of people] in all parts of the world" undertaking just such an inquiry "with all diligence and care, that any ability can perform." Hundreds of inventions and discoveries were "daily brought to light by the enforcement of wit or casual events," which proved to be of enormous use and benefit to humanity. Bodley was particularly aggrieved by Bacon's harsh characterization of the activities of physicians and alchemists. The practice of medicine, for example, flourished "with admirable remedies . . . taught by experimental effects." These experimentally derived remedies, Bodley said, already provided "the open high-way to knowledge that you recommend." There was therefore no need to do as Bacon suggested and start afresh, abandoning all supposed knowledge and the ill-advised experiments from which they were sometimes derived in a quest for new and more certain information. Such a plan "would instantly bring us to barbarism," Bodley countered, and leave the natural sciences with fewer substantive and correct theories than were already known.

Bacon was not pleased with Bodley's response to his work, and remained convinced of his position with respect to the state of current science, confident in his belief that its practitioners were ill suited to do the work that was required to amend the study of nature by establishing truths on firmer, axiomatic foundations. Still, Bacon was not eager to do the experimental work required to arrive at a more empirically grounded understanding of nature. After his death, his editor and chaplain William Rawley revealed that Bacon often spoke "complainingly . . . that his Lordship . . . who thinks he deserves to be an Architect" in the advancement of science "should be forced to be a workman and a laborer, and to dig the clay and burn the brick; and . . . to gather the straw and the stubble, all over the fields, to burn the bricks with."[81] These comments make it clear that Bacon had no interest in undertaking the necessary experimental activity that required vulgar doing and making, preferring instead the gentlemanly thinking and designing that he felt were proper activities for his new breed of natural philosopher.

Having learned that criticizing the current practice of science was no assurance of success with patrons and friends, Bacon returned to his earlier strategy of sketching out what science *could* be rather than focusing on what it was. Bacon finally managed to present a cleaned up, highly organized, and labor-extensive (as opposed to labor-intensive) science that caught the popular imagination in his posthumously published *New Atlantis*. Far away from the hustle and bustle of the London metropolis, on a fictional island called Bensalem, in a well-organized institution called Salomon's House, Bacon conjured up a way of doing the work of science that was beautifully ordered, enormously attractive, and completely impractical. There, an elite staff of thirty-six collectors, experimenters, compilers, editors, inventors, testers, and interpreters oversaw the work of "a great number of servants and attendants, men and women" in Salomon's House. Based on strict hierarchies of knowledge and power, and a complicated division of labor, here was a vision of science that would appeal to subsequent generations of gentlemen and scholars like Samuel Hartlib and Robert Boyle, who had no firsthand knowledge of Bacon or the vernacular world of London science to which he was responding. It would also appeal to a Restoration culture that had been deeply scarred by the English Civil War and the "republican experiment" of the Interregnum, and had grown fearful of the confusion that could result from a multitude of voices and approaches to a problem.[82]

Bacon achieved only limited success in his own time, and most of that was confined to the realm of politics rather than science. Several generations later he was awarded success in the field of science by seventeenth-century gentlemen interested in experimentation and the advancement of natural knowledge, and by the eighteenth- and nineteenth-century historians who saw Bacon as their logical predecessor. By then the Elizabethan foundations of Salomon's House had disappeared from view, concealed in mountains of manuscripts, buried in the rubble of war and fire, and glimpsed only fleetingly when a popular title by Plat or Hester was once again reprinted. When the Royal Society named Bacon as its founding father and "tutelary spirit," the Fellows embraced as their intellectual ancestor a gentleman who could be easily located and was their social and economic equal — one who had advanced a vision of scientific work that would not require them to "dig the clay" and "gather the straw" and delve too deeply into the messy complications of the natural world. Most members of the Royal Society wanted to be arbiters of knowledge much more than they wanted to produce it themselves, and the insatiably curious son of a London Brewer like Hugh Plat could never serve as a role model for their ideal of scientific behavior. But Bacon, who had attempted to exert his own gentlemanly authority over London's thriving world of science by sketching out the contours of an elite intellectual bureaucracy ideally suited to supervise their activity, was a perfect figurehead for their administrative aspirations.

And so today we tend to remember the orderly, imagined vistas of Salomon's House, rather than the bustling, London-based reality of Plat's *Jewell house*. Why do we remember one house and not the other? The scholarly emphasis on print culture rather than manuscript culture, the focus on singular great men rather than collaborative communities, and even our preference for a neat story of scientific development rather than a messy tale of contested knowledge and open-ended debates between humble practitioners on the streets of a busy city — all shape our historical memory of English science before the Scientific Revolution. But it is worth considering the possibility that the chief reason for our remembering Salomon's House and not Plat's *Jewell house* is that Francis Bacon himself wanted it that way. Bacon did not want us to see the rough, social foundations that lay buried in the mud and clay beneath the walls of his beautiful, orderly house of knowledge. Bacon certainly wouldn't have wanted us to value the word — much less the authority — of a Brewer's son over that of a gentleman.

Nevertheless, Plat's *Jewell house* should be remembered for its ability to convey how Elizabethan London's approach to gathering and communicating natural knowledge and doing the work of science provided a crucial developmental context for the Scientific Revolution. And the way science was practiced in Elizabethan London should also be remembered as a useful foil which gentlemanly architects — first Bacon and later the Royal Society — used to formulate their own concepts of scientific practice, scientific community, and scientific expertise.

CODA

Toward an Ethnography of Early Modern Science

The stories in this book have taken us from Lime Street to the Royal Exchange. They plunged us into Elizabethan math classes and instrument shops. We witnessed the assays of Frobisher's black rocks, and went to prison to see Clement Draper. At Hugh Plat's side we visited John Dee and Mrs. Gore. And at the end we found ourselves not in London but in an imagined institution for science called Salomon's House that was both distinct from, and eerily reminiscent of, the City. In this brief coda I share why I believe that these stories and the journey they make through Elizabethan London represent a new way of addressing one of the central problems in the history of science: what is this thing called the Scientific Revolution?

Throughout the modern age, the Scientific Revolution has been the central organizing concept in the history of science.[1] Conceptualizing it first as a radical break from medieval philosophical business as usual, early writers on the Scientific Revolution emphasized the contributions of great men of science and their visionary, pathbreaking work in such fields as physics and astronomy. Within this tradition, Isaac Newton stood at the head of a long line of geniuses, including Robert Boyle, René Descartes, Galileo Galilei, and Nicolaus Copernicus. All had, in their own way, turned the world inside out and upside down with their theories of how nature was constituted and how it worked.[2] While some scholars urged the adoption of a more cautious stance regarding this revolution in knowledge, and pointed to the significant continuities linking early modern science with what had come before, most historians (and indeed, most readers) remain attracted to the notion of radical rupture.[3]

To question the centrality of the Scientific Revolution without losing it entirely, historians have engaged in a concerted program of boundary expansion. Their efforts have reshaped the Scientific Revolution in ways that have proven to be influential for this study. First, historians have expanded the disciplinary bases of the Scientific Revolution to include not only mathematics and physics but other modes of inquiry into nature. Their work emphasizes the importance of experimenting over theorizing, and points to the significance of hands-on interactions with nature during the early modern period.[4] Second, historians have increasingly emphasized the role that social and cultural factors played in shaping science, both in its content and its practice. Studies of the places and spaces associated with science, the patronage of science, and the religious forces that helped to shape scientific debates are all part of historical efforts to situate the intellectual changes of the Scientific Revolution within a broader context. They have also led to a close examination of how and why certain kinds of knowledge flourished in particular places and times, and to studies of how scientific knowledge was produced and evaluated.[5] Third, historians of science have made a considerable effort to include a wider range of practitioners in the story of the Scientific Revolution. An increasing number of historians now put humble, vernacular efforts to understand the natural world alongside the work of great and mighty geniuses like Newton and Galileo.[6]

As historians of science expanded the parameters of the Scientific Revolution, its study became a highly contested zone of competing claims about when, where, how, and why it happened. This debate inspired one of the leading historians of the period to proclaim, "There was no such thing as the Scientific Revolution," before going on to write a whole book about it.[7] Steven Shapin hit the nail on the head with this seemingly contradictory statement. Though some have come to question the Scientific Revolution's existence, no one abandons it entirely as an analytical category. Part of our inability to move beyond what may well be a flawed central organizing principle can be explained by the striking contrast between the interest in nature in the fifteenth century and in the eighteenth. If we place these periods side by side, we cannot help concluding, like the old *New Yorker* cartoon about a scientific equation, that *something major happened* somewhere in between. Like a homeowner with a plumbing problem somewhere between the faucet and the water heater, we gamely set out to locate that major happening. But we have not achieved agreement on either the *what* or the *where*.

Instead of expanding the territorial boundaries of a central organizing concept that may itself be flawed, I decided to treat both early modern *science* and the early modern *scientist* as terms in need of further detailed study. To do so I relied on the methods of multisited ethnography, and in particular the methods of the anthropologist George Marcus. Multisited ethnographers practice a form of zen archery, circling around potential targets until a fruitful object of study emerges. Marcus encourages us to embrace contingency and chaos and resist the tendency to "analytically 'fix' by naming" our objects of study.[8] He points out that the habit results in a circumscribed scholarship because "we know what we are talking about prematurely." To resist a quickly and easily identified object of study in favor of a slowly emerging object of study, I began a four-step process adapted from an outline of ethnographic practice: *listening* to the historical sources to discover men and women who were interested in the study of nature; establishing what the anthropologist Clifford Geertz has called *rapport* with these individuals by consulting the widest possible range of source materials; *tracing the connections* between the practitioners, ideas, and communities that emerged from those sources as potentially fruitful objects of study; and finally *mapping* the twin concepts of *science* and *community* in Elizabethan London.

Anthropologists employ a variety of strategies to construct "multisited research imaginaries," including following people, following metaphors, and following intellectual property as it circulates through a given culture. In this study of scientific practice in Elizabethan London, I chose to follow people, a form of multisited ethnography that is "shaped by unexpected or novel associations among sites and social contexts suggested by life history accounts."[9] Before I could follow them, however, I had to locate them. For historians, this "fieldwork" is done in archives and libraries, and it is there that we must listen for and to our historical subjects. I began my fieldwork by assembling a list of the authors of every book written in England on a topic related to the study of nature, broadly conceived, between 1550 and 1610. As I dug into their lives and experiences, my authors introduced me to other men and women who had actively pursued natural knowledge. These men and women (who had not themselves published extant works) appeared in prefaces, letters to the reader, marginal notes printed alongside the main text, and offhand remarks in the body of the work itself.

Listening carefully to the sources enabled me to find a large, diverse, and provocative group of people to follow. My second task was to devise a way to keep track of all the individuals I met along the way, without having any clear idea about how all the information I gathered might be used. After flirting with index cards and simple lists, I decided to use a relational database — a decision that proved prescient, since by the time I sat down to write up my findings, I had gathered information on nearly eighteen hundred Elizabethans. As the data mounted, gathering prosopographical materials without a clear set of focusing questions or fixed, analytical concepts turned out to yield new insights into the practice of science in early modern England. These insights would not have risen to the surface had I restricted my inquiries from the outset to the court of Elizabeth, or a specific type of practitioner, or any of the other smaller categories that were suggested to me as eminently worthy and far more practical objects of study.[10]

As hundreds of Elizabethans entered my database, I needed to develop a deeper state of rapport with my subjects. Ethnographers struggle with this stage of their projects, and I struggled with it, too. Establishing rapport is more than just sharing a common vocabulary; it requires years of what ethnographers call "deep hanging out" with your subjects until you cross an invisible line from outsider to honorary insider. Only when an ethnographer understands inside jokes, whispered asides, and the significance of covert glances, can she or he claim to have achieved rapport. As a historian, I decided that the only way I could establish rapport was by expanding my source base and putting seemingly incommensurate sources into conversation with each other until even cryptic initials and doodles began to make sense. The research for this book involved consulting printed sources and manuscript sources. I examined intellectual history sources (science books, letters, and notebooks) and social history sources (city, church, guild, and probate records). I have read works written by well-known figures like John Dee and unknown figures like Clement Draper. I sifted through patent rolls, state papers, and diplomatic correspondence. Of course I expected to find references to science in the sources I knew best, which were intellectual history sources. I did not, however, expect to find out so much about clockmakers from parish accounts, or to be able to trace communities of midwives by tracking down how they disposed of their property. Tracing connections — between people, places, ideas, and controversies — in this wide range of sources was enormously rewarding, and revealed details about the practice of science in Elizabethan London that were fresh and surprising.

The science practiced in Elizabethan London was not revolutionary, however. The work of the practitioners I was studying did not shift paradigms, to use the vocabulary of Thomas Kuhn. Indeed, one of the larger points of this study is to provide a route for others to follow if they want to study what Kuhn called "normal science" — the intellectual work that both precedes and follows moments of enormous upheaval and change. Newton, it must be said, was an intellectual revolutionary, even if he did stand on the shoulders of giants. This study reveals that Newton also stood on the shoulders of humble, vernacular practitioners who did the more mundane (but still significant) work of consolidating and processing natural knowledge. The story of the Scientific Revolution should include Newton — but it should also include mention of humbler figures like Hugh Plat, as well as some discussion of how they did the work of science, just as Thomas Kuhn suggested.[11]

Many of the Elizabethan period's "humbler figures" clustered in the City of London, where communities of natural knowledge and practice coalesced around particular issues, geographic locations, guild memberships, and interests. Christopher Hill wrote that the "science of Elizabeth's reign was the work of merchants and craftsmen, not of dons; carried on in London, not in Oxford and Cambridge."[12] My findings lend detailed confirmation to Hill's argument. But I quickly found that no single neighborhood, discipline, occupation, or guild would have given me an accurate view of the richness that London had to offer as a site for the production of natural knowledge. Focusing on the City rather than on a smaller, institutionally defined group within it helped me understand why the work done by sociologists of science had in the early stages of my project offered up many promising methodological models that had ultimately turned out to be less useful than I had hoped. Their emphasis on institutions of science (laboratories, scientific societies, universities, even royal courts) collided with the absence of a single strong institutional framework among my London-based science communities. Instead, the Londoners who were of interest to me shared a common, broadly conceived urban culture.

As *science* and *London* emerged from the sources as central components of my project, so too did the issue of what constituted the work or *practice* of science in the period. Historians of science have been busy in recent years trying to tease out the differences between experiences and experiments, and how a hands-on knowledge of nature helped to foster modern science.[13] Londoners collected specimens, planted gardens, made medicines, cut open bodies, made and used instruments, set up distillation apparatuses, and dug into nature with wild enthusiasm. While I expected to find these hands-on practices, I did not anticipate that reading and writing should be counted so prominently among them. Adrian Johns opened up new avenues of inquiry in his pathbreaking work *The Nature of the Book: Print and Knowledge in the Making* by probing the links as well as the disjunctures that existed between print, manuscript, and experimental cultures.[14] In this study I was repeatedly struck first by the persistence of manuscript culture and second by the significant role that print played in making and breaking experimental knowledge. Johns focuses largely on the seventeenth century; I focus almost entirely on the sixteenth century. But my findings indicate that we still need to revisit the role that books — both manuscript books and printed books — played in the Scientific Revolution.

I believe that my rapport with this wide range of sources and the gradual emergence of my object of study have made it possible for me to see London science in ways that the men and women who were engaged in it would recognize and understand, even if I sometimes resort to using a different vocabulary to tell their stories and describe their work. In mapping out their world, I have used as many of their voices as possible and have tried to do justice to their complicated responses to the curious and confusing natural world. To do so I relied upon Geertz's "thick description," which has been fruitfully employed by many historians, especially those interested in the genre of microhistory and in the approach to history from below.[15] But I've also tried to keep the narrative openended and contingent, writing in the "comic mode" that Carolyn Walker Bynum discussed in her *Fragmentation and Redemption*.[16] The stories told here are not the only stories about the early modern study of nature that could be told from the archives — or even necessarily the best ones. None of them was found in a complete, whole, and smoothly narrative form. Some are fragmentary, others pieced together into a crazy quilt of texts and contexts. Despite their incompleteness and contingency, they are nevertheless *true* in the sense that they represent, singly and collectively, what some Elizabethan Londoners thought about the natural world and their place in it.

Informed by work on multisited ethnographies, I've followed a method in gathering and interpreting evidence that I believe shares some common ground with anthropologists. Though some will take issue with classifying as ethnography research on the long-dead — objects of study who died four hundred years ago cannot respond to questions or fall under the immediate, observational gaze of the researcher — I believe that my study of science in Elizabethan London is an ethnography as it discloses (as far as possible) the lived experiences of men and women in a particular time and place that is not my own. As George Marcus explains, "Ethnography is predicated upon attention to the everyday, an intimate knowledge of face-to-face communities and groups."[17] Scholars such as Donna Haraway, Bruno Latour, and Paul Rabinow have used ethnographic methods to illuminate modern science, including work on DNA, the culture of modern laboratories, and the concept of the cyborg in modern culture, and their work has deeply informed this study.[18] Like many who have admitted that an ethnography of the dead is not, strictly speaking, possible, I am nevertheless drawn to this methodological framework because I believe its combination of thick description and open-ended analysis offers up the best hope of understanding a messy, contingent, and difficult event in the history of science: the Scientific Revolution.

A subject as challenging as the Scientific Revolution is most compelling precisely because it eludes easy resolution. I have not tried to say something about the Scientific Revolution *tout court*, or to suggest that London was unique, or that social foundations of the Scientific Revolution can be found only in the Elizabethan City. Instead, I try to show how we might usefully steer the conversation about the Scientific Revolution in a new direction. By casting light on how London provided some critical, long-buried social foundations for England's seventeenth-century Scientific Revolution, I hope to encourage others to examine other urban centers more closely as sites of knowledge production. My London stories also make clear the possibility of rendering the humbler, vernacular practitioners both more visible and more audible in the historical record.

None of these findings, in and of themselves, require that we discard the Scientific Revolution in its entirety. Instead, like *science* itself, the Scientific Revolution should be considered a highly problematic term until we have taken the time to probe its complexities and contradictions more fully. At present, it is being used in the history of science as a context that must surround all efforts to understand natural knowledge during the period from 1400 to 1800 — and this book succumbs to that broad frame. Even if we believe that it doesn't exist, or find that it has outlived its usefulness, most historians of science still find they need to explain developments in natural knowledge as "leading up to it" or "resulting from it." The Scientific Revolution has become such a laden term, with so many implicit as well as explicit expectations of what is and what is not included, that we run the risk of overlooking it even when it is right in front of our eyes.

NOTES

ABBREVIATIONS

ADCL — Archdeaconry Court of London

AR — R. E. G. Kirk and Ernest F. Kirk, eds. *Returns of Aliens Dwelling in the City and Suburbs of London from the Reign of Henry VIII to That of James I.* 3 vols. Aberdeen: Aberdeen University Press, 1900–1907.

COMCL — Commissary Court of London

CPR — *Calendar of Patent Rolls*

DNB — *Dictionary of National Biography.* Oxford: Oxford University Press, 2004.

GHMS — Guildhall Library, London Manuscripts Section

Ortelius Correspondence — J. H. Hessels, ed., *Ecclesiæ Londino-Batavæ Archivum.* 4 vols. 1887; Osnabruck: Otto Zeller, 1969. Vol. 1, *Abrahami Ortelii (Geograph Antverpiensis) et Virorum Eruditorum ad Eundem et ad Jacobum Colium Ortelianum (Abrahami Ortelli Sororis Filium) Epistolæ*

PCC — Prerogative Court of Canterbury

PRO — Public Records Office

RCP Annals — Typescript translation of the *Annals* of the Royal College of Physicians, London, 2 vols.

SP — State Papers

A NOTE ABOUT "SCIENCE"

1. Ernan McMullin traces the definition of *science* through a much wider selection of texts in "Conceptions of Science in the Scientific Revolution." My efforts have been to trace the usage of the term in a particular and localized vernacular context in Elizabethan London, rather than studying its development over time.

2. Cuningham, *The cosmographical glasse,* sig. Aiir and p. 4; Ripley, *The Compound of Alchymy,* ed. Rabbards, sig. *v; Charnock, *Breviary of Philosophy,* from the copy in British Library Sloane MS 684, f. 1r; Securis, *A detection and querimonie,* sig. Biiiiv.

3. Digges, A *prognostication everlasting*, f. 1v; Baker, *The well spryng of sciences*; Dee, *General and rare memorials*, sig. Aijr; Baker, *The newe jewell of Healthe*, introduction, n.p.; Bourne, *A regiment for the sea*, sig. Aiiiv; Blagrave, *Baculum familliare*, sig. A2r; Lupton, *London and the countrey carbonadoed*, 2; Clowes, *A short and profitable treatise*, sig. Ciiv; Hood, *The use of the celestial globe*, sig. Br; Clowes, *A right frutefull and approoved treatise*, sig. A2r; Bacon, *Valerius Terminus*, in *Works*, ed. Spedding, Ellis, and Heath, 3: 223.

1. LIVING ON LIME STREET

1. Johns, *The Nature of the Book*, 74–108, discusses the workings of an early modern print shop.
2. Matthew de L'Obel's account of these events is in his *Stirpium illustrationes*, 3–4. A description of the controversy appears in L'Obel's hand in British Library MS Stowe 1069, f. 68r, and in holograph notes in his copy of the 1590 edition of the *Stirpium illustrationes* (now Magdalene College, Oxford MS 13) on ff. 6, 37. Armand Louis, *Mathieu de L'Obel*, 274, points out that no contemporary documents support L'Obel's version of events, but it is evident from subsequent editions of Gerard's work and that of L'Obel that most contemporaries were inclined to believe him. Henrey, *British Botanical and Horticultural Literature*, 45–46, asserts that Gerard's actions were not plagiarism, since he admitted that he had used Priest's translation of Dodoens and trawled through published herbals in other languages. While I tend to side with L'Obel, the important issue is not which version is true but that the different versions illuminate the rift between the Lime Street naturalists and Gerard. For a discussion of the business aspects of the publication of *The herball* in the context of the Nortons' other ventures, see Barnard, "Politics, Profits, and Idealism." On literary piracy more generally, see Johns, *The Nature of the Book*, passim.
3. *Ortelius Correspondence*, no. 338 (1 March 1607), 801–3. See ibid., no. 337 (16 December 1606), 796–800, for a song written in honor of the marriage.
4. Stow was reprinted several times in the early modern period; the most reliable edition is that of 1603. I base my sketch of Lime Street on the particulars found there and reprinted in Stow, *A Survey of London Reprinted from the Text of 1603*, 1: 150–60, 200–205.
5. Ibid., 1: 209.
6. Early modern population statistics are notoriously difficult to gather. See Finlay, *Population and Metropolis*, passim, especially 67–79 for the Stranger population. The Huguenot Society provided invaluable aids for the study of Strangers in publications dedicated to the "returns of aliens" during the early modern period. See Kirk and Kirk, *AR*; and Scouloudi, *Returns of Strangers*.
7. A few scholars have delved into the intricacies of neighborhood and society in early modern London, notably Pearl, *London and the Outbreak of the Puritan Revolution*; Boulton, *Neighborhood and Society*; and Carlin, *Medieval Southwark*. The Lime Street neighborhood seems to have been fairly stable in terms of its population; for a comparison see Boulton, "Neighborhood and Migration," 107–49. For the London-built envi-

ronment, see Schofield, "The Topography and Buildings of London," 296–321. Paster explores the imaginative possibilities of the urban experience in *The Idea of the City*.

8. For early modern natural history see Findlen, *Possessing Nature*; Ogilvie, *The Science of Describing*. For the development of natural history through the lens of community formation, see Findlen, "The Formation of a Scientific Community," 369–400. The study of botany in the university curriculum is examined in Reeds, *Botany in Medieval and Renaissance Universities*. The development of natural history in early modern England is traced in Raven, *English Naturalists*; Hoeniger and Hoeniger, *The Development of Natural History*; Hoeniger and Hoeniger, *The Growth of Natural History*.

9. For the development of these intellectual networks from humanism through the seventeenth century, see Bots and Waquet, *La république des lettres*; Jardine, *Erasmus, Man of Letters*; and Goldgar, *Impolite Learning*. For intellectual networks important in the Scientific Revolution, and their development through travel and correspondence, see Lux and Cook, "Closed Circles or Open Networks?"

10. See *Ortelius Correspondence*, no. 334 (7 January 1603/4), from Johannes Radermacher to James Cole, 787–91. I have used the spelling of Cole's name that appears in the ESTC and in most English records. He also appears variously as James Cole, Jacob Cool, Jacob Coels, and Jacobus Colius Ortelianus. Van Dorsten and Grell have studied Cole as a humanist and as a Calvinist: van Dorsten, "'I. O. C.,'" 8–20; Grell, *Dutch Calvinists*, 92n., 259, 269, 295; Grell, *Calvinist Exiles*, 24, 110, 111–13, 165, 167–68, 175–77, 179, 203, 206–7. See also the article by Grell, "Jacob Cool," in the new edition of the *DNB*.

James Cole, Sr. had already established an outpost of his silk business in the city of London in the early 1550s. By 1571 James, Elizabeth, and their young son were living in the crowded parish of St. Botolph's Bishopsgate in the northern City. They fled the Low Countries for religious reasons, along with many other members of their extended family. For the immigrant community, see Grell, *Calvinist Exiles*; Lindeboom, *Austin Friars*; and van Dorsten, *Poets, Patrons, and Professors*, along with his *Radical Arts*. While it is more accurate to describe Garret, L'Obel, and Cole as Flemish, since they came from the southern Low Countries, in Elizabethan England they were referred to as Dutch, which we more commonly associate with the northern Netherlands.

11. Grell, "Jacob Cool." For a discussion of the academic traditions of the Dutch community in England, see Grell, "The Schooling of the Dutch Calvinist Community," 45–58. *Ortelius Correspondence*, no. 57 (25 May 1575), 130–31. Books Cole received from his uncle include Ortelius, *Deorum Dearumque capita* (Antwerp, 1573), now Cambridge University Library f. 158.d.6.2; and Francesco Maurolico, *Martyrologium* (Venice, 1567), now Cambridge University Library K*.11.27. Cole's copy of Rembert Dodoens, *De frugum historia* (Antwerp, 1552), with his signature on the title page, is now Magdalene College, Oxford Goodyer 96. For Ortelius's praise of Cole's study habits, see *Ortelius Correspondence*, no. 161 (30 September 1588), 375–76: "Hij en verslyt synen tyt hier niet onnutelyck, hij studeert, hij schryft. Hij leerdt alle dage, dwelck ick geerne sie." For the range of Cole's studies see, for example, *Ortelius Correspondence*, no. 192 (25 January 1591/92), 458–62, where Cole discusses the books he has, or is seeking, concerning the immortality of the soul. These include Plato's *Axiochus*,

Epictetus's *Enchridion*, and the works of Aeneas Gazaeus and Cassiodorus. For Freherus's testimony, see *Ortelius Correspondence*, no. 313 (1 December 1597), 737–39. "Presertim dum rerum antiquarum mire et studiosum et intelligentem vidi."

12. Cole's unpublished work on ancient Roman festivals, *Fasti Triumphorum et Magistratuum Romanorum*, is now Cambridge University Library MS Gg.6.9. It is bound with his unpublished study of Greek coins, *Græca numismata externorum regum, ac populorum, descripta et exposita*, dated from Antwerp in 1588 and again from London in 1589.

   Colius, *Syntagma herbarum encomiasticum*. A second edition was published in Antwerp in 1614 by his friend Franciscus Raphelengius, the heir of the Plantin printing empire, and a third edition was published in Leiden in 1628.

   Cool, *Den staet van London in hare Groote Peste*. For more information on this publication, see Grell, "Plague in Elizabethan and Stuart London," 424–39. A modern edition, with introduction and notes by van Dorsten and Schaap, was published in 1962.

   For the paraphrase of Psalm 104, see Colius, *Paraphrasis*. Cole's treatise *Descriptio mortis* was later translated and published as *Of death a true description*.

13. Most biographical information on Ortelius is drawn from the *Ortelius Correspondence* and information contained in his *album amicorum*, now at Pembroke College, Cambridge. A modern reprint is Puraye, *Abraham Ortelius*. The best introduction to the compilation of the *Theatrum Orbis Terrarum* is Karrow, *Mapmakers of the Sixteenth Century*. Mangini, *Il "mondo" di Abramo Ortelio*, provides an analysis of the man and his times, as well as an entry point into many classic studies on the geographer.

14. *Ortelius Correspondence*, no. 192 (25 January 1590/91), 458–62. On van Meteren see Grell, *Calvinist Exiles*, 110–13, 166, 168, 179.

15. Raven, *English Naturalists*, 116, 135. For de L'Obel and his work see Greene, *Landmarks of Botanical History*, 2: 876–937; and Louis, *Mathieu De L'Obel*. Paul de L'Obel's medical formulary is now British Library MS Sloane 3252, and his *album amicorum* is British Library MS Harley 6467.

16. For James Garret see Raven, *English Naturalists*, 170, 192; Matthews, "Herbals and Formularies," 187–213; Pelling and Webster, "Medical Practitioners," 178.

17. Raven, *English Naturalists*, 192; L'Écluse, *Rariorum Plantarum Historia*, 109. See also Hunger, *Charles de L'Escluse*; Raven, *English Naturalists*, 136; Pavord, *The Tulip*, 62, 109 and Egmond et al., *Carolus Clusius*.

18. Gerard, *The herball* (1597), 117. A second, greatly expanded edition was published by Johnson in 1633. Because Johnson often expanded upon Gerard, particularly highlighting Lime Street contributions, I have used this later edition, rather than the first, as my standard text. When I cite the first edition, it is always indicated by the date (1597). See Anne Goldgar's *Tulipmania* to put Garret's early interest in tulip collection and propagation in their proper context. On Drake's Root, Johnson in Gerard, *Herball*, 1621.

19. Raven, *English Naturalists*, 164. For Penny, see Raven, *English Naturalists*, 153–71. Gunther, *Early British Botanists*, 234–35, also contains information on Penny. While Gunther must be used with caution, this information is reliable. See also Jeffers, *The Friends of John Gerard*, 26. For Camerarius and his importance in natural history, see Ashworth, "Emblematic Natural History of the Renaissance," 17–37.

20. Moffett, *The silkewormes and their flies.* For Moffett's troubles with the College, see Houliston, "Sleepers Awake," 235–46. Dawbarn recently reassessed this evidence from the perspective of patronage relationships in "New Light on Dr. Thomas Moffet," 3–22. See also *RCP Annals* 1: 31a, 31b, 32a. William Oldys's encomium is from the biographical notes contained in his introduction to Moffett, *Health's improvement*, xii. The manuscript for *The Theatre of Insects*, which contains preserved insect specimens, is British Library MS Sloane 4014. It was first published as Moffett, *Insectorum sive minimorum animalium theatrum*. It later appeared as volume 3 of Topsell, *The history of four-footed beasts*, as *The theater of insects.*

21. John Gerard, *The herball*, 274; Jeffers, *The Friends of John Gerard*, 66. For Thorius, see *Ortelius Correspondence*, no. 22 (14 June 1567), 51–52; no. 26 (12 July 1568), 61–62. For Radermacher, see AR 1: 387 and AR 2: 203 and *Ortelius Correspondence*, no. 330 (25 July 1603), no. 331 (14 August 1603), and no. 334 (7 January 1604), 772–83, 787–91. These letters recall the genesis of Abraham Ortelius's *Theatrum orbis terrarum* and reminisce about Radermacher's time in England.

22. See AR 2: 233, Lay Subsidy assessment for 1582, where James Cole, Sr., was assessed at a rate of £50. When the collectors came to the house, he actually paid the queen's servants £60. For more on the economic and social aspects of middle-class culture in London during the period, see Brooks, "Professions, Ideology, and the Middling Sort," 113–40.

23. A classic account of the murder and trial is McElwee, *The Murder of Sir Thomas Overbury*. A more recent analysis is Bellany, *The Politics of Court Scandal*. Zouche's letters to L'Obel are reprinted in Louis, *Mathieu de L'Obel*, 510–12.

24. *Ortelius Correspondence*, no. 79 (26 October 1578), 184–85; no. 90 (30 October 1579), 214–16; no. 350 (2 April 1609), 828–29. "Fratrem in magna apud Regiam Majestatem esse gratia lubens ex tuis cognovi, et Principibus placuisee viris non ultima laus est, ne tamen fidat ille nimium secundæ fortunæ, aulica enim vita splendida miseria esse solet, miseriarum officina, veritatis exilium, liberatis carcer, ni dextere ea utaris, de quibus ipsum jam monui sæpis: Ego praxin in patria exerceo, nemini addictus, meaque vivo contentus sorte. Te cum tuis quam optime valere exopto, meque tibi, nos omnes Deo, commendo."

25. Ibid., no. 144 (9 January 1585/6), 331–33. For an analysis of the contents of the letters, and their relationship to natural history, see Harkness, "Tulips, Maps, and Spiders," 184–96.

26. Ibid., no. 149 (19 January 1586/7), 345–47. Hessels's transcription has "Marvella," which I suspect is a mistranscription of "marnella," a form of wild valerian with medicinal properties. See Gerard, *The herball* (1597), 918. Because the original collection of letters was auctioned in individual lots in the early twentieth century, I have not been able to trace its current whereabouts. For the status of the sunflower, see Findlen, "Courting Nature," 66–67.

27. I base much of my framing of the Lime Street exchanges on Fumerton's brilliant study of the Elizabethan and Stuart aristocracy and the aesthetic conventions that guided their cultural expressions, *Cultural Aesthetics*, especially 1–110. For gift making and early modern culture see Davis, "Beyond the Market," 69–88; Biagioli, *Galileo, Courtier*, passim; Findlen, "The Economy of Scientific Exchange," 5–24.

28. Evans, *Rudolf II and His World*; Kaufmann, "Remarks on the Collections of Rudolf II," 22–28; Fumerton, *Cultural Aesthetics*, 67–69; Platter and Busino, *The Journals of Two Travellers*, 98; British Library MS Lansdowne 103/73 (undated); *Ortelius Correspondence*, no. 71 (4 August 1577), 167–71. A full account of Platter's adventures, with extensive footnotes, is Platter, *Beschreibung der Reisen Durch Frankreich, Spanien, England und die Niederlande*.
    Important studies on the culture of collecting include Pomian, *Collectors and Curiosities*; Findlen, *Possessing Nature*; Kaufmann, *The Mastery of Nature*; Impey and MacGregor, *The Origins of Museums*. The best-known collection of curiosities from early modern England is the Tradescant collection, which formed the basis for the Ashmolean Museum. See MacGregor, *Tradescant's Rarities*. The history of curiosity has received a great deal of scholarly attention, including Daston and Park, *Wonders and the Order of Nature*, and Evans and Marr, *Curiosity and Wonder*.

29. Will of William Martin, Barber-Surgeon, 22 January 1606/7: "Item I give to Mr. Jarret the trewe pickture of Dioscorides." This may have been James Garret, or possibly the Barber-Surgeon John Gerard. Commissary Court of London, now GHMS 9171/20, ff. 251v–253v. Gerard, *The herball*, 141, mentions Martin as someone who enjoyed doing botanical fieldwork with the physician Stephen Bredwell.

30. Platter and Busino, *The Journals of Two Travellers*, 32–36. MacGregor puts Cope's cabinet within the context of English seventeenth-century collecting patterns in "The Cabinet of Curiosities," 202–3.

31. Raven, *English Naturalists*, 154 and 159; L'Écluse, *Rariorum plantarum*, 215; James Cole's will, PCC Prob. 11/153, ff. 328b–330a. On the Quiccheberg family's connection to Moffett, see Raven, *English Naturalists*, 172. Samuel Quiccheberg's *Inscriptiones* instructs readers on how to arrange and order their cabinets of curiosities. For a sense of the range of items included in the Ortelius collection, see *Ortelius Correspondence*, no. 177 (30 March 1590), 427–29; no. 195 (10 April 1591), 469–72; no. 288 (26 April 1596), 683–85.

32. Moffett, *The theater of insects*, 924, 995, 1087, 1001; Pena and L'Obel, *Stirpium adversaria nova*, quoted in Hoeniger and Hoeniger, *Natural History*, 56. On "information overload" in the period see Blair, "Reading Strategies," 11–28, and *Coping with Information Overload in Early Modern Europe* (New Haven: Yale University Press, forthcoming); Ogilvie, "The Many Books of Nature," 29–40. On collaboration, see Johns, "The Ideal of Scientific Collaboration," 3–22.

33. Moffett, *The theater of insects*, 924, 959, 963, 983. For nature as artist, see Anne Goldgar, "Nature as Art," 324–46.

34. Johnson in Gerard, *The herball*, 434, 1610, 1618. For Penny's contributions to L'Écluse's botanical studies see, for example, L'Écluse, *Rariorum aliquot Stirpium*, 66, 87, 117, 290, 342, 362, 367, 419, 647, 657. One of the illustrations Garret sent L'Écluse of a Java pepper is reproduced in De Backer et al., *Botany in the Low Countries*, 74.

35. *Ortelius Correspondence*, no. 164 (15 May 1589), no. 314 (24 January 1597/8), no. 321 (1 June 1598), 752–53; no. 214 (5 May 1592), 512–14; no. 144 (9 January 1585/6), 331–33; Moffett, *The theater of insects*, 983, 937–38, 1008–9, 1007, 1015, 1025; Raven, *English Naturalists*, 172.

36. Moffett, *The theater of insects*, 1038, 1045, 1050, 948, 985, 1016–17, 1044, 946, 976, 979, 949, 936, 1018.

37. Ibid., 949, 994, 947; Raven, *English Naturalists*, 159; L'Obel, *Stirpium adversaria nova*, 41, 233; Gerard, *The herball*, 168; Johnson ibid., 271; *Ortelius Correspondence*, no. 353 (7 June 1610), 832–34; no. 352 (10 September 1609), pp. 831–32.

38. The itinerary for Cole's tour, as well as reactions to him and his own descriptions of what he saw are in the *Ortelius Correspondence*, no. 304 (6 June 1597), 716–17; no. 306 (2 July 1597), 719–21; no. 309 (18 October 1597), 726–29; no. 312 (21 November 1597), 735–36; no. 313 (1 December 1597), 737–39; no. 314 (24 January 1597/8), 740–41. For the cultural and intellectual significance of travel in the period see Stagl, *A History of Curiosity*.

39. Olmi, "From the Marvelous to the Commonplace"; Findlen, *Possessing Nature*, passim.

40. Moffett, *The theater of insects*, 983, 997, 990, 1045. On the traffic of zoological specimens, see George, "Alive or Dead," 250–52.

41. Moffett, *The theater of insects*, 939, 946, 1104.

42. Moffett, *The theater of insects*, 959; Matthew de L'Obel, quoted in Gerard, *The herball*, 61. For links between visual depictions of nature and the study of nature see Smith, *The Body of the Artisan*, and Long, "Objects of Art/Objects of Nature," 63–82. Freedberg traced the circulation of drawings among individuals associated with Galileo in *The Eye of the Lynx*. Janice Neri explores issues of truth and accuracy among early modern artists and their scientific illustrations in "Fantastic Observations."

43. Studies of these artists include Hodnett, *Marcus Gheeraerts the Elder*; Tahon, "Marcus Gheeraerts the Elder," 231–33; Hearn, *Marcus Gheeraerts II*; Hulton and Quinn, *The American Drawings of John White*; Hulton, *America, 1585*; Hendrix, Bocskay, and Vignau-Wilberg, *Nature Illuminated*; Hendrix, "Of Hirsutes and Insects," 373–90.

44. Moffett, *The theater of insects*, 1038, 1060.

45. Moffett, *The theater of insects*, 921, 1020; Gerard, *The herball*, 407, quoting from Pena and L'Obel's *Stirpium adversaria nova*, 105.

46. British Library MS Harley 6994, letter from Ortelius to James Cole (30 September 1586), f. 39r. This letter was not catalogued in Hessels's edition of Ortelius's correspondence and was never in the collection belonging to the Dutch Church in London. It is possible that the letter was confiscated by crown officials before reaching Cole.

47. Raven, *English Naturalists*, 168, 236; Parkinson, *Theatrum botanicum*, 613; Bauhin, *Pinax theatri botanici*, chapter 112; *Ortelius Correspondence*, no. 149 (19 January 1586/7), 345–47; no. 214 (6 May 1592), 512–14; no. 348 (3 February 1608/9), 823–25. I am indebted to Julie Hayes for helping me to decipher Peiresc's letter. For Peiresc's relationship with naturalists and other English scholars see van Norden, "Peiresc and the English Scholars," 368–89.

48. Gerard, *The herball*, 137–40, 85; Johnson ibid., 1530, 1599, 1618; L'Écluse, *Rariorum aliquot Stirpium*, 731–32.

49. Gerard, *The herball*, 61; *Ortelius Correspondence*, no. 160 (29 August 1588), 374–75, 161 (30 September 1588), 375–76; no. 164 (15 May 1589), 393–95.

50. Moffett, *The theater of insects*, 892, 926, 980.

51. *Ortelius Correspondence*, no. 86 (11 July 1579), 203–6; Bodleian Library MS Douce 363, f. 140v; Moffett, *The theater of insects*, 1081. For a discussion of the *ludens naturae* see Findlen, "Jokes of Nature," 292–331.

52. Moffett, *The theater of insects*, 1081. On interpretation of fossils see Rossi, *The Dark Abyss of Time*, especially 3–113; Rudwick, *The Meaning of Fossils*; Ashworth, "Emblematic Natural History," 17–37; Ashworth, "Natural History and the Emblematic World View," 303–33.

53. *Ortelius Correspondence*, no. 164 (15 May 1589), 393–95.

54. Ibid., no. 215 (1 July 1592), 514–17; Moffett, *The theater of insects*, 1081.

55. Findlen, *Possessing Nature*, 232–40.

56. *Ortelius Correspondence* no. 157 (15 June 1588); no. 152 (12 June 1587), 350–51.

57. Moffett, *The theater of insects*, 980, 1021.

58. Ibid., 938, 939, 976, 1001, 985.

59. Studies of *alba amicorum* include Bosters, *Alba Amicorum*, and Klose, *Corpus Alborum Amicorum*. See Egmond, "A European Community."

60. Van Meteren's *album amicorum* is now Bodleian Library MS Douce 68. See ff. 3v–4r, 5r–6, 7, 46, 56r, 60r, 61r, 71r for the individuals listed here. Paul de L'Obel's *album amicorum* is now British Library MS Harley 6467. See ff. 6r, 27r, 33r, 64r, 67r, 71r, 92r for these individuals.

61. *Ortelius Correspondence* no. 72 (24 September 1577), 169–71; no. 76 (19 August 1578), 178–79; no. 329 (2 July 1603), 770–71. Ortelius's album has been reproduced in Puraye, *Abraham Ortelius*. Camden's entry is discussed on 87–88 and is reproduced on f. 113v. Wilson's elaborately decorated entry is reproduced on f. 18r.

62. *Ortelius Correspondence*, no. 232 (16 April 1593), 556–57; no. 239 (20 September 1593), 566–67.

63. Ibid., no. 306 (2 July 1597), 719–21; no. 322 (3 June 1598), 754–55; no. 328 (2 January 1602/3), 768–69. For Gruter's relationships with scholars and intellectuals in England, see Forster, *Janus Gruter's English Years*.

64. Ibid., no. 228 (27 January 1592/3), 546–48. "Omnia mea tibi lubens in manus tradidissem. Nunc aliud cogitabo."

65. *Ortelius Correspondence*, no. 192 (25 January 1590/1), 458–62; no. 197 (12 March 1590/1), 475–76; no. 294 (October 1596), 696–97. The work in question is Garcia de Orta, *Coloquios dos simples e droges* (Goa, 1563). A translated edition by L'Écluse first appeared as *Aromatvm, et simplicivm aliqvot medicamentorvm* in 1567, followed by several other editions. See Egmond, "Correspondence and Natural History."

66. Gerard, *The herball*, 61; L'Obel, *Balsami, opobalsami, carpobalsami et xylobalsami*, 39–40.

67. Raven, *English Naturalists*, 204.

68. Gerard, *The herball*, 127. See also a reference (omitted in the 1633 edition) to a Byzantine lily that Cecil received from Constantinople and gave to Gerard to plant in his garden, Gerard, *The herball* (1597), 199,

69. Gerard, *The herball* (1597), 207, 539.

70. Gerard, *The herball*, 629, n.p.

71. Ibid., 54, 1329, 108, 132, 157, 249, 251, 480, 96, 182, 468, 216, 227, 314, 589. Lyte translated Dodoens's *Cruydeboeck* (1554) from L'Écluse's 1557 French edition in *A niewe herball* (London, 1578).

72. Gerard, *The herball*, 27, 29, 220, 229, 253, 254, 264, 277, 350, 403, 561, 562, 572, 886, 179, 218, 220, 272, 529, 596, 610, 624, 625, 631, 416, 403, 410, 429, 582, 621, 906; Johnson (quoting Pena and L'Obel's *Stirpium adversaria nova*) ibid., 135.

73. Gerard, *Catalogus arborum*. It was republished in a second edition in 1599.

74. Gerard, *The herball*, 38, 39, 79, 358, 172, 900–901, 319.

75. Ibid., 197, 443 and Gerard, *The herball* (1597), 145. This last reference to the garlic bulb was omitted in the 1633 edition.

76. Gerard, *The herball*, 218, 422, 373, 7, 27, 374.

77. The maypole was replaced after the riots and finally pulled down by Protestant reformers in 1546, Stow, *The annales*, 1: 152. There is debate among scholars about the significance that should be attached to the riot of 1517. While some scholars, notably Yungblut, emphasize the hostility between Strangers and citizens, others, such as Pettegree and Selwood, point out the ways in which the two groups sought to live and work together. See Yungblut, *Strangers Settled Here Amongst Us*; Pettegree, *Foreign Protestant Communities*; Selwood, "'English-born reputed strangers,'" 728–53. For the dynamic between Strangers and citizens interested in the natural world, see Harkness, "'Strange' Ideas and 'English' Knowledge," 137–62. Londoners did not need much excuse to resort to violence in the Elizabethan period, and one of the great challenges to City residents was finding ways to live cheek-to-jowl with all sorts of people, including foreigners. See Rappaport, *Worlds Within Worlds*; Archer, *The Pursuit of Stability*; and Ward, *Metropolitan Communities*.

78. Gerard, *The herball*, 430.

79. Ibid., 127, "The yellow Spanish daffodil does likewise deck up our London gardens, where they increase indefinitely" was all the ink Gerard was willing to spill on these hardy flowers (134).

80. Ibid., 431, 332, 348, 200. See also his reference to the Crown Imperial, 203.

81. Ibid., n.p.

82. Johnson, ibid., 164, 436, 281.

83. For the complicated means by which a work of science became a published book see Johns, *The Nature of the Book*, passim.

2. THE CONTEST OVER MEDICAL AUTHORITY

1. The spelling of Russwurin's name gave Elizabethan writers an unusually hard time, and their difficulties have been replicated by contemporary historians. In the literature of his own time as well as ours he is known as Valentyne Rawnsworm, Valentine Razewarme, and Valentine Rushworm; I selected Russwurin because that is how he spells it in British Library MS Lansdowne 101/4. Evidence surrounding the Russwurin case must be pieced together from the Patent Rolls; records of the Repertory Court of Aldermen (CLRO 7, f. 196r–v, CLRO Rep. 18, f. 211r); an intelligence report generated for William Cecil on Russwurin's activities (British Library MS Lansdowne 18/9); a

treatise on the chemical analysis of urine and ocular medicine for William Cecil by Russwurin (British Library MS Lansdowne 101/4), which is undated but which must have been written before 1587, judging from references to Cecil's mother, who died that year; and William Clowes's account of Russwurin's arrest in *A briefe and necessarie treatise.* In modern scholarship passing references to Russwurin are in Beck, *The Cutting Edge,* 186; Debus, *The English Paracelsians,* 70; Webster, "Alchemical and Paracelsian Medicine," 317; Pelling, "Appearance and Reality," 86; and Pelling, "Medical Practice in Early Modern England," 106. Russwurin's continental background and some activities in England are analyzed in Jütte, "Valentin Rösswurm," 99–112. While early modern evidence for the display of extracted bladder stones is better for Italy than England, one such testimonial survives from the Elizabethan period, now Bodleian Library MS Eng. Misc. d. 80 (R). It describes John Hubbert's removal of a bladder stone from Beatrice Shrove in Norwich in 1593. The bladder stones are still attached.

2. *RCP Annals* 2: 134b, 99b–100a, 103b, 1: 32b, 2: 110a, 133b, 62b, 79a, 86a, 92a; Barber Surgeons Company Court Minute Books, 1551–1586, GHMS 5257/1, f. 45r (10 December 1566). Scot's will, dated 22 September 1592, is PCC 73 Harrington.

3. CLRO Rep. 18, f. 211r (22 April 1574).

4. For an introduction to Paracelsian ideas and their place in chemical and medical thought, see Debus, *The Chemical Philosophy*; Pagel, *Paracelsus*; and Trevor-Roper, "The Paracelsian Movement," 149–99. The specialized literature on Paracelsus is so wide ranging that only a cursory mention of it can be made here. The spread of Paracelsian ideas is discussed in Shackelford, "Early Reception of Paracelsian Theory," 123–35. For Paracelsus's Reformation context see Moran, "Paracelsus, Religion, and Dissent," 65–79; Webster, "Paracelsus: Medicine as Popular Protest," 57–77; Webster, "Paracelsus, Paracelsianism, and the Secularization of the Worldview," 9–27. For national responses to the ideas of Paracelsus see Debus, *English Paracelsians*; Debus, *The French Paracelsians*; Kocher, "Paracelsian Medicine in England," 451–80.

5. My approach here is intended to incorporate some aspects of Jordanova's suggestion that we might fruitfully integrate the social and cultural histories of medicine; see "The Social Construction of Medical Knowledge."

6. GHMS 5257/1, f. 109r (25 [March 1577]).

7. I follow Green's revised and expanded definition of the original term coined by Pelling and Webster. I use *medical practitioner* to describe any person who, at some point in his life, would have identified himself or been identified by others as engaging in medical practice. See Pelling and Webster, "Medical Practitioners," 166, and Green, "Women's Medical Practice," 445–46.

8. *RCP Annals* 2: 70a–71a. Though none of Fairfax's advertisements are known to survive, a contemporary example of a printed advertisement for Nicholas Bowlden's surgical services (c. 1602) is Bodleian MS Ashmole 1399, f. 1. For the college's efforts to rid the market of practitioners who dispensed physic without a license, see Pelling, *Medical Conflicts.* Pelling's focus on only those practitioners who came before the college limits her view to a small percentage of the overall market in London.

9. Henry Carey was the son of William Carey and Mary Boleyn, Ann Boleyn's sister. Carey was, at the least, Elizabeth's first cousin. Some contend that he was actually her

half-brother, since Henry VIII and Mary Boleyn were intimate at the time of her marriage to Carey. For the importance of patronage to the college and its workings, see
Pelling, *Medical Conflicts*, and Dawbarn, "Patronage and Power."

10. Harold Cook was the first to draw attention to the economic realities of a medical "market," in *The Decline of the Old Medical Regime*, especially 28–69. Subsequent studies
that have emphasized its importance include Fissell, *Patients, Power, and the Poor*, especially 37–73; Gentilcore, *Healers and Healing*, especially 56–95; and Brockliss and
Jones, *The Medical World of Early Modern France*, especially 170–346.

11. Lake, "From Troynouvant to Heliogabulus's Rome and Back," 244. See also Pearl,
"Change and Stability in Seventeenth-Century London," 3–34; Archer, *The Pursuit of
Stability*.

12. Lupton, *London and the countrey*, 1; Lake, "From Troynouvant to Heliogabulus's Rome
and Back," 219.

13. Attempts to recover the patient's perspective include Porter, "The patient's view," 175–
98; Beier, *Sufferers and Healers*; Duden, *The Woman Beneath the Skin*; Fissell, *Patients,
Power and the Poor*.

14. On the contractual nature of most early modern medical arrangements see Pomata,
Contracting a Cure. For patient complaints about failed contracts brought to the College of Physicians, see Pelling, *Medical Conflicts*, 225–74.

15. *RCP Annals* 2: 78a, 106a; GHMS 5257/1, ff. 44r (24 September 1566), 58r (19 April 1569).

16. The starting point for the history of the college remains Clark, *A History of the Royal
College*. For a broader study of medicine in early modern England, consult Wear,
Knowledge and Practice in English Medicine. The classic history of the Barber-Surgeons' Company, including excerpts from its records, is Young, *The Annals of the Barber-Surgeons*.

17. See Gentilcore, *Healers and Healing*; Clouse, "Administering and Administrating
Medicine."

18. Atkinson, *St. Botolph Aldgate*, 119; Guy, "The Episcopal Licensing," 528–42.

19. Bloom and James, *Medical Practitioners*, 16. These findings suggest that Roberts's contention that the "usual system of licensing physicians and surgeons by the bishops never
really operated in London" needs to be treated with caution; Roberts, "The Personnel
and Practice of Medicine," 217.

20. DL/C/335, f. 62r. The names of the women who supported her license were Margaret
Dale, the wife of John Dale of St. Mary Aldermanbury; Anna King, the wife of William
King of St. Mildred Poultry; Thomasina Hall, the wife of Richard Hall of St. Mary
Woolchurch; Sara Hopkyns, the wife of Richard Hopkyns of St. Gile Cripplegate; Mag-
dalena Bradshaw, the wife of Richard Bradshaw of St. Anne Blackfriars; and Dorothy
Waterstone, the wife of William Waterstone of St. Mildred Poultry.

21. For Dorothy Evans, see DC/L/335 (11 May 1590). For Rosa Priest, see DC/L/332, f. 142r,
where her witnesses include Margaret Bromhead, the wife of Lewes Bromhead, and
Agatha Bailey, wife of Henry Bailey. Rosa Priest's activities as a midwife are mentioned
in the parish records of St. Mary Wolnoth when, in 1582, she intervened on behalf of a
newborn girl, Cicely Tasker, who was too weak to wait until Sunday to be brought to
church.

22. CLRO, Reps. 15, f. 156r; 18, f. 107v; 13/2, f. 506r. The college attempted, also in 1563, to shut down van Duran's practice, *RCP Annals* 1: 22b.
23. Pearl, "Change and Stability in Seventeenth-Century London," 15, suggests this was true of London more broadly.
24. *RCP Annals* 2: 7a–8a. For information on Margaret Kennix and her husband, Henry Kennix, see AR 2: 172, 250, 301; AR 3: 341. Her husband was a glover, and made denizen in 1582/83; AR 2: 301.
25. Clark, *A History of the Royal College*, 86–87.
26. Baker, *The composition or making*, 44v; Securis, *A detection and querimonie*, sig. Ciiiv.
27. Hall, *A most excellent and learned woorke of chirurgerie*, sig. +iii r–v; Clowes, *A prooved practise*, sig. A3 v. While it is possible that there was more than one female practitioner on Sea Cole Lane in the period, it is likely that Clowes refers to Margaret Kennix, the Dutch empiric protected by Elizabeth. Women were prominent healers, and could not be licensed by the college or the Barber-Surgeons' Company. Their prominence, coupled with their lack of corporate identity, may explain why they figure so largely in Clowes's diatribe.
28. Securis, *A detection and querimonie*, sig. Biiv.
29. The poem exists in two manuscript versions, one at the British Library (MS Lansdowne 241, f. 374v, John Sanderson's diary and commonplace book, 1560–1610) and one at Cambridge University Library (MS 4138, ff. 16v–18v). The two versions differ in some particulars. The Cambridge University Library copy appears to be a slightly later version of the British Library copy, based on the names of the doctors included in the poem. I am indebted to Nigel Smith for referring me to the Cambridge copy, and to Bill Sherman for supplying me with the details of the poem. Cambridge University Library MS 4138, f. 16v.
30. Ibid.
31. *RCP Annals* 2: 142a, 6b; GHMS 5257/1, f. 51r (21 October 1567); 3907/1, entries for 1591–92. Barlow's medical formularies of 1588–90 are Bodleian Library MS Ashmole 1487.
32. *RCP Annals* 1: 33a; AR 2: 411, 468; COMCL 17/381. For the importance of marriage in medical communities, see Pelling, "The Women of the Family?," 383–401.
33. GHMS 9234, f. 61v; 6836/1, f. 5r.
34. This strategy of licensed practitioners distancing themselves from empirics while adopting their techniques was common; see Lingo, "Empirics and Charlatans," 592. The historiography surrounding the adoption of Paracelsian ideas in England has split between the Kocher-Debus approach (which emphasizes the "compromise" made between practitioners who disagreed with Paracelsian theories while being attracted to his therapeutics) and the approach of Webster (who emphasizes the enormous range and importance of Paracelsian ideas that circulated during the period). I believe the evidence presented here confirms the value of Webster's approach, since the London dispute formed not simply around issues of therapies and theories, but over who should have the right to dispense Paracelsian services and remedies. Though Trevor-Roper cautions us to use the term *semi-Paracelsian* to describe doctors who used chemical medicines without embracing Paracelsus's philosophy, adding this category of analysis did not clarify this case. See Kocher, "Paracelsian Medicine in England," 451–80; De-

bus, *The English Paracelsians*; Debus, "The Paracelsian Compromise in Elizabethan England," 71–97; Trevor-Roper, "Court Physicians and Paracelsianism," 79–94; Webster, "Alchemical and Paracelsian Medicine," 301–34. For a review of the debates and recent historiography consult Smith, "Paracelsus as Emblem," 314–22.

35. British Library MS Lansdowne 18/9, 29 August 1573; 101/4, undated. Russwurin was by no means the only German immigrant in London, even though their numbers were smaller than the Flemish and Walloon immigrants. All three groups were called "Dutch" in contemporary accounts. The use of *Allmaigne* is unusual for Elizabethans. For the German immigrant presence in the early modern period, see Esser, "Germans in Early Modern Britain."

36. British Library MS Lansdowne 18/9, 29 August 1573. For Peter Turner see Munk, *The Roll of the Royal College*, 1: 84. Turner was locked in disputes with the College of Physicians over his right to practice medicine in London for many years, *RCP Annals* 2: 1b, 13a, 18a, 19a–b. For early modern ideas about the eyes and their treatment, see Sorsby, "Richard Banister," 42–55.

37. British Library MS Lansdowne 101/4, undated.

38. Ibid.

39. For Cecil's relationship with his Galenic physicians, see Harkness, "Nosce Teipsum," 171–92.

40. Russwurin signs his letter to Cecil as "Medicus Spagiricus opt[halamista]," British Library MS Lansdowne 101/4. The *ars spagyrica* were those branches of chemical distillation that had to do not with alchemical transformation but with Paracelsian medicines. Tycho Brahe refused to engage in alchemy but was deeply interested in the healing power of the ars spagyrica; see Christianson, *On Tycho's Island*, 91. We do not know whether Russwurin treated Cecil's mother, Jean, but she had gone blind by 12 December 1574, according to her letter to her son, British Library MS Lansdowne 104/61. For Russwurin's denization, see Webster, "Alchemical and Paracelsian Medicine," 305, where he cites *CPR* 6: 261, 25 February 1574. Russwurin was made denizen shortly before he ran afoul of Mrs. Currance.

41. The surgeons included two Englishmen (Robert Clarke and Richard Wistowe), as well as foreigner James Markady. On 16 March 1567/68 Parke was ordered by the company to remove this flattering self-description and "to sette his signe as other Surgeons do w[i]thout any Superscription." GHMS 5257/1, f. 52r. For the significance of street signs, see Garrioch, "Shop Signs and Social Organization," 20–48.

42. The College punished Cornet by seizing "his feigned and unwholesome remedies" and throwing them into a bonfire in Westminster's market. *RCP Annals* 1: 8a. In taking to the streets and erecting a temporary stall in such a well-trafficked part of the City, Russwurin invited comparisons between his activities and those of other itinerant peddlers, who often provoked civic censure along with other "masterless men"; Beier, "Social Problems in Elizabethan London." For a discussion of charlatans and their mobility, see Porter, *Quacks*. Even so, many itinerant charlatans did remain sedentary for some cures. See Gentilcore, "'Charlatans, Mountebanks, and Other Similar People,'" 297–314; Lingo, "Empirics and Charlatans," passim. For the Royal Exchange and its place at the heart of London, see Saunders, "Reconstructing London."

43. Clowes, A *Briefe and necessarie treatise*, 10r, 11r.

44. CLRO Rep. 18, f. 196r–v.

45. Clowes, A *Briefe and necessarie treatise*, 11r–v; CLRO Rep. 18, ff. 196r–v, 211r. The doctors consulted were Dutch physician Peter Symons and Italian physician Julio Borgarucci.

46. Clowes, A *briefe and necessarie treatise*, 12r in Henry E. Huntington Library copy, HEH 29002. He is also mentioned in marginalia on f. 24r. Eamon, in his *Science and the Secrets of Nature*, 254–55, refers to Hester in association with the Italian practitioner Fioravanti. Debus also refers to Hester in *English Paracelsians*, especially 64–69.

47. Clowes, A *briefe and necessarie treatise*, 12r–v; British Library MS Lansdowne 101/4, ff. 12r, 14r. Though undated, Russwurin's claims of mistreatment by London practitioners are in keeping with the tenor of the events of 1574, f. 8r. For the Paracelsian emphasis on the chemical analysis of urine, see Debus, *The Chemical Philosophy* 1: 59, 109–10. For the medical implications of tartar in Paracelsian medicine, see ibid., 1: 107.

48. Munk, *The Roll of the Royal College*, 1: 104–6. For military medicine in the period, see Cruickshank, *Elizabeth's Army*, 174–87.

49. GHMS 5257/1, ff. 82r (6 October 1573), 95v (7 February 1574/75), 104v ([28] February 1575/76), 113v (14 February 1577/88). For Goodrich, see Hessels, *Ecclesiæ Londino-Batavæ Archivum*, 2: 820 and *RCP Annals* 1: 8a.

50. Pelling and Webster, "Medical Practitioners," 172.

51. On medical publishing see Slack, "Mirrors of Health and Treasures of Poor Men," 237–73; Furdell, *Publishing and Medicine*; Johns, *The Nature of the Book*. Much of my subsequent analysis on the rhetorical strategies employed against Russwurin was shaped by Harley, "Rhetoric and the Social Construction of Sickness."

52. Harris traces the idea of venomous, foreign infection and the danger it posed to England, as well as its relationship to Paracelsian medicine, in *Foreign Bodies and the Body Politic*.

53. Baker, *The composition or making*. The preface to the reader is dated 15 March 1574, shortly before Clowes and Baker took the Russwurin case to the Aldermen.

54. Baker, *The composition or making*, sig. Ciir–v.

55. Arrizabalaga, Henderson, and French, *The Great Pox*; Pelling, "Appearance and Reality," 95–105; Kevin Siena, *Venereal Disease, Hospitals, and the Urban Poor*.

56. A single copy survives of Hester's A *true and perfect order to distil oils* at Glasgow University. No copy of Coxe's *Treatise on the Making and Use of Divers Oils* survives. The work is mentioned in Maunsell, A *first part of the catalogue of English printed books*, pt. 2, 6. For Coxe's apology for dabbling with the occult, see his *Short treatise*. He also published an almanac, A *prognostication*. For a broader discussion of Coxe, see Jones, "Defining Superstitions," 187–203.

57. Webster, "Alchemical and Paracelsian Medicine," 333.

58. Baker, *The newe jewell of health*, sigs. *iiiv, *ivr.

59. In addition to Clowes's work on syphilis and Baker's work on the medical uses of mercury, these works included Baker, *Guidos questions*, and Hester, A *joyfull jewell*, Fioravanti's plague treatise translated by Thomas Hill.

60. Clowes, A *short and profitable treatise*, sigs. Civ, Ciir.

61. This was the name given to him by Gabriel Harvey, who inscribed it in the margins of Hester's advertisement. The advertisement is not dated, but is thought to have been produced sometime between 1585 and 1588. See the British Library's copy of John Hester's *These oiles, waters, extractions*. For more on Hester, see Kocher, "John Hester, Paracelsian," 621–38.

62. John Hester, *A short discours*, and John Hester, *A compendium of the rationall secretes*. For Fioravanti and his importance in the "book of secrets" tradition, see Eamon, *Science and the Secrets of Nature*.

63. Hester, *A hundred and fourtene experiments and cures*, letter of dedication to Sir Walter Raleigh, n.p. Hester's interests bring into question Roberts's conclusion that "Paracelsian medicine hardly touched the apothecaries"; "The Personnel and Practice of Medicine," 226.

64. Hester, *A compendium of the rationall secretes*, sig. iiijv.

65. This indicates that Hester's translation was based on the 1582 Lyon imprint of the work.

66. Hester, *A hundred and fourtene experiments and cures*, 6, 7. On the importance of distillation in the literature on chemical medicines, see Multhauf, "The Significance of Distillation in Renaissance Chemistry," 329–46.

67. Bostocke, *The difference between the auncient phisicke*, sigs. Bir, [Bvir]. For more on Bostocke, see Harley, "Rychard Bostok of Tandgridge, Surrey," 29–36.

68. Clowes, *A briefe and necessarie treatise*, 9r–v.

69. Ibid., 9r–11v.

70. Ibid., 12r, 13r. Jonathan Gil Harris brilliantly connects concerns about foreigners to contagion in *Foreign Bodies and the Body Politic*, passim.

71. Webster pointed out that foreign medical practitioners threatened men like Clowes because they attracted patients, reputedly knew efficacious cures, and had achieved a certain level of public recognition in the marketplace; "Alchemical and Paracelsian Medicine," 301–34. Though historians often label negative reactions to foreigners as xenophobia, that appears to be too broad a term to help us understand the different reactions the Russwurin case engendered. See Yungblut, *Strangers Settled Here Amongst Us*, and Dillon, *Language and Stage in Medieval and Renaissance England*. Joseph Ward, through a juxtaposition of Thomas Dekker's *The Shoemaker's Holiday* and a letter sent from the Weavers' Company to elders of the French church, suggested we should avoid "leaping to the conclusion that xenophobia was the essential characteristic of life in early modern London." See Ward, "Fictitious Shoemakers, Agitated Weavers," 85.

72. Pickering, "Epistle," f. 59r.

73. Baker, *The whole worke of that famous chiru[r]gion*, sigs. iiv–iiir.

74. I. W., *The copie of a letter*, sig. Ar–v.

75. Clowes, *A prooved practise for all young chirurgians*, "Epistle to the Reader," n.p.; sig. Ar–A2v.

76. Read, *A most excellent and compendious method*, sigs. Aiir–Aiiir. For a broader view of Read's effort see Cook, "Against Common Right and Reason," 301–22. The opening wedge may well have been made when the queen successfully lobbied the College of Physicians to grant the surgeon John Banister a license to practice medicine; *RCP An-*

nals 2: 63. Later, the earl of Essex requested that an unlicensed empiric he went to for medical services, Leonard Poe, be granted a similar license based on the Banister precedent; ibid., 104b–105a.

3. EDUCATING ICARUS AND DISPLAYING DAEDALUS

1. Baker's advertisement survives in a single copy in the Society of Antiquaries, London. For information on Baker, see McConnell, "Baker, Humphrey," *DNB*. London records suggest that her dates of activity are too conservative. Vestry Minute Books for St. Bartholomew by the Exchange list him as a parishioner in 1586 (GHMS 4384/1, 89) and show assessments for Baker as late as 1593 (132).
2. Record, *The ground of artes*, sig. Bi, v. His life and work are surveyed in F. M. Clark, "New Light on Robert Recorde," 50–70. Keith Thomas's article on numeracy was the first to define and characterize mathematical literacy in the period, "Numeracy in early modern England," 103–32.
3. Ibid., sig. [Aviir]; Digges, *A prognostication everlasting*, sig. Air.
4. Studies of the interest in mathematics include classic works by Taylor, *The Mathematical Practitioners*, and Feingold, *The Mathematicians' Apprenticeship*. More recently, Bennett, Clucas, Johnston, and Cormack have extended and deepened the lines of analysis opened up by Taylor and Feingold. See Bennett, "The 'Mechanics' Philosophy and the Mechanical Philosophy"; Clucas, "'No Small Force'"; Johnston, "Mathematical Practitioners and Instruments"; Johnston, "Making Mathematical Practice"; Cormack, "'Twisting the Lion's Tail'"; Cormack, *Charting an Empire*. These studies of English mathematics are complemented by works done by other scholars on the roles fashioned by practitioners in the Scientific Revolution, especially Westman, "The Astronomer's Role"; Hall, "The Scholar and the Craftsman"; Biagioli, "The Social Status of Italian Mathematicians"; and Biagioli, *Galileo, Courtier*. Shapin's "'A Scholar and a Gentleman'" provides a reconsideration of the practitioner's role, while extending the analysis beyond a single disciplinary group. Though Tudor and Stuart mathematicians themselves used the term *mathematical practitioners* to collectively describe those interested in mathematics and their applications, its use has been criticized by Ash, *Power, Knowledge, and Expertise*, 140–41, who asserts, "It may well be doubted whether they themselves would have, or even could have, self-identified as any sort of community; therefore the question arises whether it is most useful or informative for historians to treat them as one." While I am deeply indebted to the work of these scholars, as this chapter shows, here I am less interested in the role of *mathematicians* than in the roles of *mathematics* and *literacy* in the City.
5. Bacon, *The wisedome of the Ancients*, 92–95.
6. Hill, "'Juglers or Schollers?'"
7. Dee, "Compendious Rehearsal," 5–6.
8. For the tendency to confuse mathematics and magic, see Zetterberg, "The Mistaking of 'The Mathematics' for Magic," 83–97.
9. I derived these statistics from consulting the *English Short Title Catalogue* from 1558 through 1603. Using the subject headings, I compiled a list of titles that included works

on both theoretical mathematics and applied mathematics (almanacs and works on astrology, astronomy, navigation, cosmology, surveying, maps and mapmaking, military science, and mensuration–the application of geometry to measurements). As some of these subject headings are inconsistently or inadequately noted in the *English Short Title Catalogue*, I also did an annual survey of published titles so as to include any works that contained a high degree of mathematical content but had not been given a mathematical subject heading.

10. Record, *The pathway to knowledg*, sig. ir. It is difficult to gauge the popularity of any early modern title. Still, publishers found that reprinting titles that would still sell was a good source of profit; see Johns, *The Nature of the Book*, 454–57. I see frequent reprints as an indication that consumers would indeed buy the books if they were offered, and thus as a marker of popularity.

11. Meskens noted the same gap between the high incidence of arithmetic publications and the relatively low number of geometry publications in late-sixteenth-century Antwerp and concluded that for "merchants it was arithmetic that was important, not Euclid or Sacrobosco." Meskens, "Mathematics Education," 152, 155.

12. London GHMS 1568, f. 18r (1549).

13. See, for example, Bourne, *An almanac and prognostication*, with a second edition in 1567, and Digges, *A prognostication everlasting*, also with a second edition in 1567. Digges also published a general prognostication in 1553. A comprehensive study of almanacs in early modern English culture is Capp, *Astrology and the Popular Press*.

14. See Davis, "Sixteenth-Century French Arithmetics and Business Life."

15. Record, *The ground of artes*, sig. Bvv; Anonymous, *An introduction for to learne to recken*, sig. Aiir–v.

16. Digges, *A booke named Tectonicon* [unsig., Preface to the Reader].

17. Johnston, "Mathematical Practitioners and Instruments"; Hood, "Thomas Hood's Inaugural Address," 98. On this idea see also Smith, *The Body of Nature*, passim.

18. See Jones, "Gemini, Thomas," *DNB*, for a list of extant instruments and their present locations.

19. Record's pedagogical strategies are discussed in Johnson and Larkey, "Robert Recorde's Mathematical Teaching."

20. For the reception of Copernican ideas, see Gingerich, *The Book Nobody Read*.

21. Tartaglia also provided a preface, which listed all the branches of mathematics. See Tartaglia, *Euclide Megarense reassettato*. Commandino's translation of Euclid into Latin was published in 1572, and a translation of the work into Italian followed in 1575.

22. Gaukroger draws attention to the fact that Dee's "Mathematical Preface" was "ignored by contemporaries and successors alike" and "was not even mentioned by natural philosophers such as Bacon and Boyle." Gaukroger, *Francis Bacon*, 23. While the Billingsley-Dee edition of Euclid's *Elements* may not have been influential within the circle of elite natural philosophers, and was never reissued, the ideas contained within were popularized in cheaper and more widely available English vernacular publications.

23. Billingsley, "The Translator to the Reader," in Euclid, *The elementes of geometrie*, n.p.

24. For Dee's interest in mathematics see Clulee, *John Dee's Natural Philosophy*, 143–76,

and Harkness, *John Dee's Conversations with Angels,* 91–97. Bourne, *A book called the treasure for travseilers,* sig. ***iir–[***iiir].

25. Dee, "Mathematical Preface," unsig., sigs. ciiiiv–div.
26. Ibid., sig. bjv; Harvey quoted in Stern, *Gabriel Harvey,* 167; Bourne, *Inventions or devises,* 98. Harvey's copy of Blagrave is British Library C.60.0.7.
27. Ferguson, *The Articulate Citizen,* 363.
28. For Smith's biography see Dewar, *Sir Thomas Smith,* and Archer, "Smith, Sir Thomas," *DNB.*
29. Smith, *Discourse of the Commonweal,* 12, 28.
30. Ibid., 124, 88.
31. Dee, "Mathematical Preface," sig. Aiiiir.
32. Inscription on the funeral monument of Elizabeth Withypool Lucar (d. 1537), St. Laurence Pountney Church, London, courtesy of Judith Bennett. The monument was probably constructed in the Elizabethan period based on its stylistic elements. Ferguson, *Dido's Daughters,* discusses the problems associated with assessing female literacy in the period.
33. Smith, *The Body of the Artisan,* passim.
34. Cuningham, *The cosmographical glasse,* sig. Aiir.
35. Baker, *The well spryng of sciences,* sig. Aiiir, 128r; Bourne, *A booke called the treasure for travseilers,* sig. [**ivr].
36. Baker, *The well spryng of sciences,* 78v; Bourne, *The art of shooting,* sig. Aivr.
37. Baker, *The well spryng of sciences,* sig. Aiiiv–Aiiiir.
38. Van Egmond, *The Commercial Revolution;* Meskens, "Mathematics Education," 137–55. For the European promotion of mathematical education, see Keller, "Mathematics, Mechanics," 350–52. Rose's study *The Italian Renaissance of Mathematics* remains the classic starting point for historians interested in Italian developments, which inspired the rest of Europe. For Florence, see Goldthwaite, "Schools and Teachers of Commercial Arithmetic."
39. Simon's *Education and Society* remains the starting point for those interested in early modern education in England. For the history of mathematics education, including mention of the St. Olave's curriculum, see Howson, *A History of Mathematics Education in England.* See also Alexander, *The Growth of English Education,* especially, 203; Watson, *The Beginning of the Teaching of Modern Subjects,* especially 304; and Woodbridge, "Introduction," 1.
40. British Library MS Lansdowne 98/1.
41. Johnson, "Gresham College," 426. Two studies of mathematics in the early history of the college are Feingold, "Gresham College and London Practitioners," and Clucas, "'No Small Force.'" Larry Stewart has connected later public lectures on natural philosophy with the rise of technological literacy; *The Rise of Public Science.*
42. Buck, *The Third Universitie,* in Stow, *The Annales,* 965. For a discussion of the options for instruction in mathematical geography in London, see Cormack, "The Commerce of Utility," 311–17.
43. Mellis's additions can be found in Record, *The grounde of artes* (London, 1582) and subsequent editions; Mellis, "To the Right worshipfull," sig. Aiiv.

44. Mellis, "To the Reader," in Oldcastle, *A Briefe Instruction*, sig. A3r.

45. Johnson, ed., "Thomas Hood's Inaugural Address," 103; George Buck, *The Third Universitie*, 981. Scholars who have argued for the importance of the mathematics lecture as a means of educating Londoners include Taylor, *The Mathematical Practitioners*. Feingold has made a persuasive case that the lectures were not well attended and were discontinued due to lack of interest and support; Feingold, *The Mathematicians' Apprenticeship*, 171–76.

46. Mellis, "To the Right worshipfull," sig. Aiir.

47. British Library MS Lansdowne 121/13 (undated); British Library MS Lansdowne 29/20 (2 March 1579). The work on water, like the work on statics, is now lost, but aspects are probably included in *A regiment for the sea* and *A booke called the treasure for travelers*, the latter of which includes a section on statics.

48. British Library MS Lansdowne 121/13; Bourne, *Inventions or devises*, sig. ¶3; Bourne, *A book called the treasure for travelers*, sig. *iiir and 2.

49. Bourne, *A booke called the treasure for travelers*, sigs. *iiv, *iiiv.

50. Norman, *The newe attractive*, sigs. Aiiiv, Biv.

51. CLRO, Rep. 15, f. 598v.

52. Turner, *Elizabethan Instrument Makers*, 3. In the following pages, my dates of activity for the instrument makers can vary from those in other scholarly works since they are based on information drawn from social history sources rather than on extant examples of craftsmanship. In addition to Turner, classic reference works on instrument makers include Hind, *Engraving in England*; Loomes, *The Early Clockmakers*; Clifton, *Dictionary of British Scientific Instrument Makers*.

53. One of Cole's *compendia* is in the National Maritime Museum, London, Greenwich Hospital Collection, signed and dated 1569. For a full catalogue of Cole's work, and biographical details, see Ackermann, *Humphrey Cole*. Vallin's musical chamber clock is in the Ilbert Collection of the British Museum. For more on Vallin see Drover and Lloyd, *Nicholas Vallin*. An example of Vallin's craftsmanship can be seen and heard at http://www.youtube.com/watch?v=EcmyZjuRsrw.

54. The history of the Clockmakers' Company is outlined in Atkins and Overall, *Some Account of the Worshipful Company of Clockmakers*, and White, *The Clockmakers of London*. Joyce Brown discovered that the Grocers' Company was home to many early clockmakers; see her *Mathematical Instrument Makers*, which includes references to earlier makers. M. A. Crawforth extended Brown's analysis and examined other London guilds, discovering genealogies of instrument-making masters and apprentices in both the Broderers' and Joiners' companies; "Instrument Makers in the London Guilds," 319–77.

55. Only Geminus and the Vallins are prominent in the historical literature; see Turner, *Elizabethan Instrument Makers*, 12–20, and Drover and Lloyd, *Nicholas Vallin*, passim, for biographical details and examples of their craftsmanship. Information on other immigrant instrument makers is scattered in the London sources, including the unpaginated parish registers of St. Anne Blackfriars, GHMS 4510/1. In addition to the registers, see for Mary, AR 1:192, 298; for Michael Noway, AR 2: 34, 212, 276, 318, 410, and the will of Nicholas Vallin, ADCL 5/58; for Francis Noway, GHMS 4508, AR 2: 179, 253, 410; for

Roian, COMCL 16/251, AR 2: 357, AR 3: 50; for Tiball, AR 2: 357, AR3: 5; for Daunte-
ney, AR 2: 180–81, AR3: 397. Michael Noway's watch is preserved in the collection of
the Clockmakers' Company.

56. Like the immigrant clockmakers, many of these individuals must be traced through social history records rather than through apprenticeship documents or by consulting
extant specimens of craftsmanship. For Blunte, GHMS 10342 and GHMS 6419, f. 112v;
for Modye, GHMS 10342 and ADCL 4/201v; for Brome, GHMS 10342; for Islberye,
GHMS 10342; for Gawnt, C66/1032, no. 298 and COMCL 19/65; for Stevens, ADCL 3/
225; for Thomas, GHMS 9220 and GHMS 9221; for Hearne, the matrimonial enforcement case in the Consistory Court, DL/C/214, ff. 164–65, 180–88, and GHMS 9221; for
Francis, GHMS 9223, f. 30v; for White, GHMS 9223, f. 58v, and ADCL 4/226.

57. Loomes, *The Early Clockmakers*, 269. Examples of Grynkin's craftsmanship are in the
British Museum, the Fitwilliam Museum, and the Victoria and Albert Museum; For
Humfrey Cole, see Turner, *Elizabethan Instrument Makers*, 20–25; Brown, "Mathe-
matical Instrument Makers," passim.

58. For Geminus, see Turner, *Elizabethan Instrument Makers*, 12–20; for Ryther and examples of his work for Dudley, see Turner, *Elizabethan Instrument Makers*, 27–29; for
Whitwell, see Turner, *Elizabethan Instrument Makers*, 29–31.

59. GHMS 577/1, ff. 3r, 10v, 12v, 15r, 16v, and 20v; GHMS 1046/1, ff. 59v–60r; GHMS 4524/
1, ff. 42v–43v.

60. British Library Add. GHMS 12222, 63 (St. Giles Cripplegate); GHMS 2593/1, f. 9r (St.
Lawrence Jewry), GHMS 4352/1, f. 30r (St. Margaret Lothbury), GHMS 4165/1, f. 2r
(St. Peter Cornhill).

61. GHMS 9235/2, pt. 1, f. 11v (St. Botolph Aldgate), GHMS 1279/2, f. 132r (St. Andrew
Hubbard), GHMS 1046/1 (St. Antholin Budge Row), f. 14v.

62. Annual maintenance contracts of 4 shillings were routine in Elizabethan London, up
from the 3 shillings, 4 pence that Henry VIII's clockmaker, Bruce Awsten, received
from St. Andrew Hubbard around 1540; see GHMS 1279/2, f. 72v. St. Andrew by the
Wardrobe paid its clocksmith that amount annually for his services; GHMS 2088, f. 8r.
St. Andrew by the Wardrobe was one of the ten poorest churches in all of London, but
it still managed to raise £9 over two years to buy a new clock, which it took scrupulous
care of over the next several decades; GHMS 2088, ff. 8v, 11r.

63. GHMS 1454, roll 91. The clock was made by John Williamson (fl. 1580–92/93), who
also repaired the St. Lawrence Jewry clock (GHMS 2953/1, f. 64r). He may have been a
member of the Carpenters' Company; see GHMS 2953/1, f. 64v. The Poultry's clock
may have been linked to the elaborate ship on the top of St. Mildred's church, which
has been taken as a weathervane but might have been part of a noteworthy timepiece;
see Prockter and Taylor, *The A to Z of Elizabethan London*, 51. Stow mentions neither
a clock nor a weathervane in his description of the church in *A Survey of London*.

64. GHMS 1002/1A, f. 246r.

65. British Library MS Cotton Vespasian F.VI, f. 270r (29 January 1572/73), Sir Thomas
Smith to Francis Walsingham.

66. The decisions made by Baker and other educators and authors to cater to the interests

of their audience can be seen as early evidence for the development of a mathematical and instrumental public culture. An analogy can be made between their efforts in London and the later development of a public culture of chemistry as described by Golinski, *Science as Public Culture*.

67. Johnston, "Mathematical Practitioners," especially 324–27. To compare with later efforts to lecture about machines, see Morton, "Concepts of Power," 63–78.

68. Bourne, *The art of shooting*, sig. Aiiiv; Borough, *Discourse on the variation of the cumpas*, sig. *iiiv.

69. Harvey quoted in Stern, *Gabriel Harvey*, 167.
    Harvey's marginal notes have received a great deal of attention. See especially Stern, *Gabriel Harvey*. Anthony Grafton made a persuasive case that these marginalia, previously dated 1580, were more likely to have been written in 1590. See Grafton, "Geniture Collections, Origins, and Use of a Genre," note 16. Harvey's annotations in Gaurico have been crucial to the work done by a number of scholars, including Taylor, *The Mathematical Practitioners*; Johnston, "Making Mathematical Practice"; and Popper, "The English Polydaedali."

70. Harvey quoted in Stern, *Gabriel Harvey*, 168, 202. Harvey's copy of Gaurico is Bodleian Library 4°Rawl.61. In Stern, Robert Norman has been mistranscribed as "Robert Norton." The note on Kynvin appears in Harvey's copy of Blagrave's *Mathematical Jewel*, now British Library C.60.0.7.

71. Blagrave, *The mathematical jewel*, title page; Cyprian Lucar, *A treatise named Lucarsolace*, sig. Aiiv.

72. Hood's astrological diagrams—four of them, approximately forty-two centimeters square each, mounted on old boards—are now British Library Add. MS 71494, 71495. Formerly held by the Patent Office, they were transferred to the Science Museum before their addition to the British Library collections. Stephen Johnston, in an unpublished conference paper, discusses this "paper instrument" and its implications in "The Astrological Instruments of Thomas Hood," (July 1998), http://www.mhs.ox.ac.uk/staff/saj/hood-astrology/. For the connections between mathematics and astrology see Dunn, "The True Place of Astrology," 151–63.

73. Lucar, *A treatise named Lucarsolace*, sig. Aiiv; Lucar, *A treatise named Lucar Appendix*, 1; Blagrave, *The mathematical jewel*, title page; Hood, *The use of the two mathematical instrumentes*, title page.

74. Tebeaux, *The Emergence of a Tradition*; Hood, *The use of the celestial globe*, 42r.

75. Blagrave, *Baculum familliare*, title page, 3, 17. On precision, see Moran, "Princes, Machines, and the Valuation of Precision."

76. Bourne, *A regiment for the sea*, unsig.; Wright, *Certaine errors*, sig. I3r.

77. Lucar, *A treatise named Lucar Appendix*, 2, 102, 10; Blagrave, *Astrolabium uranicum*, sig. Cr. On the significance of advertising for instrument makers of the period, see Bryden, "Evidence from advertising."

78. Hood, *The use of the two mathematical instrumentes*, sig. Ciiv; Lucar, *A treatise named Lucar Appendix*, 56; Hood, *The making and use of the geometricall instrument*, end page; Hood, *The use of the celestial globe*, sig. [A4]r.

79. Wright, *Certaine errors*, sig. A2r, sig. [M4v], sig. [Gg2r].
80. Stevin, *The haven-finding Art*, sig. A3r; Wright in Stevin, *The haven-finding art*, sig. B3r; Blagrave, *The mathematical jewel*, sig. iiv; Lucar, *A treatise named Lucar Appendix*, 23.

4. "BIG SCIENCE" IN ELIZABETHAN LONDON

1. SP Domestic 12/111/2 (15 January 1577), signed statement of conversations had between "Giovanni Battista" and William Cecil, Lord Burghley. These documents have not been linked to Giovan Battista Agnello by scholars of the Frobisher voyages, yet the handwriting is identical to other Agnello documents. Michael Lok reveals that his formal relationship with Agnello began a few days later on 18 January 1577. For Agnello, see Harkness, *John Dee's Conversations with Angels*, 204. From 1547 to 1549, "J. B. Agnelli and Company" was authorized by royal officials to import gold bullion for use in the royal mint; see PRO E 101/3/9. Agnello's alchemical treatise, first published in Italian in 1567, was translated into English by R[obert] N[apier], *A revelation of the secret spirit*. Agnello is mentioned in Chaplin, *Subject Matter*, 48, 50, 57. For an introduction to alchemy during the period, consult Moran, *Distilling Knowledge*.
2. The classic biographies of Cecil are Read, *Mr. Secretary Cecil*, and Read, *Lord Burghley*. For his medical concerns, see Harkness, "*Nosce teipsum*," passim.
3. For an introduction and case studies, see Galison and Hevly, *Big Science*. On the elusiveness of a consistent definition for Big Science see Hevly, "Reflections on Big Science and Big History." As the notes in this chapter show, much of the conceptual framework I use here is drawn from this pathbreaking collection of essays. They emphasize that Big Science indicates not simply a difference in the scale of research but changes in scientific work necessitated by new internal and external contexts.
4. For the need to embed such conceptual terms in appropriate contexts and recognize that they often emerge before the language used to denote them, see Skinner, "Language and Social Change," and Long, *Openness, Secrecy, Authorship*. While I recognize the complications that stem from using Big Science as an analogy, it seemed to me that the notion existed *conceptually* if not in actual usage. My use of the term is similar to Long's useful, though anachronistic, employment of *intellectual property* to describe the relationship between author and artifact in the medieval craft tradition; Long, *Openness, Secrecy, Authorship*, 5.
5. Carr, *Select Charters*, 21; McDermott, "The Company of Cathay," 173.
6. There are a few notable exceptions, the most important being the crown-subsidized plan to rebuild Dover Harbor; see Ash, *Power, Knowledge, and Expertise*, 55–86.
7. These questions are adapted from Galison, "The Many Faces of Big Science," 2: "How does a coordinated effort change direction when new scientific results arrive? Who controls research when sponsors and scientists have different intentions? How does the organization of the research team reflect the wider culture in which it is embedded?"
8. British Library MS Lansdowne 101/16 (1 June 1595).
9. Scholars have emphasized the economic significance of these projects, while I emphasize their place in Elizabethan science. See, for example, Heal and Holmes, "The Economic Patronage of William Cecil." I am also attempting to extend the discussions

of the relationships between science and technology in the eighteenth century and the period of the Industrial Revolution; see Musson and Robinson, *Science and Technology in the Industrial Revolution;* Jacob, *The Cultural Meaning of the Scientific Revolution;* Inkster, *Science and Technology in History.*

10. Galison, "The Many Faces of Big Science."

11. Shapin, "The House of Experiment," passim; Sherman, *John Dee,* passim; Harkness, "Managing an Experimental Household," passim.

12. Galison, "The Many Faces of Big Science," 1. Modern Big Science required the collaboration of hundreds of practitioners, including scientists, lab managers, and technicians. As a form of work, modern Big Science differed sharply from the dominant nineteenth-century model of scientific practice, under which physicists and chemists worked in smaller laboratories with small teams of two or three assistants. See Nye, *Before Big Science.* In the case of Elizabethan Big Science, I am making a similar claim that there was a difference in scale between the projects Cecil promoted and the technical projects of the Middle Ages.

13. For Dee's patronage problems at the court of Elizabeth see Sherman, *John Dee,* passim.

14. I use the vocabulary developed by Pumfrey and Dawbarn in "Science and Patronage in England, 1570–1625," who argue that the English court prized utility, while European courts often favored ostentatious display. After Robert Westman's influential article pointed to the significance of patronage in early modern science ("The Astronomer's Role in the Sixteenth Century"), many historians have explored particular case studies. Useful European comparisons to English patronage are Biagioli, *Galileo, Courtier;* Evans, *Rudolf II and His World;* Findlen, *Possessing Nature;* Lux, *Patronage and Royal Science;* Moran, "German Prince-Practitioners"; Eamon, "Court, Academy, and Printing House"; Shackelford, "Paracelsianism and Patronage"; Smith, *The Business of Alchemy.*

15. Pumfrey and Dawbarn, "Science and Patronage in England," 151; Goodman, "Philip II's Patronage," 65.

16. Kettering, *Patrons, Brokers, and Clients,* 3–11. Kettering does much to complicate early modern patronage, which is often explored only as a straightforward relationship between patrons and clients. Ash's *Power, Knowledge, and Expertise,* while adding the category of "expert mediator" to the mix, retains the emphasis on the two poles of patron and client. I do not assert here that Elizabethan England was unique in the way it undertook patron-client relationships, and strongly suspect that careful scrutiny of court records and correspondence will reveal that complicated chains of influence and entreaty colored the dispensation of patronage in other early modern courts and cities.

17. Cecil's role in the restoration of the coinage is discussed in Read, *Mr. Secretary,* 194–97.

18. Challis, *The Tudor Coinage;* Challis, *A New History of the Royal Mint;* Goldman, "Eloye Mestrelle"; Symonds, "The Mint of Queen Elizabeth."

19. Smith, *Discourse,* 3; Fuller, *Worthies,* 387, quoted in Thirsk, *Economic Policies and Projects,* 53, n. 6.

20. See, for example, British Library MS Lansdowne 6/9 (1563), MS Lansdowne 10/6 (15 March 1566/7), MS Lansdowne 10/18 (10 April 1568).

21. British Library MS Lansdowne 12/58 (1570), MS Lansdowne 12/59 (27 January 1569/70),

MS Lansdowne 18/61 (1574), MS Lansdowne 101/14 (11 May 1591), MS Lansdowne 683/10 (1576), SP Domestic 12/8/13 (1559), SP Domestic 12/46/64 (3 May 1568).

22. For Mistrell see *CPR* 3 Elizabeth, C 66/972 (1561), *CPR* 11 Elizabeth, C 66/1053 (1569), British Library MS Lansdowne 26/22 (4 December 1578), MS Lansdowne 48/15 (1586) and Goldman, "Eloye Mestrelle," passim.

23. Discussions of the English system include Hulme, "The History of the Patent System"; Hulme, "The History of the Patent System . . . A Sequel"; Price, *The English Patents of Monopoly*; Gough, *The Rise of the Entrepreneur*; Thirsk, *Economic Policy and Projects*. My focus here is less on the development of the modern patent system (Hulme and Price), or on the development of consumer goods (Thirsk), than on the ways in which these documents shed light on Elizabethan Big Science. Thirsk, for example omitted from her study "patents for engines that would dredge, drain, grind, raise water, pipe a water-supply, and refine pit coal . . . since they did not directly result in the production of consumer goods." Thirsk, *Economic Policy*, 57, n. 20.

24. Thirsk, *Economic Policy*, 53; Yungblut, *Strangers Here Among Us*, 95–113.

25. Heal and Homes identified the same thoroughness and evaluative tendencies in Cecil's handling of a wider range of economic projects; "The Economic Patronage of William Cecil," 208, 220.

26. Harkness, "Strange Ideas and English Knowledge," 28, 30; Pumfrey and Dawbarn, "Science and Patronage," especially 139–43.

27. British Library MS Lansdowne 77/59 (1 October 1594). Jentill may have been a member of a distinguished family of Italian immigrants that included the jurist and law professor Alberico Gentili, but I have not been able to definitively prove the connection.

28. *CPR* 31 Elizabeth, C 66/1333(8 February 1588/89); SP Domestic 12/125/50 (July 1578); Everitt, "Background to History."

29. For the complicated relationship between proprietary knowledge and science, see the important work of Long, especially *Openness, Secrecy, Authorship*. She traces how the more open culture of technical knowledge in antiquity and the early Middle Ages gave way, amid urbanism and competition beginning in the thirteenth century, to a discourse of protection and secrecy. Her approach differs from Eamon, *Science and the Secrets of Nature*, who sees the Middle Ages as a time of secrecy, in contrast to the new openness of early modern culture.

30. SP Domestic 12/125/48 (July 1578).

31. See, for example, Cecil's annotations on Leonard Engelbreght of Aachen's proposal for making saltpeter, British Library MS Lansdowne 24/54 (1577). For an overview of the literature on the structure of patronage relationships, see Eisenstadt and Roniger, "Patron-Client Relations," especially 49–50.

32. British Library MS Lansdowne 29/16 (29 June 1579); Carr, *Select Charters*, lviii–lix.

33. British Library MS Harley 6991/56 (28 January 1574/75); Rabb, *Enterprise and Empire*, 28–39.

34. *CPR* 4 Elizabeth, C 66/985 (26 May 1562). Letters patent quoted from Carr, *Select Charters*, lviii.

35. *CPR* 2 Elizabeth, C 66/960 (3 October 1560).

36. CLRO Rep. 15, ff. 498r–v, 502r. Aconcio received an annuity from Elizabeth on 27 Feb-

ruary 1559/60; *CPR* 2 Elizabeth, C 66/948. He was made denizen on 8 October 1561; ibid., 3 Elizabeth, C 66/968. His twenty-year patent for engineering machines and furnaces is ibid., 7 Elizabeth, C 66/1017. Aconcio was the author of one of the first books published on the subject of toleration, *Stratagematum Satanæ*.

37. Carr, *Select Charters*, xiv–xv, xxx, xxii, xlviii–xlix.

38. Ibid., xvii–xviii.

39. British Library MS Lansdowne 105/44 (undated); *CPR* 10 Elizabeth, C 66/1049; SP Domestic 12/1/56 (December 1558).

40. *CPR* 9 Elizabeth, C 66/1040 (8 September 1567). Patent quoted in Carr, *Select Charters*, lviii. SP Domestic, 12/16/30 (13 March 1561); Ash, *Power, Knowledge, and Expertise*, 19–54.

41. For Morice, see *Analytical Index to the Series of Records Known as the Remembrancia*, 550, 551, 553. He obtained a five hundred–year lease of London Bridge's first arch in 1581, which was followed by leases of the other four arches. See Carr, *Select Charters*, cxxiii.

42. Thirsk, *Economic Policy*, 87.

43. For example, letters patent for the Mineral and Battery Works, quoted in Carr, *Select Charters*, 17, and the letters patent issued to Pratt and van Herwick for furnaces, British Library MS Lansdowne 105/44 (undated).

44. There are parallels between Cecil's reliance on men like Waad and Herle and the use of professional informants to enforce Elizabeth's 1563 Statute of Artificers. See Davies, *The Enforcement of English Apprenticeship*, 40–62.

45. *CPR* 7 Elizabeth, C 66/1012 (29 January 1564/5); SP Domestic 12/83/12 (18 November 1571). Subsequent documents relating to Franckard's proposals are SP Domestic 12/83/13, 12/83/16, and 12/83/21. Information on Waad drawn from Hicks, "Waad, Armagil," *DNB*. To put Waad in the context of Cecil's other servants, consult Barnett, *Place, Profit, and Power*.

46. British Library MS Lansdowne 101/17 (4 March 1596, with additions dated 26 April 1596). For further information on the Dover project, see Ash, *Power, Knowledge, and Expertise*.

47. See, in contrast, Kettering, *Patrons, Brokers, and Clients*, 5.

48. Lupton, *London and the countrey carbonadoed*, 1.

49. GHMS 4069/1, ff. 27r, 29r, e.g.

50. Fisher, *London and the English Economy*, 185–98.

51. For points of comparison, see Moran, "German Prince-Practitioners"; Rose, *The Italian Renaissance of Mathematics*; Middleton, *The Experimenters*; Evans, *Rudolf II and His World*. For London see Werner and Berlin, "Developing an Interdisciplinary Approach?"

52. British Library MS Lansdowne 12/7 (11 August 1569); McDermott, "The Company of Cathay," 147; Ramsay, *The City of London in International Politics*.

53. An analogy can be made here to the California industrialists who provided funds for Lawrence's pre–World War II cyclotron research; Seidel, "The Origins of the Lawrence Berkeley Laboratory."

54. Thirsk, *Economic Policy*, 26; McDonnell, *Medieval London Suburbs*, 72–118.

55. Accounts of the Ordnance Office, January 1580/1–December 1581, British Library MS Cotton Julius F.I, f. 112r. For the Artillery Garden see Walton, "The Bishopsgate Artillery Garden."

56. For Honricke, see Moens, *The Marriage, Baptismal, and Burial Records*, 33; British Library MS Lansdowne 4/47; SP 12/125/50. For Smith, see British Library MS Harley 286/85, MS Sloane 3682, MS Add. 12, 503, MS Cotton Titus B.5.262; GHMS 9234/1. For Bonetto, see AR 2: 222, 252, 404; GHMS 4508/1. For Lane, see British Library MS Sloane 2192, 2228, MS Lansdowne 24/30; GHMS 4399/1, 162; SP Domestic 12/200/56 (30 April 1587). For Helwiss, see British Library MS Harley 7009, MS Lansdowne 101/13.

57. British Library MS Royal 18.A.21, f. 2v; MS Harley 7009, f. 104. Kynvin's letters patent for a position as gunner in the Tower of London are in *CPR* 31 Elizabeth, C 66/1330 (22 October 1589).

58. British Library MS Lansdowne 121/14 (undated).

59. See, for example, SP 12/88/7 (4 June 1572), Lane's plan for military readiness; SP Additional 21/79 (July 1572), Lane's method for levying soldiers; SP 12/92/25 (August 1573), criticisms of Lane's suit for forfeitures of bonds; British Library MS Lansdowne 19/80 (15 January 1574/75), Lane's military plan regarding the Turks; British Library MS Lansdowne 39/27 (9 July 1583), Lane's project to prosecute Strangers.

60. SP Domestic 12/200/56 (30 April 1587); SP Domestic 12/206/12 (6 December 1587), SP 12/209/118 (30 April 1588), SP 12/216/12 (9 September 1588), SP 12/216/29 (17 September 1588), SP 12/217/2 (1 October 1588).

61. Wheeler, *A Treatise of Commerce*, 363; SP Domestic 12/40/12 (22 June 1566), 12/127/57 (December 1578), 12/36/44 (April 1565).

62. SP Domestic 12/161/11 (10 June 1583).

63. Ibid., 12/40/12 (22 June 1566).

64. Galison, "The Many Faces," 2; Carr, *Select Charters*, 18–19; McDermott, "The Company of Cathay," 150. Garrard and Heyward were made the first governors of the Society of the Mineral and Battery Works, although there were noble members and privy councilors among the members. A full description of the Company is Donald, *Elizabethan Monopolies*.

65. SP Domestic 12/36/40 (20 April 1565).

66. Rabb, *Enterprise and Empire*, 4, 52–53; Wheeler, *A Treatise of Commerce*, 373; McDermott, "Michael Lok," 139; McDermott, "The Company of Cathay," 151.

67. Sir Humphrey Gilbert, in Peckham, *A True reporte*, page before f. 1, quoted in Rabb, *Enterprise and Empire*, 21.

68. Galison, Hevly, and Lowen, "Controlling the Monster," 46–77; McDermott, "The Company of Cathay," 157–58, 163.

69. McDermott, "The Company of Cathay," 160–65.

70. Ibid., "The Company of Cathay," 171–72.

71. For the European interest in alchemy and mining, see Moran, "German Prince-Practitioners," 261; Goodman, "Philip II's Patronage," 55; Marín, *Felipe II y la alquimia*; Smith, *The Business of Alchemy*.

72. See, for example, Smith, *The Business of Alchemy*; Clericuzio, "Agricola e Paracelso"; and Beretta, "Humanism and Chemistry."

73. For a work dedicated to Elizabeth I credited to Lannoy, see British Library Sloane MS 1744, ff. 4r–8v; British Library MS Add. 35831, ff. 236–40. The history of the Society for the New Art is in Dewar, *Sir Thomas Smith*, 149–55. The patent is reprinted in Strype, *The Life of the Learned Sir Thomas Smith*, 282–86.

74. Wood, "Custom, Identity, and Resistance."

75. SP Domestic, 12/36/12 (7 February 1564/5), 12/36/13 (9 February 1564/5).

76. For Waad's reports on Lannoy, see SP 12/39/39 (7 March 1565/6), 12/32/3 (7 August 1565), Add. 13/23.2 (July 1566), 12/40/28 (15 July 1566), 12/40/32 (19 July 1566), 12/40/53 (26 August 1566), 12/42/70 (28 May 1567).

77. SP Domestic, 12/37/3A (12 August 1565), 12/39/39 (7 March 1565/6). For more on Princess Cecilia's visit, see Bell and Seaton, *Queen Elizabeth and a Swedish Princess*. Lannoy's treatises and copies of letters to Elizabeth appear in contemporary alchemical collections, such as Sloane 3654, ff. 4r–6v; 1744, ff. 4r–8v. Lannoy was not the first alchemist to approach Elizabeth through Cecil. For an earlier example, see Pritchard, "Thomas Charnock's Book."

78. The last reference to Lannoy in the State Papers relates to a command to report to court, SP Domestic 12/42/70 (28 May 1567). I have not been able to uncover Lannoy's fate.

79. Donald, *Elizabethan Copper*; Donald, *Elizabethan Monopolies*. A general survey is in Tylecote, *A History of Metallurgy*.

80. SP Domestic 12/36/49 (12 May 1565). I cannot agree with Donald: "Although extensive correspondence exists between William Humfrey and Cecil it is found, on examination, to be without substance and not worth quoting." Donald, *Elizabethan Monopolies*, vii. The correspondence is indeed voluminous, and it contains a great deal of information on the management of Elizabethan projects.

81. Carr, *Select Charters*, 21, 20; British Library MS Harley 6991/62 (7 March 1574/75).

82. See, for example, SP 70/146/428 (8 February 1572/73).

83. SP 70/146/431 (8 February 1572/73).

84. Accounts of Frobisher's discoveries are in Collinson, *The Three Voyages of Martin Frobisher*; Best, *The Three Voyages of Martin Frobisher*; Hogarth, Boreham, and Mitchell, *Martin Frobisher's Northwest Venture*. A recent biography is McDermott, *Martin Frobisher*. For Lok and his family, see McDermott, "Michael Lok." For a discussion of the metallurgical work, see Hogarth, "Mining and Metallurgy"; Allaire, "Methods of Assaying Ore."

85. Collinson, *Three Voyages*, 92; SP Colonial 35.

86. SP Domestic 112/25 (22 April 1577).

87. Mandosio, "La place de l'alchemie"; Halleux, "L'alchimiste et l'essayeur"; SP Domestic, 122/62 (February 1577/78); Collinson, *Three Voyages*, 175.

88. SP Domestic 122/62; Collinson, *Three Voyages*, 175–76.

89. Other metals could include lead, litharge, charcoal, saltpeter, fluorite, borax, soda, iron sulphide, and copper sulphides. Agricola, *De re metallica*, 401.

90. Hogarth, Boreham, and Mitchell, *Martin Frobisher's Northwest Venture*, 170–71.
91. SP Domestic, 118/36, 118/54; Collinson, *Three Voyages*, 192, 194–95.
92. SP Domestic, 118/43, 122/62; Collinson, *Three Voyages*, 194, 176; Donald, "Burchard Kranich," 308.
93. Agricola, *De re metallica*, 401.
94. Ibid., 456; SP Domestic, 122/62.
95. Hogarth, Boreham, and Mitchell, *Martin Frobisher's Northwest Venture*, 170.
96. SP Dom, 122/62, 122/61; Collinson, *Three Voyages*, 176–78, 181.
97. SP Dom 122/62; Collinson, *Three Voyages*, 179–80.
98. British Library MS Lansdowne 12/7 (11 August 1569).
99. Marcus, Mueller, and Rose, *Elizabeth I*, 341.
100. Carr, *Select Charters*, lxv.

5. CLEMENT DRAPER'S PRISON NOTEBOOKS

1. British Library MS Sloane 3686, ff. 70v–71r.
2. I have attributed fifteen manuscript notebooks or partial manuscript notebooks to Draper. British Library Draper notebooks are Sloane 95 (ff. 98r–126v), 317, 320 (c. 1578–80), 1423, 3657 (c. 1591–96/97), 3686 (c. 1581), 3687 (c. 1600), 3688 (c. 1590–94), 3689, 3690 (c. 1583–93), 3691, 3692 (c. 1583), 3707, and 3748 (c. 1597–1606). A fragment of Draper's notebooks survives as Bodleian Library Ashmole 1394, item 6 (c. 1591). Additional Draper correspondence survives in the Henry E. Huntington Manuscripts at the Huntington Library and in Draper's lengthy letter to Queen Elizabeth, SP Domestic 12/243. I am indebted to Celeste Chamberland for bringing British Library MS Sloane 95 to my attention. Some Sloane manuscripts passed through the hands of Robert Kellam, who annotated them; thus many Draper notebooks are attributed to Kellam in the catalogue and given a seventeenth-century date. Internal evidence, including frequent self-referential statements, confirms that they were written by Draper.
3. On communities, see Anderson, *Imagined Communities*, especially chapter 1. For another case study of how notebooks can be used to gain entrée into a world of lived experience, see Kassell, "How to Read Simon Forman's Casebooks," and her *Medicine and Magic in Elizabethan London*.
4. Minshull, *Essayes and characters*, n.p.
5. Edmund Neville's alchemical writings are British Library MS Harley 853, written when he was in the Tower under suspicion of conspiring against Elizabeth. For Raleigh's activities in the Tower, see Shirley, "The Scientific Experiments of Sir Walter Ralegh." Shirley's contention that Raleigh was engaged in an elaborate program of experimentation has been recently tempered by scholars, notably Clucas, "Thomas Harriot and the Field of Knowledge."
6. Johnston, "Recorde, Robert," *DNB*; Wallis, "Bullein, William," *DNB*.
7. Will of Thomas Draper (27 February 1544/45), PCC Prob. 11/30/327.
8. GHMS 16,981, f. 39v (26 April 1564): "Item for the oath of Clement Draper of Mr. Alderman [Christopher] Draper."

9. Will of Christopher Draper (21 July 1580), PCC Prob. 11/63/248; British Library MS Sloane 320, ff. 1–32r. Draper and Clitherowe were in partnership with another Londoner, Roger Clark.

10. Bettey, "The Production of Alum and Copperas," 91. For the controversy see Bettey, "A Fruitless Quest for Wealth." For the issuance of writs for Draper's arrest, see SP Domestic, Supplement 46/32, f. 224.

11. SP Domestic 12/177; Bettey, "A Fruitless Quest for Wealth," 6.

12. Muldrew brilliantly analyzes this culture of debt and obligation in its economic and social terms in *The Economy of Obligation*.

13. Cross, *The Puritan Earl*, 90–93.

14. Henry E. Huntington Library MS Ha 5366 (3 October 1583); MS Ha 2363 (28 June 1584), Clement Draper to Francis Walsingham; MS Ha 2364 (18 October 1586), Clement Draper to Thomas Sampson. That these letters ended up in the earl's possession is some indication of the serious threat the earl posed to Draper.

15. Dobb surveys London prisons during the Elizabethan and Jacobean periods in "London Prisons." Byrne's *Prisons and Punishments of London* takes a longer chronological approach, while Watson's "The Compter Prisons of London" focuses on the institutions that housed minor debtors, petty offenders, drunks, and vagrants. Blatcher explains the workings of the court of the King's Bench before Draper's arrest and sheds light on the complicated procedures by which someone came into the prison in *The Court of the King's Bench*.

16. This neat division of crime and punishment was the ideal, but in Elizabethan times it was no longer possible to segregate prisoners accused of certain crimes from the rest of the prison population. Folger Shakespeare Library MS L.b. 202 (11 March 1581/2), for example, confirms that the King's Bench prison held two priests, one condemned for praemunire or assertion of papal supremacy, and one for treason. Later in the decade (7 March 1585/86), the King's Bench housed several recusants, as did the Marshalsea, the Clink, and the White Lion. See Folger Shakespeare Library MS L.b.239.

17. Taylor, *Works*, 292–93; Browner, "Wrong Side of the River."

18. The Rules surrounding the Fleet Prison are best known, but the King's Bench also had such an area. Salgado, *The Elizabethan Underworld*, 175.

19. British Library MS Lansdowne 29/28 (17 March 1579/80); Dobb, "London Prisons," 90. That prisons did not function as penal institutions should not suggest that early modern England lacked a penal culture. Indeed, punishments for many offenses were harsh and publicly administered. See Devereaux and Griffiths, *Penal Practice and Culture*.

20. Salgado, *Elizabethan Underworld*, 171–72; Middleton, *The roaring Girle*, III.iii, sig. G3v; Mynshull, *Essayes and characters*, n.p. See also Fennor, *The miseries of a jaile* for a hair-raising tale of the author's time in the Wood Street Compter.

21. Elizabeth Draper, daughter of Clement and Elizabeth Draper, was christened in the parish of All Hallow's the Less, London on 7 December 1583. Conjugal visits and even marriages took place in prisons all over London. Edmund Neville, for example, was married to Jane Smythe in the Tower of London on 7 January 1587/88; Loomie, "Neville, Edmund," *DNB*.

22. British Library MS Sloane 3686, f. 18r; 3687, f. 88r. Giles Farnaby may have been a composer who was educated at Oxford and apprenticed to the Joiners' Company. See Marlow, "The Life and Music of Giles Farnaby"; Owen, "Giles and Richard Farnaby."

23. British Library MS Sloane 3690, f. 111v; 3692, f. 50r; 95, f. 103v.

24. Ibid., 95, f. 105v; 3688, f. 53v. Brooke may have been a member of the Ironmongers' Company. Clement's uncle Christopher Draper gave a legacy to John Brooke, Ironmonger, in his will of 1581. See PCC Prob. 11/63.

25. British Library MS Sloane 3689, f. 1v; 3690, f. 112r; 3688, ff. 3r, 150v. I have not been able to locate another reference to *scala celie.*

26. Ibid., 3692, f. 61r; 3690, f. 118r–v; 3688, f. 150v.

27. Ibid., 3748, f. 8r; 3690, f. 112r–v; 3688, f. 65r; 3748, ff. 115r–116v.

28. Ibid., 3690, f. 80v; 3748, ff. 57v, 11v–12r, 97r–v; 3687, f. 68r.

29. Ibid., 3687, ff. 44r, 71v–72r; 3657, f. 26v; 3689, f. 2r. The Verzelini family came to England in the 1570s. Specimens of glass made by Giacamo Verzelini are in the Fitzwilliam, Victoria and Albert, and British museums, as well as in the Fine Arts Museum in San Francisco. The funeral brasses for him and his wife, Elizabeth, are on view at Downe Parish Church in the London borough of Bromley. Verzelini's glass factory was in the old Crutched Friars Church.

30. For information on Gans (alternatively spelled Ganz and, in Elizabethan times, Gannes), see Feuer, *Jews in the Origins of Modern Science;* Grassl, "Joachim Gans of Prague"; Quinn, *The Roanoke Voyages,* 907; Abrahams, "Joachim Gaunse." Passing reference is made to Gans in Chaplin, *Subject Matter,* 20, 68. None of these accounts includes any reference to Gans's relationship with Draper or their shared chemical interests. Further information on the Roanoke assays is in Hume, "Roanoke Island," 20, and Hume, *The Virginia Adventure,* 76–88. For Jews in England during the period, see Prior, "A Second Jewish Community"; Katz, *Jews in the History of England.*

31. Donald, *Elizabethan Copper,* 76; SP 12/152/88, 12/152/89 (March 1582); SP Domestic 12/226/40, 12/226/40.1, 12/226/40.2 (17 September 1589); British Library MS Sloane 3748, ff. 26v–30v. Grassl surveyed the extant records from the Marshalsea and Tower for 1589 and did not find Gans's name among the inmates; Grassl, "Joachim Gans of Prague," 8. He suggested that Gans translated Ecker's treatise on saltpeter, a key military supply, for Francis Walsingham during his incarceration. Gans's translation of Ecker is Hatfield House MS Cecil Paper 276.5. A transcription of the undated letter accompanying the translation addressed to Walsingham is in Grassl, "Joachim Gans of Prague," 14–15.

32. Some of the most important studies on early modern figures and their reading habits are Oakeshott, "Sir Walter Ralegh's Library"; Stern, *Gabriel Harvey;* Anthony Grafton, "Kepler as Reader"; William Sherman, *John Dee;* Ann Blair, *The Theater of Nature.* On the book trade in England, consult Blayney, *The Bookshops of Paul's Cross Churchyard;* Johns, *The Nature of the Book.*

33. Davis, "Beyond the Market"; Darnton, *The Literary Underground,* 182; Darnton, *The Kiss of Lamourette,* 111.

34. British Library MS Sloane 95, ff. 98r–126v; 3686, f. 94r.

35. See, for example, ibid., 3748, ff. 52r, 53v, 58r–v; 3657, f. 44v. For Padden's translation see ibid., 3748, ff. 62v–63r. Draper recopies this recipe ibid. 3687, f. 55r.

36. Ibid., 3657, f. 27r–v; 3748, f. 77v.

37. On commonplace books and miscellanies, see Marotti, *Manuscript, Print, and the English Renaissance Lyric*. Ann Blair has undertaken an important examination of how commonplacing functioned in natural philosophy; "Humanist Methods in Natural Philosophy." Chronological note taking and a secondary process of commonplacing often went hand in hand. Commonplace books were popular in printed formats; see Moss, *Printed Commonplace-Books*. An English example is Hugh Plat, *The Floures of Philosophie*. Collective experimental notebooks were kept as well; for an example see Middleton, *The Experimenters*, 359–82. Receipt or recipe books were the most popular form of notebook that has survived. See Leong, "Medical Remedy Collections."

38. Dear, *Discipline and Experience*, 12–13; Smith, *The Body of the Artisan*, 17–20.

39. Martin, *The History and Power of Writing*; Spiller, *Science, Reading, and Renaissance Literature*; Preston, *Thomas Browne*; Patterson, *Reading Holinshed's Chronicles*; Woolf, *Reading History*; Sharpe, *Reading Revolutions*; Sherman, *John Dee*; Cunningham, "Virtual Witnessing."

40. This notebook entry, headed "Epistola," may be part of Draper's attempt to write his own alchemical master text, or it may simply be Draper's copy of another person's account of alchemy that was circulating in manuscript; see British Library MS Sloane 3657, ff. 45r–53r. The work has references to printed alchemical texts, so it must date after c. 1450, and has common turns of phrase that indicate that its language of composition was English. I have not been able to definitively determine the author of the work, and therefore it is not possible to conclusively attribute the work to Draper. The continuation of the text may be ibid., 3657, ff. 34r–37v, which begins, "Heare begynnethe the makinge of the great Elixer of ph[ilosoph]or's Stone."

41. Ibid., 3689, f. 36r. See also ibid., 3688, f. 109r.

42. Ibid., 3686, f. 34r.

43. Shapin and Schaffer, *Leviathan and the Airpump*, 20–24, 60. Shapin and Shaffer modified and extended van Leeuwen's arguments about vicarious witnessing in *The Problem of Certainty*. Dear discusses the role that reporting experiments played in the early Royal Society in "*Totius in verba*," 145–61. Shapiro has taken up many of these issues in *Probability and Certainty*. Cunningham points out that virtual witnessing was evident in the work of Draper's contemporary William Gilbert; "Virtual Witnessing," 209.

44. British Library MS Sloane 3690, f. 88r–v.

45. Ibid., 3657, f. 37r.

46. Ibid., 3748, ff. 52r, 53v, 58r–v, f. 62v–63r.; 95, ff. 98r–126v; 3688, f. 24v–34v. Draper copies the recipe for tincture of antimony ibid., 3687, f. 55r.

47. Ibid., 3687, ff. 1r, 23v–39v; 3686, f. 40v; 3690, f. 26v; 3688, f. 107v–108r; 3748, f. 82r.

48. Ibid., 3686, ff. 105v–106r, ff. 107v–110v; 3690, ff. 112v–115v; 320, f. 117v. This transcriptive habit was common; Love, *Scribal Publication*.

49. British Library MS Sloane 3688, f. 44v.

50. Ibid., 3691, ff. 68v–79r; 3686, f. 94r; 3688, ff. 92r, 96r, 101r–102r; 3692, f. 35r–49v, f. 35r. See also Plat, *The jewell house*, 1: 47, 2: 29.

51. British Library MS Sloane 3691, f. 95v; 3688, f. 109r–v; 3686, f. 85r–v. For the Ripley Scrolls and a reproduction of the Huntington Library's example, see Dobbs, *Alchemical Death and Resurrection*, 16–24.

52. British Library MS Sloane 3867, f. 71v; 3686, f. 82r; 3688, ff. 102r, 104r.

53. Ibid., 3686, ff. 91v–92r; 3688, ff. 95r, 137v–139r, 141v–142r, 143v–144r. There are also copious illustrations in Draper's copies of late medieval treatises, ibid., 3707.

54. Ibid., 3688, f. 8r.

55. Ibid., 3692, f. 25v.

56. Ibid., 3689, f. 36r; 3688, f. 53v; 3748, f. 2r.

57. Ibid., 3686, ff. 90v, f. 91v; 3657, f. 44r.

58. British Library MS Sloane 3686, f. 29r; 3687, f. 66r–v.

59. See, for example, a recipe for the manufacturing of the white stone with aqua fortis, where Draper noted this was the second time Gans had taught him the procedure; ibid., 3478, f. 18v.

60. Ibid., 3748, f. 15r–v.

61. Ibid., 3688, ff. 4v–5r; 3686, f. 91r–v; 3690, f. 105v.

62. In this respect Draper has much in common with medieval textual communities; Stock, *Listening for the Text*; Clanchy, *From Memory to Written Record*. Sharpe has argued that the Medievalist concept of interpretive communities should be less confined socially than it has been traditionally, and not limited to manuscript culture; *Reading Revolutions*, 60.

6. FROM THE JEWEL HOUSE TO SALOMON'S HOUSE

1. See Prest, *The Inns of Court*, and Raffield, *Images and Cultures of Law*.

2. Biographical information has been drawn from Sydney Lee's article on Plat in the *DNB*, from his writings and manuscripts, and from Mullett, "Hugh Plat." Passing reference to Plat is in Thirsk, *Economic Policy and Projects*, 99, and Martin, *Francis Bacon*, 56.

3. Plat was a prolific recorder of his face-to-face encounters. I have attributed twenty-three notebooks or partial notebooks to him: British Library MSS Sloane 2170, 2171, 2172, 2175 (ff. 71v–86r), 2176, 2177 (Plat family papers), 2189, 2194, 2195, 2197, 2203, 2209, 2210, 2212, 2216, 2223, 2244, 2245, 2246, 2247, 2249, 2272, 3574. The *DNB* attributes British Library MS Sloane 3690 to Hugh Plat, but it is in the hand of Clement Draper. Given his large manuscript output and his relatively small publication output, it is not clear whether Plat deserves a central place among the professors of secrets studied by Eamon. Instead, I see Plat as a bridge between the book-of-secrets tradition and the seventeenth-century figures Eamon describes as "Baconian intelligencers." Eamon, *Science and the Secrets of Nature*, 311–14 (which discuss Plat), 322–32 (on Baconian intelligencers). Eamon does place Plat in a middling spot in his analyses, seeing him as a representative of the move toward scientific virtuosity, while admitting that he "was perhaps too practical to

qualify as a typical virtuoso." More important to my analysis is the fact that he was not an aristocrat but an upwardly mobile urban elite.

4. Plat, *The jewell house*, sig. B2v.

5. Raffield, *Images and Cultures of Law*; Bacon, *Gesta Grayorum*.

6. Bacon, *The New Atlantis*, in *Critical Edition*, 457–89.

7. Glanvill, *Scepsis Scientifica*, quoted in Vickers, notes to *The New Atlantis*, 789.

8. British Library MS Sloane 2210, ff. 113r–v, 106r–v, 107v–108r, 111v, 116r. Plat's networks of consultants are comparable to those established by Robert Hooke a century later; Iliffe, "Material Doubts." Points of comparison between Plat and Hooke do not end there. Like Plat's, Hooke's boundless enthusiasms, according to Iliffe, may have "prevented him from concentrating on the kinds of projects which would have elevated him above ordinary mortals," 286.

9. British Library MS Sloane 2210, ff. 139v, 118v, 66v, 71r, 166v, 52v; 2216, ff. 18r, 142r, 37r, 151r; 2172, f. 12r; 2245, ff. 30v–31r; 2189, f. 29v.

10. Plat, *The jewell house*, sig. B2v. This democratic approach to nature strongly resonates with that of Hooke; see Iliffe, "Material Doubts," 287.

11. Plat, "Diverse new sorts of soyle," *The jewell house*, 3. Plat seems appreciative of the "craft secrets" tradition—see Long, *Openness, Secrecy, Authorship*, 78–89–all the while the dispersal of this information was gradually eroding a sense of proprietary craft knowledge in favor of a general sphere of "natural knowledge."

12. Bennett, *Ale, Beer, and Brewsters*, discusses the nature of the work and the social meanings that were attached to it.

13. British Library MS Sloane 2189, f. 23v; 2210, f. 70v.

14. Ibid., 2216, f. 130r; 2210, ff. 164v, 166v; 2189, f. 28v.

15. Ibid., 2210, f. 79r, 45v, 154v; 2216, f. 99v.

16. Ibid., 2209, ff. 6r, 9r; 2216, f. 113r; 2210, f. 76v.

17. Ibid., 2210, ff. 136v, 124v, 145r, 79r; 2209, f. 3r; 2216, ff. 63r, 124v; 2189, ff. 38v, 41r.

18. See, for example, ibid., 2170; 2175; ff. 1–51; 2176; 2194; 2195. These collections show that Charnock, Paracelsus, Norton, and Ripley were strong influences.

19. Eamon, *Science and the Secrets of Nature*, passim.

20. Shapin, *A Social History of Truth*.

21. Pearl, "Change and Stability," 27.

22. British Library MS Sloane 2210, f. 45r; 2245, f. 9v.

23. Ibid., 2210, ff. 26r–28r; Dee, *The Diaries*, 47. For Dee's home, see Harkness, "Managing an Experimental Household." Another of Plat's teachers was Godfrey Mosanus. In August 1608 Plat wrote to the Landgrave of Hessen that the father of his physician, James Mosanus, "did first teach me to close a helm and body together" in a chemical laboratory; see British Library MS Sloane 2172, ff. 18r–19v. Godfrey, like his son, was a physician, and after immigrating to England was pursued repeatedly by London's College of Physicians between 1581 and 1587 for practicing without their permission. *RCP Annals* 2: 1b, 25a, 29a, 58b.

24. For a discussion of Paracelsus's ideas about the homunculus, see Newman, *Promethean Ambitions*, 164–237, and his "The Homunculus and His Forebears."

25. Rappaport, Archer, and Ward have evinced telling evidence surrounding urban contention and the City's ability to quench it. Rappaport, *Worlds within Worlds*, 201–15; Archer, *The Pursuit of Stability*, especially chapter 3; Ward, *Metropolitan Communities*, 92–98.

26. Shapin, *Social History of Truth*, 107–14. For a discussion of this blend of collaboration and competition, see also Barry, "Bourgeois Collectivism?"

27. Plat, *The jewell house*, 30. Plat's notebook account in British Library MS Sloane 2210, f. 68v, does not include the detail that he immediately recorded the experiment.

28. Plat, *The jewell house*, sig. B3v–[B4r]; British Library MS Sloane 2172, f. 18v.

29. British Library MS Sloane 2210, ff. 91v, 174v; 2216, ff. 123r, 33v; 2189, f. 47r. See Paula Findlen, "Jokes of Nature," passim.

30. British Library MS Sloane 2210, ff. 35r, 38v, 163r, 169v, 175v.

31. See, for example, British Library MS Sloane 2216, ff. 53r ("I have heard Baylie the dyer affirm . . . "), 102r ("I have been credibly informed that if peter, sal niter, sal armoniac and arsenic be boiled . . . "), 114r ("I had at 7 months end by mixing some new red wine with the aforesaid liquor . . . and here I do observe that after these wines have once fined of themselves . . . "); 2245, f. 79r ("I have also found by experience . . . ").

32. Plat's attitude was common in the period. See Schaffer, "Self Evidence"; Shapin, "The Philosopher and the Chicken"; Harkness, "'Nosce teipsum.'"

33. British Library MS Sloane 2209, ff. 4r–v, 27r, 8v.

34. Ibid., 2210, ff. 123r, 172r, 179r–v; 2216, f. 18r.

35. Ibid., 2216, ff. 53r, 55v, 79r; 2189, ff. 8v–9r.

36. Ibid., 2245, ff. 2r–3r, 40v. Plat leaves instructions to refer back to the method on f. 3r.

37. Ibid., f. 11r–v; 2209, f. 7v.

38. Ibid., 2209, f. 6v. Also, ibid., 2245, f. 80r–v ("I doo also suppose that if the right redd wardens an excellent quodoniate & an excellent gelly may also be made according to this resceit").

39. Ibid., 2245, f. 84r; 2216, f. 102r ("this secrete is utterly false"); 2212, f. 10r ("this secret I can not yet finde trew in experience").

40. Ibid., 2209, f. 19r. For his medicines, see ibid., f. 15r. For the medicinal uses of mumia, see Dannenfeldt, "Egyptian Mumia"; Cook, "Time's Bodies," 230–32. Plat discusses the corpses most suitable for mumia in British Library, MS Sloane 2249, f. 5r. His oath to preserve the secrets surrounding mumia is ibid., 2246, f. 60v. On proprietary medicines and print culture, see Isaac, "Pills and Print." "Anthony's Pill" may relate to Francis Anthony's sensational cure-all, potable gold; see Anthony, *The apologie, or defence*.

41. British Library, MS Sloane 2209, ff. 17r, 19r–v.

42. For the Elizabethan period, Slack's classic work *The Impact of Plague* is indispensable.

43. The genealogy of the recipe is confusing. Plat asserted that the medicine was devised by a Dr. Siringe in Milan in 1579, and later mentions that a Signor Romero began to distribute the medicine in London; British Library MS Sloane 2209, f. 21r–v. Plat notes on f. 18v to "see diverse agues cured with my defensative cake," which appears a few pages later on ff. 22r–25v.

44. Ibid., 2209, ff. 22r–25v.

45. Ibid., 2189, f. 1v.

46. Ibid., 2172, ff. 13r–14v.

47. Plat, *A discoverie*, sig. A2r–v. *Delightes for ladies* was reprinted twelve times in the seventeenth century and can be described as one of the period's best sellers.

48. Plat, *The jewell house*, sig. [B4v]. To compare Plat's emphasis with other examples from the "books of secrets" tradition, see Eamon, *Science and the Secrets of Nature*, especially, 139, 313. For the role of print culture in early modern science see Long, "The Opennness of Knowledge"; Johns, *The Nature of the Book*.

49. Plat, *Certaine philosophicall preparations*, n.p.

50. Plat, "Diverse chimicall conclusions," *The jewell house*, 20.

51. Ibid., 37; Plat, *A discoverie*, sig. A2r.

52. Plat, *The jewell house*, 34–35, 93–94; Plat, *Sundrie new and artificiall remedies*, sig. Bv–B2r.

53. Plat, *The jewell house*, title page and sig. A3r; "Diverse chimicall conclusions," ibid., 6; Plat, "Secrets in Distillation," *Delightes for ladies*, experiment no. 16, n.p.; "The Art of Preserving," ibid., experiment no. 63, n.p.

54. Plat, "Diverse chimicall conclusions," *The jewell house*, 8–9.

55. On emerging notions of authorship see Long, "Invention, Authorship"; Saunders, *Authorship and Copyright*; Rose, *Authors and Owners*; Long, *Openness, Secrecy, Authorship*; Loewenstein, *The Author's Due*; Loewenstein, *Ben Jonson*. Much of this work responds, often critically, to Michel Foucault's "What Is an Author?"

56. For Oldenburg as editor, see Bazerman, *Shaping Written Knowledge*, 128–50. For his correspondence networks, see *The Correspondence of Henry Oldenburg*; Hall, "Henry Oldenburg."

57. Saunders, "The Stigma of Print." Historians and critics have been gradually advancing a more nuanced interpretation.

58. Plat, *The jewell house*, 52, 54, 62; "Diverse new soyles," ibid., 31; British Library MS Sloane 2210, f. 58v.

59. Plat, *A discoverie*, sig. A2v.

60. The literature on Francis Bacon is vast, and it is impossible to give a full survey of it here. My discussion of Bacon is intended not to provide an overview of his thought but only to highlight specific points of commonality and difference between him, Plat, and London science, especially in his earlier works. Here, I mention only those studies that have been influential in making this comparison. Two classic sources are *The Works*, ed. Spedding, Ellis, and Heath, and *The Letters and Life*, ed. Spedding. Recently Rees and Jardine undertook a new edition of the works for Oxford University Press. Two of the most recent biographies of Bacon are Jardine and Stewart, *Hostage to Fortune*, and Zagorin, *Francis Bacon*. For his place in the history and philosophy of science, consult Jardine, *Francis Bacon*; Rossi, *Francis Bacon*; Webster, *The Great Instauration*; Weinberger, *Science, Faith, and Politics*; Urbach, *Francis Bacon's Philosophy of Science*; Pérez-Ramos, *Francis Bacon's Idea of Science*; Briggs, *Francis Bacon*; Martin, *Francis Bacon*; Peltonen, *The Cambridge Companion*; Solomon, *Objectivity in the Making*; Gaukroger, *Francis Bacon*.

61. Martin, *Francis Bacon*, passim. Gaukroger expands on this point in *Francis Bacon*, 16–18.

62. Bacon, *The Advancement of Learning*, 32. Images of a feminine Nature serving as a courtesan appear in *Valerius Terminus*, in *Works*, ed. Spedding, Ellis, and Heath, 3: 222: "And therefore knowledge that tendeth but to satisfaction is but as a courtesan, which is for pleasure and not for fruit or generation."

63. Esler, *The Aspiring Mind*.

64. Letter to William Cecil, 1592, in Bacon, *Letters and Life*, 1: 108–9.

65. Bacon, *Valerius Terminus*, in *Works*, ed. Spedding, Ellis, and Heath, 3: 223.

66. Ibid., 222, 223.

67. Bacon, *The Advancement of Learning*, 88, 64–65, 31–32; Plat, *The jewell house*, sig. A2r.

68. Bacon, *Filium Labyrinthi, sive Formula Inquisitionis*, in *Works*, ed. Spedding, Ellis, and Heath, 3: 504. For Bacon's chemical interests see Jardine and Stewart, *Hostage to Fortune*, 506–8.

69. Bacon, *The Refutation of Philosophies*, 120.

70. Bacon, *The Advancement of Learning*, 30; Jardine, *Francis Bacon*, 76–108.

71. Bacon, *The Instauratio Magna*, 131, 71.

72. Bacon, *The Advancement of Learning*, 91, 92, 119, 31.

73. Bacon's notebook is British Library MS Add. 27278. For an analysis, see Michael Kiernan, "Introduction: Bacon's Programme for Reform," in Bacon, *The Advancement of Learning*, xxxiv–xxxvi.

74. British Library MS Add. 27278, ff. 13v, 11r.

75. Bacon, *The Advancement of Learning*, 101; Bacon, *Valerius Terminus*, in *Works*, ed. Spedding, Ellis, and Heath, 3: 246–47.

76. Bacon, *The Advancement of Learning*, 26; Bacon, *Sylva sylvarum*, sig. Av.

77. For example, ibid., 12, 19, 30.

78. Ibid., sig. Ar.

79. On this point see Johnson, *Astronomical Thought*, 296. The lack of contemporary response to *The Advancement of Learning* is often commented upon by Bacon scholars. See, for example, Gaukroger, *Francis Bacon*; Martin, *Francis Bacon*; Rossi, *Francis Bacon*.

80. Excerpts quoted here and subsequently are from Thomas Bodley to Francis Bacon, 29 February 1608/[9], in Bacon, *Works*, ed. Montagu, 12: 83–90.

81. Bacon, *Sylva Sylvarum*, sig. A2r. Bacon's ambivalence about experimental work was shared by successive generations of gentlemen, including many members of the Royal Society. See Pumfrey's "Who Did the Work?"

82. Bacon, *The New Atlantis*, 486–87. For Bacon's belief that his work would be appreciated only in "ages to come," see Jardine and Stewart, *Hostage to Fortune*, 473–78.

CODA

1. Excellent surveys are Porter, "The Scientific Revolution"; Lindberg and Westman, *Reappraisals of the Scientific Revolution*; Porter and Teich, *The Scientific Revolution*.

2. For example, Burtt, *The Metaphysical Foundations*; Crombie, *Augustine to Galileo*; Hall, *The Scientific Revolution*; Jones, *Ancients and Moderns*; Hall, "On the Historical Singularity of the Scientific Revolution."

3. For example, Duhem, *The Aim and Structure of Physical Theory*; Schmitt, "Towards a Reassessment of Renaissance Aristotelianism"; Schmitt, *Aristotle and the Renaissance*.

4. For example, Schaffer, "Glass Works"; Hall, *Promoting Experimental Learning*; Wilson, *The Invisible World*; Kuhn, "Mathematical versus Experimental Traditions."

5. Much of the literature can be traced back to pioneering work of Marxist historians of science: Jacob, *The Newtonians and the English Revolution*; Jacob, *Robert Boyle and the English Revolution*; Jacob, "Restoration Ideologies and the Royal Society"; Jacob, *The Radical Enlightenment*; Jacob, *The Cultural Meaning of the Scientific Revolution*. Recently, historians adopted sociological methodologies, tracing their work back to Zilsel, "The Sociological Roots of Science," and Merton, *Science, Technology, and Society*. A valuable overview of this approach is Golinski, *Making Natural Knowledge*.

6. For example, Hall, "The Scholar and the Craftsman"; Field and James, *Humanists, Scholars, Craftsmen*.

7. Shapin, *The Scientific Revolution*, 1. Dobbs suggested that the Scientific Revolution as a category of analysis had outlived its usefulness; "Newton as Final Cause."

8. Marcus, *Ethnography*, 187–88.

9. Ibid., 90–95; Fischer, "The Uses of Life Histories."

10. For prosopographic methods, see Bulst and Genet, *Medieval Lives and the Historian*; Mathesen, "Medieval Prosopography and Computers"; Goudriaan et al., *Prosopography and Computer*. For early modern prosopography see Hans, *New Trends in Education*; Stone, "Prosopography"; de Ridder-Symoens, "Prosopographical Research in the Low Countries." In the history of science see, for example, Shapin and Thackeray, "Prosopography as a Research Tool"; Hunter, *The Royal Society and Its Fellows*; Westfall, "Science and Technology."

11. Kuhn, *The Structure of Scientific Revolutions*.

12. Christopher Hill, *Intellectual Origins*, 15.

13. Charles B. Schmitt, "Experience and Experiment"; Shapin and Schaffer, *Leviathan and the Air-Pump*; Dear, "Narratives, Anecdotes, and Experiments"; Garber, "Experiment, Community"; Shapin, *A Social History of Truth*; Dear, *Discipline and Experience*.

14. Eisenstein, *The Printing Press as an Agent of Change*; Rider, "Literary Technology and Typographic Culture"; Johns, *The Nature of the Book*.

15. For example, Davis, *The Return of Martin Guerre*; Ginzburg, *The Cheese and the Worms*.

16. Bynum, *Fragmentation and Redemption*.

17. Marcus, *Ethnography*, 83.

18. For example, Haraway, *Primate Visions*; Haraway, *Simians, Cyborgs, and Women*; Latour, *Science in Action*; Rabinow, *French DNA*.

BIBLIOGRAPHY

MANUSCRIPTS

Archdeaconry Court of London
  MS 3
  MS 4
  MS 5
Bodleian Library
  MS Ashmole 1394
  MS Ashmole 1399
  MS Ashmole 1487
  MS Douce 68
  MS Douce 363
  MS Eng. Misc. d. 80 (R)
British Library
  MS Additional 12222
  MS Additional 12503
  MS Additional 27278
  MS Additional 35831
  MS Additional 71494
  MS Additional 71495
  MS Cotton Julius F.I
  MS Cotton Titus B.V
  MS Cotton Vespasian F.VI
  MS Douce 68
  MS Harley 286
  MS Harley 853
  MS Harley 6467
  MS Harley 6991
  MS Harley 6994

MS Harley 7009
MS Lansdowne 4
MS Lansdowne 6
MS Lansdowne 10
MS Lansdowne 12
MS Lansdowne 18
MS Lansdowne 19
MS Lansdowne 21
MS Lansdowne 24
MS Lansdowne 26
MS Lansdowne 27
MS Lansdowne 29
MS Lansdowne 39
MS Lansdowne 42
MS Lansdowne 43
MS Lansdowne 48
MS Lansdowne 69
MS Lansdowne 75
MS Lansdowne 77
MS Lansdowne 80
MS Lansdowne 98
MS Lansdowne 101
MS Lansdowne 103
MS Lansdowne 104
MS Lansdowne 105
MS Lansdowne 121
MS Lansdowne 241

British Library
  MS Lansdowne 683
  MS Royal 18.A
  MS Sloane 95
  MS Sloane 317
  MS Sloane 320
  MS Sloane 684
  MS Sloane 1423
  MS Sloane 1744
  MS Sloane 2170
  MS Sloane 2171
  MS Sloane 2172
  MS Sloane 2175
  MS Sloane 2176
  MS Sloane 2177
  MS Sloane 2189
  MS Sloane 2192
  MS Sloane 2194
  MS Sloane 2195
  MS Sloane 2197
  MS Sloane 2203
  MS Sloane 2209
  MS Sloane 2210
  MS Sloane 2212
  MS Sloane 2216
  MS Sloane 2223
  MS Sloane 2228
  MS Sloane 2244
  MS Sloane 2245
  MS Sloane 2246
  MS Sloane 2247
  MS Sloane 2249
  MS Sloane 2272
  MS Sloane 3252
  MS Sloane 3574
  MS Sloane 3654
  MS Sloane 3657
  MS Sloane 3682
  MS Sloane 3686
  MS Sloane 3687
  MS Sloane 3688
  MS Sloane 3689
  MS Sloane 3690
  MS Sloane 3691
  MS Sloane 3692
  MS Sloane 3707
  MS Sloane 3748
  MS Sloane 4014
  MS Stowe 1069
Cambridge University, University Library
  MS 4138
  MS Gg.6.9
Commissary Court of London
  MS 19
Corporation of London Record Office
  MS Rep. 13
  MS Rep. 15
  MS Rep. 17
  MS Rep. 18
Folger Shakespeare Library
  MS L.b.202
  MS L.b.239
Guildhall Library, London
  MS 577/1
  MS 1002/1A
  MS 1046/1
  MS 1279/2
  MS 1454
  MS 1568/1
  MS 2088
  MS 2593/1
  MS 2953/1
  MS 3907/1
  MS 4069/1
  MS 4165/1
  MS 4524/1
  MS 4352/1
  MS 4384/1
  MS 4399/1
  MS 4508/1
  MS 4510/1
  MS 5257/1
  MS 6419/1
  MS 6836/1
  MS 9171/20
  MS 9220
  MS 9221
  MS 9223

Guildhall Library, London
  MS 9234/1
  MS 9235/2
  MS 10,342
  MS 16,981
Henry E. Huntington Library
  MS Ha 2363
  MS Ha 2364
  MS Ha 5366
London Metropolitan Archives
  MS DL/C/214
  MS DL/C/332
  MS DL/C/335
Oxford University, Magdalene College
  MS Goodyer 13
  MS Goodyer 96
Prerogative Court of Canterbury
  Prob. 11/30
  Prob. 11/63
  Prob. 11/73
  Prob. 11/153
Public Records Office
  C 66/948
  C 66/960
  C 66/968
  C 66/972
  C 66/985
  C 66/1012
  C 66/1017
  C 66/1032
  C 66/1040
  C 66/1049
  C 66/1053
  C 66/1330
  C 66/1333
  E 101/3/9

SP Additional 13/23.2
SP Additional 21/79
SP Domestic 12/1
SP Domestic 12/8
SP Domestic 12/16
SP Domestic 12/32
SP Domestic 12/36
SP Domestic 12/37
SP Domestic 12/39
SP Domestic 12/40
SP Domestic 12/42
SP Domestic 12/46
SP Domestic 12/83
SP Domestic 12/88
SP Domestic 12/92
SP Domestic 12/111
SP Domestic 12/118
SP Domestic 12/122
SP Domestic 12/125
SP Domestic 12/127
SP Domestic 12/152
SP Domestic 12/161
SP Domestic 12/177
SP Domestic 12/200
SP Domestic 12/206
SP Domestic 12/209
SP Domestic 12/216
SP Domestic 12/217
SP Domestic 12/226
SP Domestic 12/243
SP Domestic 122/62
SP Domestic Supplement 46/32
SP 70/146
Royal College of Physicians, London, MS
  Annals

PRIMARY SOURCES

Aconcio, Iacopo. *Stratagematum Satanæ libri octo quos Iacobus Acontius vir summi iudicij nec minoris pietatis, annis abhinc pène [sic] 70 primum edidit & sereniss. Q Reginæ Elizabethæ inscripsit.* Oxford, 1631.

Agnello, Giovan Battista. *A revelation of the secret spirit Declaring the most concealed secret of alchymie.* Trans. R[obert] N[apier]. London, 1623.

Agricola, Georg. *De re metallica*. Trans. and ed. Herbert Clark Hooever and Lou Henry Hoover. London: Mining Magazine, 1912.

Analytical Index to the Series of Records Known as the Remembrancia, 1579–1664. London: E. J. Francis, 1878.

Anonymous. *An introduction for to learne to recken wyth the pen or with the counters, according to the true rule of algorisme, in whole numbers or in broken Newly over seene and corrected*. London, 1566.

Anthony, Francis. *The apologie, or defence of a verity heretofore published concerning a medicine called aurum potabile*. London, 1616.

Bacon, Francis. *The Advancement of Learning. The Oxford Francis Bacon*, vol. 4. Ed. Michael Kiernan. Oxford: Clarendon, 2000.

———. *Francis Bacon: A Critical Edition of the Major Works*. Ed. Brian Vickers. Oxford: Oxford University Press, 1996.

———. *Gesta Grayorum*. London, 1688.

———. *The Instauratio Magna: Part II Novum organum. The Oxford Francis Bacon*, vol. 11. Ed. Graham Rees and Maria Wakeley. Oxford: Clarendon, 2004.

———. *The Letters and Life of Francis Bacon*. 7 vols. Ed. James Spedding. London, 1861–1874.

———. *The Refutation of Philosophies*. In Farrington, *The Philosophy of Francis Bacon*, 103–33.

———. *Sylva sylvarum: or A naturall historie in ten centuries*. London, 1627.

———. *The wisedome of the Ancients*. Trans. Arthur Gorges. London, 1619.

———. *The Works of Francis Bacon*. 17 vols. Ed. Basil Montagu. London: William Pickering, 1830.

———. *The Works of Francis Bacon*. 7 vols. Ed. James Spedding, Robert Leslie Ellis, and Douglas Denon Heath. London, 1859–64.

Baker, George. *The composition or making of the moste excellent and pretious oil, called oleum magistrale*. London, 1574.

———, ed. *Guidos questions newly corrected*. London, 1579.

———, ed. *The newe jewell of health wherein is contayned the most excellent secretes of phisicke and philosophie, devided into fower books*. London, 1576.

———, ed. *The whole worke of that famous chiru[r]gion Maister John Vigo*. Trans. Thomas Gale. London, 1586.

Baker, Humfrey. *Such as are desirous, eyther themselves to learne, or to have theyr children or servants instructed*. London, ca. 1590.

———. *The well spryng of sciences which teacheth the perfect worke and practise of arithmeticke*. London, 1568.

Bauhin, Gaspard. *Pinax theatri botanici*. Basel, 1623.

Bell, James, and Ethel Seaton, eds. *Queen Elizabeth and a Swedish Princess: Being an Account of the Visit of Princess Cecilia of Sweden to England in 1565*. London: Haslewood, 1926.

Best, George. *The Three Voyages of Martin Frobisher*. 2 vols. Ed. Vilhjamlur Stefansson and Eloise McCaskill. London: Argonaut, 1938.

Billingsley, Henry, trans. *The elementes of geometrie*. London, 1570.

Blagrave, John. *Astrolabium uranicum generale*. London, 1596.

——. *Baculum familiare, catholicon sive generale. A booke of the making and use of a staffe, newly invented by the author, called the familiar staffe*. London, 1590.

——. *The mathematical jewel*. London, 1585.

Bloom, J. Harvey, and R. Rutson James, eds. *Medical Practitioners in the Diocese of London, Licensed under the Act of 3 Henry VIII, C. 11, an Annotated List, 1529–1725*. Cambridge: Cambridge University Press, 1935.

Borough, William. *Discourse on the variation of the cumpas*. Annexed to Norman, *The newe attractive*.

Bostocke, Richard. *The difference between the auncient phisicke . . . and the latter phisicke*. London, 1585.

Bourne, William. *An almanac and prognostication for three years*. London, 1564.

——. *The arte of shooting in great ordnaunce*. London, 1587.

——. *A booke called the treasure for traveilers*. London, 1578.

——. *Inventions or devises*. London, 1578.

——. *A regiment for the sea*. Ed. Thomas Hood. London, 1592.

Buck, George. *The Third Universitie of England*, in John Stow, *The Annales, or a generall chronicle of England*. London, 1615.

Carr, Cecil T., ed. *Select Charters of Trading Companies*. New York: Burt Franklin, 1970.

Clowes, William. *A briefe and necessarie treatise, touching the cure of the disease called morbus Gallicus, or lues venerea, by unctions and other approoved waies of curing*. London, 1585.

——. *A prooved practise for all young chirurgians, concerning burnings with gunpowder*. London, 1588.

——. *A right frutefull and approoved treatise, for the artificiall cure of that malady called in Latin Struma*. London, 1602.

——. *A short and profitable treatise touching the cure of the disease called Morbus Gallicus by unctions*. London, 1579.

Cole, James (see also Colius, Jacobus, and Cool, Jacob). *Of death a true description*. London, 1629.

Colius, Jacobus. *Descriptio mortis, & Præparation contra eandem*. Middelburg, 1624.

——. *Paraphrasis, ofte verklaringe ende verbredinge van den CIIII psalm de Propheten Davids*. Middelburg, 1618.

——. *Syntagma herbarum encomiasticum*. Leiden, 1606.

Cool, Jacob. *Den staet van London in hare Groote Peste*. Middelburg, 1606.

——. *Den staet van London in hare Groote Peste*. Ed. J. A. van Dorsten and K. Schaap. Leiden, 1962.

Coxe, Francis. *A prognostication made for y[e] yeere of our Lorde God 1566 declaryng the chau[n]ge, full, & quarters of the moone, w[ith] other, accustomable matters, seruing all England*. London, 1566.

——. *A short treatise declaringe the detestable wickednesse, of magicall sciences as necromancie. coniurations of spirites, curiouse astrologie and such lyke*. London, 1561.

Cuningham, William. *The cosmographical glasse*. London, 1559.

Dee, John. "Compendious Rehearsal." In *Autobiographical Tracts of Dr. John Dee, Warden of the College of Manchester*, ed. James Crossley. Chetham Society (Old Series) 24 (1851): 1–45.

———. *The Diaries of John Dee.* Ed. Edward Fenton. Oxford: Day, 1998.

———. *General and rare memorials pertaining to the perfect arte of navigation.* London, 1577.

———. "Mathematical Preface." In Billingsley, *Elementes of geometrie.*

Digges, Leonard. *A booke named Tectonicon.* London, 1556.

———. *A prognostication everlasting of right good effect.* London, 1564.

Dodoens, Rembert. *De frugum historia.* Antwerp, 1552.

———. *A niewe herball, or historie of plantes.* Trans. Henry Lyte. London, 1578.

Fennor, William. *The miseries of a jaile: or A true description of a prison.* London, 1610.

Gerard, John. *Catalogus arborum, fructium ac plantarum tam indigenarum quam exoticarum, in horto Johannes Gerardi.* London, 1596.

———. *The herball or Generall historie of plantes.* London, 1597.

———. *The herball or Generall historie of plantes. Gathered by John Gerarde of London Master in Chirurgerie very much enlarged and amended by Thomas Johnson citizen and apothecary of London.* London, 1633.

Glanvill, Joseph. *Scepsis scientifica.* London, 1665.

Hall, John, trans. *A most excellent and learned woorke of chirurgerie, called Chirurgia parva Lanfranci, Lanfranke of Mylayne his briefe.* London, 1565.

Hessels, J. H., ed. *Ecclesiæ Londino-Batavæ Archivum.* 4 vols. 1887; Osnabruck: Otto Zeller, 1969.

Hester, John. *A compendium of the rationall secretes, of the worthie knight and most excellent doctour of phisicke and chirurgerie, Leonardo Phioravante Bolognese devided into three bookes.* London, 1582.

———. *A hundred and fourtene experiments and cures of the famous phisition Philippus Aureolus Theophrastus Paracelsus.* London, 1583.

———. *A joyfull jewell.* London, 1579.

———. *A short discours Of the excellent doctour and knight, maister Leonardo Phioravanti Bolognese uppon chirurgerie.* London, 1580.

———. *These oiles, waters, extractions, or Essences[,] saltes, and other compositions; are at Paules wharfe ready made to be solde.* [London], [c. 1585–88].

———. *A true and perfect order to distill oyles out of al manner of spices, seedes, rootes and gummes.* London, 1575.

Hood, Thomas. *The making and use of the geometricall instrument, called a sector.* London, 1598.

———. *The use of the celestial globe in plano, set foorth in two hemispheres.* London, 1590.

———. *The use of the two mathematical instrumentes the crosse staffe, and the Jacobes staffe.* London, 1596.

Hulton, Paul, ed. *America, 1585: The Complete Drawings of John White.* Chapel Hill: University of North Carolina Press, 1986.

Johnson, Francis R., ed. "Thomas Hood's Inaugural Address as Mathematical Lecturer of the City of London (1588)." *Journal of the History of Ideas* 3 (1942): 94–106.

Kirk, R. E. G., and Ernest F. Kirk, eds. *Returns of Aliens Dwelling in the City and Suburbs of London from the Reign of Henry VIII to That of James I.* 3 vols. Aberdeen: Aberdeen University Press, 1900–1907.

L'Écluse, Charles de. *Nederlandsch kruidkundige, 1526–1609.* Ed. Friedrich Wilhelm Tobias Hunger. 2 vols. The Hague: M. Nijhoff, 1927–43.

———. *Rariorum aliquot Stirpium, per Pannonium, Austriam, & vicinas quasdam Provincias observatarum Historia.* Antwerp, 1583.

———. *Rariorum Plantarum Historia.* Antwerp, 1601.

L'Obel, Matthew de (see also Pena, Pierre). *Balsami, opobalsami, carpobalsmi et xylobalsami, cum suo cortice, explanatio.* London, 1598.

———. *Nova stirpium adversaria perfacilis vestigatio.* Antwerp, 1576.

———. *Stirpium adversaria nova, perfacilis vestigatio.* London, 1605.

———. *Stirpium illustrationes.* London, 1655.

Lucar, Cyprian. *A treatise named Lucar Appendix.* Annexed to Tartaglia, *Three bookes of colloquies.*

———. *A treatise named Lucarsolace.* London, 1590.

Lupton, Donald. *London and the countrey carbonadoed and quartered into severall characters.* London, 1632.

Madge, Sidney, ed. *Inquisitiones Post Mortem Relating to the City of London Returned into the Court of Chancery.* Part 2, 1561–77. London: London and Middlesex Archaeological Society, 1901.

Marcus, Leah, Janel Mueller, and Mary Beth Rose, eds. *Elizabeth I: Collected Works.* Chicago: University of Chicago Press, 2000.

Maunsell, Andrew. *The first part of the catalogue of English printed books.* London, 1595.

Maurolico, Francesco. *Martyrologium.* Venice, 1567.

Mellis, John. "To the Reader." In Hugh Oldcastle, *A briefe instruction and maner how to keepe bookes of accompts.* London, 1588.

———. "To the Right worshipfull M. Robert Forth Doctor of Law." In Record, *The grounde of artes.* London, 1582.

Middleton, Thomas. *The roaring Girle.* London, 1611.

Minshull, Geffray. *Essayes and characters of a prison and prisoners.* London, 1618.

Moens, W. J. C., ed. *The Marriage, Baptismal, and Burial Registers, 1571 to 1874, and Monumental Inscriptions of the Dutch Reformed Church, Austin Friars, London.* Lymington, UK: privately printed, 1884.

Moffett, Thomas. *Healths improvement: or, Rules comprizing and discovering the nature, method and manner of preparing all sorts of foods used in this nation.* London, 1746.

———. *Insectorum sive minimorum animalium theatrum.* London, 1634.

———. *The silkewormes and their flies.* London, 1599.

———. *The theater of insects.* Vol. 3 of Topsell, *The history of four-footed beasts.*

Munk, William. *The Roll of the Royal College of Physicians of London.* 4 vols. London: Harrison, 1878.

Norman, Robert. *The newe attractive.* London, 1581.

Oldcastle, Hugh. *A briefe instruction and maner how to keepe bookes of accompts.* London, 1588.

Oldenburg, Henry. *The Correspondence of Henry Oldenburg*. Ed. Marie Boas Hall and Rupert Hall. 11 vols. Madison: University of Wisconsin Press, 1965–77.

Orta, Garcia de. *Aromatum, et simplicium aliquot medicamentorum apud Indos nascentium historia*. Trans. Charles de L'Écluse. Antwerp, 1567.

———. *Coloquios dos simples e drogas e coisas medicinais da India e de algumas frutas*. Goa, 1563.

Ortelius, Abraham. *Deorum Dearumque capita*. Antwerp, 1573.

———. *Theatrum orbis terrarum*. Antwerp, 1570.

Parkinson, John. *Theatrum botanicum: The theater of plants. Or, An herball of a large extent*. London, 1640.

Peckham, George. *A true reporte, of the late discoveries . . . of the new-found Landes: By . . . Sir Humfrey Gilbert*. London, 1583.

Pena, Pierre, and Mathias de L'Obel. *Stirpium adversaria nova*. London, 1571.

Plat, Hugh. *Certaine philosophicall preparations of foode and beverage for sea-men, in their long voyages*. London, [1607].

———. *Delightes for ladies*. London, 1602.

———. *A discoverie of certaine English wants*. London, 1595.

———. *The Floures of Philosophie*. London, 1572.

———. *The jewell house of art and nature*. London, 1594.

———. *Sundrie new and artificiall remedies against famine*. London, 1596.

Platter, Thomas. *Beschreibung der Reisen durch Frankreich, Spanien, England und die Niederlande, 1595–1600*. 2 vols. Ed. I. Teil. Basel: Schwabe, 1968.

Platter, Thomas, and Horatio Busino. *The Journals of Two Travellers in Elizabethan and Early Stuart England*. London: Caliban, 1995.

Prockter, Adrian, and Robert Taylor, eds. *The A to Z of Elizabethan London*. Kent: Harry Margary, 1979.

Puraye, Jean, ed. *Abraham Ortelius: Album Amicorcum*. Amsterdam: Van Gendt, 1969.

Quiccheberg, Samuel. *Inscriptiones vel tituli theatri amplissimi*. Munich, 1565.

Read, John, trans. *A most excellent and compendious method of curing woundes in the head, and in other partes of the body*. London, 1588.

Record, Robert. *The ground of artes*. London, 1558.

———. *The ground of artes*. London, 1582.

———. *The pathway to knowledg containing the first principles of geometrie*. London, 1551.

Ripley, George. *The compound of alchymy*. Ed. Ralph Rabbards. London, 1591.

Scouloudi, Irene. *Returns of Strangers in the Metropolis, 1593, 1627, 1635, 1639: A Study of an Active Minority*. London: Huguenot Society, 1985.

Securis, John. *A detection and querimonie of the daily enormities and abuses co[m]mitted in physick concernyng the thre[e] partes therof: that is, the physitions part, the part of the surgeons, and the arte of poticaries*. London, 1566.

Smith, Thomas. *A Discourse of the Commonweal of This Realm of England*. Ed. Mary Dewar. Charlottesville: University Press of Virginia for the Folger Shakespeare Library, 1969.

Sotheby and Company. *Catalogue of the Highly Important Correspondence of Abraham Ortelius (1528–98) together with Some Earlier and Later Letters Presented by Ortelius's*

Nephew, Jacob Cole, to the Dutch Church in London. London: Charles F. Ince and Sons, 1955.

Stevin, Simon. *The haven-finding art.* Trans. Edward Wright. London, 1599.

Stow, John. *The annales, or a generall chronicle of England.* London, 1615.

———. *A survay of London.* London, 1598.

———. *A Survey of London Reprinted from the Text of 1603.* 2 vols. Ed. Charles Lethbridge Kingsford. Oxford: Clarendon, 1908.

Tartaglia, Niccolò. *Euclide Megarense reassettato et alla integrite ridotto per Niccolo Tartalea.* Venice, 1543.

———. *Three bookes of colloquies concerning the arte of shooting.* Trans. Cyprian Lucar. London, 1588.

Taylor, John. *Works.* London: Spenser Society, 1869.

Topsell, Edward. *The history of four-footed beasts and serpents.* London, 1658.

———. *The History of Four Footed Beasts and Serpents and Insects.* 2 vols. New York: Da Capo, 1967.

W., I. *The copie of a letter sent by a learned physician to his friend.* London, 1586.

Wheeler, John. *A Treatise of Commerce.* Ed. George Burton Hotchkiss. New York: New York University Press, 1931.

Wright, Edward. *Certaine errors in navigation.* London, 1599.

SECONDARY SOURCES

Abrahams, Israel. "Joachim Gaunse: A Mining Incident in the Reign of Queen Elizabeth." *Transactions of the Jewish Historical Society of England* 4 (1903): 83–101.

Ackermann, Silke, ed. *Humphrey Cole: Mint, Measurement, and Maps in Elizabethan England.* British Museum Occasional Paper, no. 126. London: British Museum, 1998.

Alexander, Michael Van Cleave. *The Growth of English Education 1348–1648: A Social and Cultural History.* University Park: Penn State University Press, 1990.

Allaire, Bernard. "Methods of Assaying Ore and Their Application in the Frobisher Ventures." In *Meta Incognita: A Discourse of Discovery, Martin Frobisher's Arctic Expeditions, 1576–1578.* ed. Thomas H. B. Symons, 477–504. Hull, Quebec: Canadian Museum of Civilization, 1999.

Ames-Lewis, Francis, ed. *Sir Thomas Gresham and Gresham College: Studies in the Intellectual History of London in the Sixteenth and Seventeenth Century.* Aldershot: Ashgate, 1999.

Anderson, Benedict. *Imagined Communities: Reflections on the Origin and Spread of Nationalism.* London: Verso, 1991.

Archer, Ian. *The Pursuit of Stability: Social Relations in Elizabethan London.* Cambridge: Cambridge University Press, 1991.

———. "Smith, Sir Thomas (1513–1577)." In *Oxford Dictionary of National Biography.* Oxford: Oxford University Press, 2004 [http://www.oxforddnb.com/view/article/25906, accessed 26 October 2006].

Arrizabalaga, John, John Henderson, and Roger French. *The Great Pox: The French Disease in Renaissance Europe.* New Haven: Yale University Press, 1997.

Ash, Eric. *Power, Knowledge, and Expertise in Elizabethan England*. Baltimore: Johns Hopkins University Press, 2004.

Ashworth, William B. "Emblematic Natural History of the Renaissance." In *Cultures of Natural History*, ed. Nicholas Jardine, James Secord, and Emma Spary, 17–37. Cambridge: Cambridge University Press, 1996.

——. "Natural History and the Emblematic World View." In Lindberg and Westman, *Reappraisals of the Scientific Revolution*, 303–33.

Atkins, S. E., and W. H. Overall. *Some Account of the Worshipful Company of Clockmakers*. London, 1881.

Atkinson, A. G. B. *St. Botolph Aldgate: The Story of a City Parish*. London: Grant Richards, 1898.

Baillie, G. H. *Watchmakers and Clockmakers of the World*. London: Methuen, 1929.

Barnard, John. "Politics, Profit, and Idealism: John Norton, the Stationers' Company, and Sir Thomas Bodley." *Bodleian Library Record* 17 (2002): 385–408.

Barnett, Richard C. *Place, Profit, and Power: A Study of the Servants of William Cecil, Elizabethan Statesman*. Chapel Hill: University of North Carolina Press, 1969.

Barry, Jonathan. "Bourgeois Collectivism? Urban Association and the Middling Sort." In Barry and Brooks, *The Middling Sort of People*, 84–112.

Barry, Jonathan, and Christopher Brooks, eds. *The Middling Sort of People: Culture, Society, and Politics in England, 1550–1800*. New York: St. Martin's, 1994.

Bazerman, Charles. *Shaping Written Knowledge: The Genre and Activity of the Experimental Article in Science*. Madison: University of Wisconsin Press, 1998.

Beck, R. Theodore. *The Cutting Edge: Early History of the Surgeons of London*. London: Lund Humphries, 1974.

Beier, A. L. "Social Problems in Elizabethan London." *Journal of Interdisciplinary History* 9 (1978): 203–21.

Beier, L. M. *Sufferers and Healers: The Experience of Illness in Seventeenth-Century England*. London: Routledge, 1987.

Bellany, Alastair. *The Politics of Court Scandal in Early Modern England: News Culture and the Overbury Affair, 1603–1660*. Cambridge: Cambridge University Press, 2002.

Bennett, James A. "The 'Mechanics' Philosophy and the Mechanical Philosophy." *History of Science* 24 (1986): 1–28.

Bennett, Judith M. *Ale, Beer, and Brewsters in England: Women's Work in a Changing World, 1300–1600*. Oxford: Oxford University Press, 1996.

Beretta, Marco. "Humanism and Chemistry: The Spread of Georgius Agricola's Metallurgical Writings." *Nuncius* 12 (1997): 17–47.

Bettey, J. H. "A Fruitless Quest for Wealth: The Mining of Alum and Copperas in Dorset, c. 1568–1617." *Southern History* 23 (2001): 1–9.

——. "The Production of Alum and Copperas in Southern England." *Textile History* 13 (1982): 91–98.

Biagioli, Mario. *Galileo, Courtier: The Practice of Science in the Culture of Absolutism*. Chicago: University of Chicago Press, 1993.

——. "The Social Status of Italian Mathematicians, 1400–1600." *History of Science* 27 (1989): 41–95.

Blair, Ann. "Humanist Methods in Natural Philosophy: The Commonplace Book." *Journal of the History of Ideas* 53 (1992): 541–51.

————. "Reading Strategies for Coping with Information Overload ca. 1550–1700." *Journal of the History of Ideas* 64 (2003): 11–28.

————. *The Theater of Nature: Jean Bodin and Renaissance Science.* Princeton: Princeton University Press, 1997.

Blatcher, Marjorie. *The Court of the King's Bench, 1450–1550: A Study in Self Help.* London: Athlone, 1978.

Blayney, Peter W. M. *The Bookshops of Paul's Cross Churchyard.* Occasional Papers of the Bibliographical Society, no. 5. London: Bibliographical Society, 1990.

Bosters, C., ed. *Alba Amicorum: Viif Eeuwen Vriendschap op Papier Gezet: Het Album Amicorum en Het Poëziealbum in de Nederlanden.* The Hague: CIP-Gegevens Koninklijke, 1990.

Bots, Hans, and Françoise Waquet. *La république des lettres.* Paris: Belin — De Boeck, 1977.

Boulton, Jeremy. "Neighborhood and Migration in Early Modern London." In Clark and Souden, *Migration and Society in Early Modern England,* 107–49.

————. *Neighborhood and Society: A London Suburb in the Seventeenth Century.* Cambridge: Cambridge University Press, 1987.

Briggs, John C. *Francis Bacon and the Rhetoric of Nature.* Cambridge: Harvard University Press, 1989.

Brockliss, Laurence, and Colin Jones. *The Medical World of Early Modern France.* Oxford: Clarendon, 1997.

Brooks, Christopher. "Professions, Ideology, and the Middling Sort in the Late Sixteenth and Early Seventeenth Centuries." In Barry and Brooks, *The Middling Sort of People,* pp. 113–40.

Brown, Joyce. *Mathematical Instrument-Makers in the Grocers' Company, 1688–1800.* London: Science Museum, 1979.

Browner, Jessica A. "Wrong Side of the River: London's Disreputable South Bank in the Sixteenth and Seventeenth Century." *Essays in History* 36 (1994) [http://etext.lib.virginia .edu/journals/EH/EH36/browner1.html, accessed 26 October 2006].

Bryden, D. J. "Evidence from Advertising for Mathematical Instrument Making in London, 1556–1714." *Annals of Science* 49 (1992): 301–36.

Bulst, Neithard, and Jean-Philippe Genet, eds. *Medieval Lives and the Historian: Studies in Medieval Prosopography.* Kalamazoo, 1986.

Burke, Peter. "The Language of Orders in Early Modern Europe." In *Social Orders and Social Classes in Europe Since 1500: Studies in Social Stratification,* ed. M. L. Bush, 1–12. New York: Longman, 1992.

Burtt, E. A. *The Metaphysical Foundations of Early Modern Science.* New York: Doubleday, 1924.

Bynum, Caroline Walker. *Fragmentation and Redemption.* New York: Zone, 1992.

Byrne, Richard. *Prisons and Punishments of London.* London: Harrap, 1989.

Capp, Bernard. *Astrology and the Popular Press: English Almanacs, 1500–1800.* New York: 1979.

Carlin, Martha. *Medieval Southwark.* London: London and Hambledon, 1996.

Challis, C. E. *The Tudor Coinage*. Manchester: Manchester University Press, 1978.

———, ed. A *New History of the Royal Mint*. Cambridge: Cambridge University Press, 1992.

Chamberland, Celeste C. "With a Lady's Hand and a Lion's Heart: Gender, Honor, and the Occupational Identity of Surgeons in London, 1580–1640." Ph.D. diss., University of California, 2004.

Chaplin, Joyce E. *Subject Matter: Technology, the Body, and Science on the Anglo-American Frontier, 1500–1676*. Cambridge: Harvard University Press, 2001.

Christianson, John Robert. *On Tycho's Island: Tycho Brahe and His Assistants, 1570–1601*. Cambridge: Cambridge University Press, 2000.

Clanchy, M. T. *From Memory to Written Record: England, 1066–1307*. Cambridge: Harvard University Press, 1979.

Clark, F. M. "New Light on Robert Recorde." *Isis* 8 (1926): 50–70.

Clark, George. A *History of the Royal College of Physicians of London*. 2 vols. Oxford: Clarendon, 1964.

Clark, Peter, and David Souden. *Migration and Society in Early Modern England*. Totowa, NJ: Barnes and Noble, 1988.

Clericuzio, Antonio. "Agricola e Paracelso: Mineralogia e iatrochemica nel Rinascimento." *Nuova civiltà delle machine* 12 (1994): 113–21.

Clifton, Gloria. *Dictionary of British Scientific Instrument Makers, 1550–1851*. London: Zwemmer/National Maritime Museum, 1995.

Collinson, Richard. *The Three Voyages of Martin Frobisher*. London: Hakluyt Society, 1867.

Clouse, Michele. "Administering and Administrating Medicine: Philip II and the Medical World of Early Modern Spain." Ph.D. diss., University of California, 2004.

Clucas, Stephen. "'No Small Force': Natural Philosophy and Mathematics in Thomas Gresham's London." In Ames-Lewis, *Sir Thomas Gresham and Gresham College*, 146–73.

———. "Thomas Harriot and the Field of Knowledge in the English Renaissance." In *Thomas Harriot: An Elizabethan Man of Science*, ed. Robert Fox, 93–136. Aldershot: Ashgate, 2000.

Clulee, Nicholas. *John Dee's Natural Philosophy: Between Science and Religion*. London: Routledge, 1988.

Cook, Harold J. "Against Common Right and Reason: The College of Physicians Versus Dr. Thomas Bonham." *American Journal of Legal History* 29 (1985): 301–22.

———. *The Decline of the Old Medical Regime in Stuart London*. Ithaca: Cornell University Press, 1986.

———. "Good Advice and Little Medicine: The Professional Authority of Early Modern English Physicians." *Journal of British Studies* 33 (1994): 1–31.

———. "Time's Bodies: Crafting the Preparation and Preservation of Naturalia." In Smith and Findlen, *Merchants and Marvels*, 223–47.

———. The *Trials of an Ordinary Doctor: Joannes Groenevelt in Seventeenth-Century London*. Baltimore: Johns Hopkins University Press, 1994.

Cormack, Lesley. *Charting an Empire*. Chicago: University of Chicago Press, 1997.

————."The Commerce of Utility: Teaching Mathematical Geography in Early Modern England." *Science and Education* 15 (2006): 305–22.

————. "'Twisting the Lion's Tail': Practice and Theory in the Court of Henry, Prince of Wales." In Moran, *Patronage and Institutions*, 67–84.

Cottret, Bernard. *The Huguenots in England: Immigration and Settlement c. 1550–1700*. Cambridge: Cambridge University Press, 1991.

Crawforth, M. A. "Instrument Makers in the London Guilds." *Annals of Science* 44 (1987): 319–77.

Crombie, A. C. *Augustine to Galileo: The History of Science, A.D. 400–1650*. London: Falcon, 1952.

Cross, Claire. *The Puritan Earl: The Life of Henry Hastings Third Earl of Huntington, 1536–1595*. New York: St. Martin's, 1966.

Cruickshank, C. G. *Elizabeth's Army*. Oxford: Oxford University Press, 1966.

Cunningham, Richard. "Virtual Witnessing and the Role of the Reader in a New Natural Philosophy." *Philosophy and Rhetoric* 34 (2001): 207–24.

Dannenfeldt, Karl H. "Egyptian Mumia: The Sixteenth-Century Experience and Debate." *Sixteenth-Century Journal* 16 (1985): 163–80.

Darnton, Robert. *The Kiss of Lamourette: Reflections in Cultural History*. New York: Norton, 1995.

————. *The Literary Underground of the Old Regime*. Cambridge: Harvard University Press, 1982.

Daston, Lorraine, and Katharine Park. *Wonders and the Order of Nature, 1150–1750*. New York: Zone, 2001.

Davies, Margaret Gay. *The Enforcement of English Apprenticeship: A Study in Applied Mercantilism, 1563–1642*. Cambridge: Harvard University Press, 1956.

Davis, Natalie Zemon. "Beyond the Market: Books as Gifts in Sixteenth-Century France." *Transactions of the Royal Historical Society*, 5th series, 33 (1983): 69–88.

————. *The Return of Martin Guerre*. Cambridge: Harvard University Press, 1983.

————. "Sixteenth-Century French Arithmetics and Business Life." *Journal of the History of Ideas* 21 (1960): 18–48.

Dawbarn, Frances. "New Light on Dr. Thomas Moffet: The Triple Roles of an Early Modern Physician, Client, and Patronage Broker." *Medical History* 47 (2003): 3–22.

————. "Patronage and Power: The College of Physicians and the Jacobean Court." *British Journal of the History of Science* 31 (1998): 1–19.

Dear, Peter. *Discipline and Experience: The Mathematical Way in the Scientific Revolution*. Chicago: University of Chicago Press, 1995.

————. "Narratives, Anecdotes, and Experiments: Turning Experience into Science in the Seventeenth Century." In *The Literary Structure of Scientific Argument*, ed. Peter Dear, 135–63. Philadelphia: University of Pennsylvania Press, 1991.

————. "*Totius in verba*: Rhetoric and Authority in the Early Royal Society." *Isis* 76 (1985): 145–61.

De Backer, W., Francine de Nave, D. Imhof, et al. *Botany in the Low Countries (end of the Fifteenth Century — ca. 1650)*. Antwerp: Plantin-Moretus Museum, 1993.

Debus, Allen G. *The Chemical Philosophy: Paracelsian Science and Medicine in the Six-teenth and Seventeenth Centuries.* 2 vols. New York: Science History Publications, 1977.

―――. *The English Paracelsians.* Cambridge: Cambridge University Press, 1965.

―――. *The French Paracelsians: The Chemical Challenge to Medical and Scientific Tra-dition in Early Modern France.* Cambridge: Cambridge University Press, 1991.

―――. "The Paracelsian Compromise in Elizabethan England." *Ambix* 8 (1960): 71–97.

Devereaux, Simon, and Paul Griffiths, eds. *Penal Practice and Culture, 1500–1900: Pun-ishing the English.* New York: Palgrave Macmillan, 2004.

Dewar, Mary. *Sir Thomas Smith: A Tudor Intellectual in Office.* London: Athlone, 1964.

Dillon, Janette. *Language and Stage in Medieval and Renaissance England.* Cambridge: Cambridge University Press, 1998.

Dobb, Clifford. "London Prisons." *Shakespeare Survey* 17 (1964): 87–100.

Dobbs, Betty Jo Teeter. *Alchemical Death and Resurrection: The Significance of Alchemy in the Age of Newton.* Washington, DC: Smithsonian Institution Libraries, 1990.

―――. "Newton as Final Cause and First Mover." In *Rethinking the Scientific Revolution,* ed. Margaret J. Osler, 25–40. Cambridge: Cambridge University Press, 2000.

Donald, M. B. "Burchard Kranich (c. 1515–1578), Miner and Queen's Physician, Cornish Mining Stamps, Antimony, and Frobisher's Gold." *Annals of Science* 6 (1950): 308–52.

―――. *Elizabethan Copper: The History of the Mines Royal, 1568–1608.* London: Perga-mon, 1955.

―――. *Elizabethan Monopolies: The History of the Company of Mineral and Battery Works.* London: Oliver and Boyd, 1961.

Drover, C. B., and H. A. Lloyd. *Nicholas Vallin, 1565–1603: Connoisseur Year Book.* Lon-don, 1955.

Duden, Barbara. *The Woman Beneath the Skin: A Doctor's Patients in Eighteenth-Century Germany.* Trans. Thomas Dunlap. Cambridge: Harvard University Press, 1991.

Duhem, Pierre. *The Aim and Structure of Physical Theory.* Princeton: Princeton University Press, 1991.

Dunn, Richard. "The True Place of Astrology among the Mathematical Arts of Late Tudor England." *Annals of Science* 51 (1994): 151–63.

Eamon, William. "Court, Academy, and Printing House: Patronage and Scientific Careers in Late Renaissance Italy." In Moran, *Patronage and Institutions,* 25–50.

―――. *Science and the Secrets of Nature.* Princeton: Princeton University Press, 1994.

Earle, Peter. "The Middling Sort in London." In Barry and Brooks, *The Middling Sort of People,* pp. 141–58.

Egmond, Florike. "Correspondence and Natural History in the Sixteenth Century: Cul-tures of Exchange in the Circle of Carolus Clusius." In *Correspondence and Cultural Exchange in Early Modern Europe,* ed. Francisco Bethencourt and Florike Egmond. Cambridge, Cambridge University Press, forthcoming.

―――. "A European Community of Scholars: Exchange and Friendship among Early Modern Natural Historians." In *Finding Europe: Discourses on Margins, Communities, Images,* ed. Anthony Molho and Diogo Ramada Curto. Oxford: Berghahn, forthcom-ing.

Egmond, Florike, Paul Hoftijzer, and Robert Vissers, eds. *Carolus Clusius in a New Context: Cultural Histories of Renaissance Natural Science.* Amsterdam: Edita, 2006.

Eisenstadt, S. N., and Louis Roniger. "Patron-Client Relations as a Model of Structuring Social Exchange." *Comparative Studies in Society and History* 22 (1980): 42–77.

Eisenstein, Elizabeth. *The Printing Press as an Agent of Change: Communication and Cultural Transformations in Early-Modern Europe.* New York: Cambridge University Press, 1979.

Esler, Anthony. *The Aspiring Mind of the Elizabethan Younger Generation.* Durham: Duke University Press, 1966.

Esser, Raingard. "Germans in Early Modern Britain." In *Germans in Britain Since 1500*, ed. Panikos Panayi, 17–27. London: Hambledon, 1996.

Evans, R. J. W. *Rudolf II and His World: A Study in Intellectual History, 1576–1612.* Oxford: Oxford University Press, 1973.

Evenden, D. A. "Gender Differences in the Licensing and Practice of Female and Male Surgeons in Early Modern England." *Medical History* 42 (1998): 194–216.

Everitt, C. W. F. "Background to History: The Transition from Little Physics to Big Physics in the Gravity Probe B Relativity Gyroscope Program." In Galison and Hevly, *Big Science*, 212–35.

Farrington, Benjamin. *The Philosophy of Francis Bacon: An Essay on Its Development from 1603 to 1609 with New Translations of Fundamental Texts.* Chicago: University of Chicago, 1964.

Feingold, Mordechai. "Gresham College and London Practitioners: The Nature of the English Mathematical Community." In Ames-Lewis, *Sir Thomas Gresham and Gresham College*, 174–88.

———. *The Mathematicians' Apprenticeship: Science, Universities, and Society in England, 1560–1640.* Cambridge: Cambridge University Press, 1984.

Ferguson, A. B. *The Articulate Citizen and the English Renaissance.* Durham: Duke University Press, 1965.

Ferguson, Margaret. *Dido's Daughters: Literacy, Gender, and Empire in Early Modern England.* Chicago: University of Chicago Press, 2002.

Feuer, Lewis S. *Jews in the Origins of Modern Science and Bacon's Scientific Utopia: The Life and Work of Joachim Gause, Mining Technologist and First Recorded Jew in English-Speaking North America.* Cincinnati: American Jewish Archives, 1987.

Field, J. V., and Frank A. J. L. James, eds. *Renaissance and Revolution: Humanists, Scholars, Craftsmen, and Natural Philosophers in Early Modern Europe.* Cambridge: Cambridge University Press, 1993.

Findlen, Paula. "Controlling the Experiment: Rhetoric, Court Patronage and the Experimental Method of Francesco Redi." *History of Science* 31 (1993): 35–64.

———. "Courting Nature." In *Cultures of Natural History*, ed. Nicholas Jardine, James Secord, Emma Spary, 57–74. Cambridge: Cambridge University Press, 1996.

———. "The Economy of Scientific Exchange in Early Modern Italy." In Moran, *Patronage and Institutions*, 5–24.

———. "The Formation of a Scientific Community: Natural History in Sixteenth-Century Italy." In *Natural Particulars: Nature and the Disciplines in Renaissance Europe*, ed. Anthony Grafton and Nancy Siraisi, 369–400. Cambridge: MIT Press, 1999.

———. "Jokes of Nature and Jokes of Knowledge: The Playfulness of Scientific Discourse in Early Modern Europe." *Renaissance Quarterly* 43 (1990): 292–331.

———. *Possessing Nature: Museums, Collecting, and Scientific Culture in Early Modern Italy*. Berkeley: University of California Press, 1994.

Finlay, Roger. *Population and Metropolis: The Demography of London, 1580–1625*. Cambridge: Cambridge University Press, 1981.

Fischer, M. J. "The Uses of Life Histories." *Anthropological Humanities Quarterly* 16 (1991): 24–27.

Fisher, F. J. *London and the English Economy, 1500–1700*. Ed. P. J. Corfield and N. B. Harte. London: Hambledon, 1990.

Fissell, Mary E. *Patients, Power, and the Poor in Eighteenth-Century Bristol*. Cambridge: Cambridge University Press, 1991.

Forbes, Thomas. *Chronicle from Aldgate: Life and Death in Shakespeare's London*. New Haven: Yale University Press, 1971.

Forster, Leonard. *Janus Gruter's English Years: Studies in the Continuity of Dutch Literature in Exile in Elizabethan England*. London: Oxford University Press, 1967.

Foucault, Michel. "What Is an Author?" In *Language, Counter-Memory, Practice: Selected Essays and Interviews*, trans. Donald F. Bouchard and Sherry Simon. Ithaca: Cornell University Press, 1977.

Freedberg, David. *The Eye of the Lynx: Galileo, His Friends, and the Beginnings of Modern Natural History*. Chicago: University of Chicago Press, 2003.

Fumerton, Patricia. *Cultural Aesthetics: Renaissance Literature and the Practice of Social Ornament*. Chicago: University of Chicago Press, 1991.

Furdell, Elizabeth Lane. *Publishing and Medicine in Early Modern England*. Rochester, NY: University of Rochester Press, 2002.

Galison, Peter, and Bruce Hevly, eds. *Big Science: The Growth of Large-Scale Research*. Stanford: Stanford University Press, 1992.

———. "The Many Faces of Big Science." In Galison and Hevly, *Big Science*, 1–17.

Galison, Peter, Bruce Hevly, and Rebecca Lowen. "Controlling the Monster: Stanford and the Growth of Physics Research, 1935–1962." In Galison and Hevly, *Big Science*, 46–77.

Garber, Daniel. "Experiment, Community, and the Constitution of Nature in the Seventeenth Century." *Perspectives on Science* 3 (1995): 173–201.

Garrioch, David. "Shop Signs and Social Organization in Western European Cities, 1500–1900." *Urban History* 21 (1994): 20–48.

Gaukroger, Stephen. *Francis Bacon and the Transformation of Early-Modern Philosophy*. Cambridge: Cambridge University Press, 2001.

Gentilcore, David. "'Charlatans, Mountebanks, and Other Similar People': The Regulation and Role of Itinerant Practitioners in Early Modern Italy." *Social History* 20 (1995): 297–314.

———. *Healers and Healing in Early Modern Italy*. Manchester: Manchester University Press, 1998.

George, Wilma. "Alive or Dead: Zoological Collections in the Seventeenth Century." In Impey and MacGregor, *The Origins of Museums*, 245–55.

Gingerich, Owen. *The Book Nobody Read: Chasing the Revolutions of Nicolaus Copernicus*. New York: Penguin, 2005.

Ginzburg, Carlo. *The Cheese and the Worms: The Cosmos of a Sixteenth-Century Miller*. Baltimore: Johns Hopkins University Press, 1980.

Goldgar, Anne. *Impolite Learning: Conduct and Community in the Republic of Letters, 1680–1750*. New Haven: Yale University Press, 1995.

———. "Nature as Art: The Case of the Tulip." In Smith and Findlen, *Merchants and Marvels*, 324–46.

———. *Tulipmania: Money, Honor, and Knowledge in the Dutch Golden Age*. Chicago: University of Chicago Press, 2007.

Goldman, P. H. J. "Eloye Mestrelle and the Introduction of the Mill and Screw Press into English Coinage, circa 1561–1575." *Spink's Numismatic Circular* 82 (1974): 422–27.

Goldthwaite, Richard A. "Schools and Teachers of Commercial Arithmetic in Renaissance Florence." *Journal of European Economic History* 1 (1972–73): 418–33.

Golinski, Jan. *Making Natural Knowledge: Constructivism and the History of Science*. Cambridge: Cambridge University Press, 1998.

———. *Science as Public Culture*. Cambridge: Cambridge University Press, 1992.

Goodman, David. "Philip II's Patronage of Science and Engineering." *British Journal of the History of Science* 16 (1983): 49–66.

Goudriaan, Koen, Kees Mandemakers, Joachim Reitsma, and Peter Stabel, eds. *Prosopography and Computer: Contributions of Medievalists and Modernists on the Use of Computer in Historical Research*. Lewen: Garant, 1995.

Gough, J. W. *The Rise of the Entrepreneur*. London: B. T. Batsford, 1969.

Grafton, Anthony. "Geniture Collections, Origins, and Use of a Genre." In *Books and the Sciences in History*, ed. Marina Frasca-Spada and Nicholas Jardine, 49–68. Cambridge: Cambridge University Press, 2000.

———. "Kepler as Reader." *Journal of the History of Ideas* 53 (1992): 561–72.

Grassl, Gary C. "Joachim Gans of Prague: America's First Jewish Visitor." *Review of the Society for the History of Czechoslovak Jews* 1 (1987): 53–90.

———. "Joachim Gans of Prague: The First Jew in English America." *American Jewish History* 86 (1998): 195–217.

Graves, Michael A. R. *Burghley: William Cecil, Lord Burghley*. New York: Longman, 1998.

Green, Monica. "Women's Medical Practice and Health Care in Medieval Europe." *Signs* 14 (1989): 434–73.

Greene, Edward Lee. *Landmarks of Botanical History*. 2 vols. Stanford: Stanford University Press, 1983.

Grell, Ole Peter. *Calvinist Exiles in Tudor and Stuart England*. Aldershot: Scolar, 1996.

———. *Dutch Calvinists in Early Stuart London: The Dutch Church in Austin Friars, 1603–1642*. Leiden: E. J. Brill, 1989.

———. "Plague in Elizabethan and Stuart London: The Dutch Response." *Medical History* 34 (1990): 424–39.

———. "The Schooling of the Dutch Calvinist Community in London, 1550 to 1650." *De zeventiende eeuw*, 2, no. 2 (1986): 45–58.

Gunther, R. T. *Early British Botanists and Their Gardens*. Oxford: Oxford University Press, 1922.

Guy, John R. "The Episcopal Licensing of Physicians, Surgeons and Midwives." *Bulletin of the History of Medicine* 56 (1982): 528–42.

Hall, A. Rupert. "On the Historical Singularity of the Scientific Revolution of the Seventeenth Century." In *The Diversity of History: Essays in Honour of Sir Herbert Butterfield*, ed. J. H. Elliott and H. G. Koenigsberger, 199–222. London: Routledge, 1970.

———. "The Scholar and the Craftsman in the Scientific Revolution." In *Critical Problems in the History of Science*, ed. Marshall Clagett, 3–23. Madison: University of Wisconsin Press, 1959.

———. *The Scientific Revolution, 1500–1800*. Boston: Beacon, 1954.

Hall, Marie Boas. "Henry Oldenburg and the Art of Scientific Communication." *British Journal for the History of Science* 2 (1965): 277–90.

———. *Promoting Experimental Learning: Experiment and the Royal Society, 1660–1727*. Cambridge: Cambridge University Press, 1991.

Halleux, Robert. "L'alchimiste et l'essayeur." In *Die Alchemie in der europäischen Kultur- und Wissenschaftsgeschichte*, ed. Christoph Meinel, 277–91. Wiesbaden: Otto Harrassowitz, 1986.

Haraway, Donna. "Situated Knowledges: The Science Question in Feminism and the Privilege of Partial Perspective." *Feminist Studies* 14 (1988): 575–99.

Harkness, Deborah E. *John Dee's Conversations with Angels: Cabala, Alchemy, and the End of Nature*. Cambridge: Cambridge University Press, 1999.

———. "Managing an Experimental Household: The Dees of Mortlake and the Practice of Natural Philosophy." *Isis* 88 (1997): 247–62.

———. *"Nosce Teipsum:* Curiosity, the Humoural Body, and the Culture of Therapeutics in Sixteenth- and Early Seventeenth-Century England." In *Curiosity and Wonder from the Renaissance to the Enlightenment*, ed. R. J. W. Evans and Alexander Marr, 171–92. Aldershot: Ashgate, 2006.

———. "Strange Ideas and English Knowledge: Natural Science Exchange in Elizabethan London." In Smith and Findlen, *Merchants and Marvels*, 137–62.

———. "Tulips, Maps, and Spiders: The Cole-Ortelius-Lobel Family and the Practice of Natural Philosophy in Early Modern London." In *From Strangers to Citizens: Foreigners and the Metropolis, 1500–1800*, ed. Randolph Vigne and Charles Littleton, 184–96. London: Huguenot Society and Sussex Academic Press, 2001.

Harley, David. "Rhetoric and the Social Construction of Sickness and Healing." *Social History of Medicine* 12 (1990): 407–35.

———. "Rychard Bostok of Tandgridge, Surrey (c. 1530–1605), M.P., Paracelsian Propagandist, and Friend of John Dee." *Ambix* 47 (2000): 29–36.

Harris, Jonathan Gil. *Foreign Bodies and the Body Politic: Discourses of Social Pathology in Early Modern England*. Cambridge: Cambridge University Press, 1998.

———. *Sick Economies: Drama, Mercantilism, and Disease in Shakespeare's England*. Philadelphia: University of Pennsylvania Press, 2004.

Heal, Felicity, and Clive Holmes. "The Economic Patronage of William Cecil." In *Patronage, Culture, and Power: The Early Cecils*, ed. Pauline Croft, 199–229. New Haven: Yale University Press, 2002.

Hearn, Karen. *Marcus Gheeraerts II: Elizabethan Artist in Focus*. London: Tate Publishing, 2002.

Hendrix, Lee. "Of Hirsutes and Insects: Joris Hoefnagel and the Art of the Wondrous." *Word and Image* 11 (1995): 373–90.

Hendrix, Lee, with Georg Bocskay, and Thea Vignau-Wilberg, eds. *Nature Illuminated: Flora and Fauna from the Court of Emperor Rudolf II*. Los Angeles: J. Paul Getty Museum, 1997.

Henrey, Blanche. *British Botanical and Horticultural Literature Before 1800: Comprising a History and Bibliography of Botanical and Horticultural Books Printed in England, Scotland, and Ireland from the Earliest Times until 1800*. 3 vols. London: Oxford University Press, 1975.

Hevly, Bruce. "Reflections on Big Science and Big History." In Galison and Hevly, *Big Science*, 357–63.

Hicks, Michael. "Waad, Armagil (c. 1510–1568)." In *Oxford Dictionary of National Biography*. Oxford: Oxford University Press, 2004 [http://www.oxforddnb.com/view/article/28363, accessed 26 October 2006].

Hill, Christopher. *Intellectual Origins of the English Revolution*. Oxford: Clarendon, 1965.

Hill, Katherine. "'Juglers or Schollers?': Negotiating the Role of a Mathematical Practitioner." *British Journal for the History of Science* 31 (1998): 253–74.

Hind, Arthur M. *Engraving in England in the Sixteenth and Seventeenth Centuries: A Descriptive Catalogue with Introductions*. 3 vols. Cambridge: University Press, 1952–64.

Hodnett, Edward. *Marcus Gheeraerts the Elder of Bruges, London, and Antwerp*. Utrecht: Haentjens Dekker and Gumbert, 1971

Hoeniger, F. D., and J. F. M. Hoeniger. *The Development of Natural History in Tudor England*. Charlottesville: University of Virginia Press for the Folger Shakespeare Library, 1969.

———. *The Growth of Natural History in Stuart England: From Gerard to the Royal Society*. Charlottesville: University of Virginia Press for the Folger Shakespeare Library, 1969.

Hogarth, D. D. "Mining and Metallurgy of the Frobisher Ores." In *Archaeology of the Frobisher Voyages*, ed. William W. Fitzhugh and Jacqueline S. Olin, 137–51. Washington, DC: Smithsonian Institution Press, 1993.

Hogarth, D. D., P. W. Boreham, and John G. Mitchell. *Martin Frobisher's Northwest Venture, 1576–1581: Mines, Minerals, and Metallurgy*. Hull, Quebec: Canadian Museum of Civilization, 1994.

Holbrook, Mary, et al. *Science Preserved: A Directory of Scientific Instruments in Collections in the United Kingdom and Eire*. London: HMSO, 1992.

Houliston, V. H. "Sleepers Awake: Thomas Moffet's Challenge to the College of Physicians of London, 1584." *Medical History* 33 (1989): 235–46.

Howson, Geoffrey. *A History of Mathematics Education in England*. Cambridge: Cambridge University Press, 1982.

Hulme, E. Wyndham. "The History of the Patent System Under the Prerogative and at Common Law." *Law Quarterly Review* 12 (1896): 141–54.

———. "The History of the Patent System Under the Prerogative and at Common Law: A Sequel." *Law Quarterly Review* 16 (1900): 44–56.

Hulton, Paul, and David Beers Quinn, eds. *The American Drawings of John White, 1577–1590.* Chapel Hill: University of North Carolina Press, 1964.

Hume, Ivor Noel. "Roanoke Island: America's First Science Center." *Colonial Williamsburg* 16 (1994): 14–28.

———. *The Virginia Adventure, Roanoke to James Towne: An Archaeological and Historical Odyssey.* New York: Knopf, 1994.

Hunter, Michael. *The Royal Society and Its Fellows, 1660–1700: The Morphology of an Early Scientific Institution.* Chalfont St. Giles: British Society for the History of Science, 1982.

Iliffe, Rob. "Material Doubts: Hooke, Artisan Culture, and the Exchange of Information in 1670s London." *British Journal for the History of Science* 28 (1995): 285–318.

Impey, Oliver, and Arthur MacGregor, eds. *The Origins of Museums: The Cabinet of Curiosities in Sixteenth- and Seventeenth-Century Europe.* 1985; London: Stratus, 2001.

Inkster, Ian. *Science and Technology in History: An Approach to Industrial Development.* New Brunswick, NJ: Rutgers University Press, 1991.

Isaac, Peter. "Pills and Print." In *Medicine, Mortality, and the Book Trade*, ed. Robin Myers and Michael Harris, 25–47. New Castle, DE: Oak Knoll, 1998.

Jacob, James R. "Restoration Ideologies and the Royal Society." *History of Science* 18 (1980): 25–38.

———. *Robert Boyle and the English Revolution: A Study in Social and Intellectual Change.* New York: Burt Franklin, 1977.

Jacob, Margaret C. *The Cultural Meaning of the Scientific Revolution.* New York: Knopf, 1988.

———. *The Newtonians and the English Revolution, 1689–1720.* Ithaca: Cornell University Press, 1976.

———. *The Radical Enlightenment: Pantheists, Freemasons, and Republicans.* London: Allen and Unwin, 1981.

Jardine, Lisa. *Erasmus, Man of Letters: The Construction of Charisma in Print.* Princeton: Princeton University Press, 1993.

———. *Francis Bacon: Discovery and the Art of Discourse.* Cambridge: Cambridge University Press, 1974.

Jardine, Lisa, and Alan Stewart. *Hostage to Fortune: The Troubled Life of Francis Bacon, 1561–1626.* London: Victor Gollancz, 1998.

Jeffers, Robert H. *The Friends of John Gerard (1545–1612), Surgeon and Botanist: Biographical Appendix.* Falls Village, CT: Herb Grower, 1969.

Johns, Adrian. "The Ideal of Scientific Collaboration: The 'Man of Science' and the Diffusion of Knowledge." In *Commercium Litterarium: Forms of Communication in the Republic of Letters, 1600–1750*, ed. Hans Bots and Francoise Waquet, 3–22. Amsterdam: APA — Holland University Press, 1994.

———. *The Nature of the Book: Print and Knowledge in the Making.* Chicago: University of Chicago Press, 1998.

Johnson, Francis R. *Astronomical Thought in Renaissance England: A Study of the English Scientific Writing, 1500 to 1645.* Baltimore: Johns Hopkins University Press, 1937.

———. "Gresham College: Precursor of the Royal Society." *Journal of the History of Ideas* 1 (1940): 413–38.

Johnson, Francis R., and S. V. Larkey. "Robert Recorde's Mathematical Teaching and the Anti-Aristotelian Movement." *Huntington Library Bulletin* 7 (1935): 59–87.

Johnston, Stephen. "The Astrological Instruments of Thomas Hood." July 1998, http://www.mhs.ox.ac.uk/staff/saj/hood-astrology/.

———. "Making Mathematical Practice: Gentlemen, Practitioners, and Artisans in Elizabethan England." Ph.D. diss., University of Cambridge, 1994.

———. "Mathematical Practitioners and Instruments in Elizabethan England." *Annals of Science* 48 (1991): 319–34.

———. "Recorde, Robert (c. 1512–1558)." In *Oxford Dictionary of National Biography.* Oxford: Oxford University Press, 2004 [http://www.oxforddnb.com/view/article/23241, accessed 26 October 2006].

Jones, Norman. "Defining Superstitions: Treasonous Catholics and the Act Against Witchcraft of 1563." In *State, Sovereigns, and Society in Early Modern England: Essays in Honour of A. J. Slavin,* ed. Charles Carlton, Robert L. Woods, Mary L. Robertson and Joseph L. Block, 187–203. New York: Palgrave, 1997.

Jones, Peter Murray. "Gemini, Thomas (*fl.* 1540–1562)." In *Oxford Dictionary of National Biography.* Oxford: Oxford University Press, 2004 [http://www.oxforddnb.com/view/article/10513, accessed 26 October 2006].

Jones, Richard Foster. *Ancients and Moderns: A Study of the Rise of the Scientific Movement in Seventeenth-Century England.* New York: Dover, 1962.

Jordanova, Ludmilla. "The Social Construction of Medical Knowledge." *Social History of Medicine* 8 (1995): 361–81.

Jütte, Robert. "Valentin Rösswurm: Zur Sozialgeschicte des Paracelsismus im 16. Jarhundert." In *Resultate und Desiderate der Paracelsus-Forschung,* ed. Peter Dilg and Hartmut Rudolph, 99–112. Stuttgart: Franz Steiner Verlag, 1993.

Karrow, Robert J., Jr. *Mapmakers of the Sixteenth Century and Their Maps: Bio-Bibliographies of the Cartographers of Abraham Ortelius, 1570.* Chicago: Speculum Orbis, 1993.

Kassell, Lauren. "How to Read Simon Forman's Casebooks: Medicine, Astrology, and Gender in Elizabethan London." *Social History of Medicine* 12 (1999): 3–18.

———. *Medicine and Magic in Elizabethan London: Simon Forman, Astrologer, Alchemist, and Physician.* Oxford: Clarendon, 2005.

Katz, David S. *Jews in the History of England, 1485–1850.* Oxford: Oxford University Press, 1994.

Kaufmann, Thomas DaCosta. *The Mastery of Nature: Aspects of Art, Science, and Humanism in the Renaissance.* Princeton: Princeton University Press, 1993.

———. "Remarks on the Collections of Rudolf II: The *Kunstkammer* as a Form of *Representatio.*" *Art Journal* 38 (1978): 22–28.

Keller, Alexander. "Mathematics, Mechanics, and the Origins of the Culture of Mechanical Invention." *Minerva* 23 (1985): 348–61.

Kettering, Sharon. *Patrons, Brokers, and Clients in Seventeenth-Century France.* Oxford: Oxford University Press, 1986.

Kiernan, Michael. "Introduction: Bacon's Programme for Reform." In Bacon, *Advancement of Learning*, xxxiv–xxxvi.

Klose, Wolfgang. *Corpus Alborum Amicorum.* Stuttgart: CAAC, 1988.

Kocher, Paul. "John Hester, Paracelsian (fl. 1576–93)." In *Joseph Quincy Adams Memorial Studies*, ed. James G. McManaway, Giles E. Dawson, and Edwin E. Willoughby, 621–38. Washington, DC: Folger Shakespeare Library, 1948.

———. "Paracelsian Medicine in England: The First Thirty Years (ca. 1570–1600)." *Journal of the History of Medicine* 2 (1947): 451–80.

Kuhn, Thomas. "Mathematical Versus Experimental Traditions in the Development of Physical Science." In *The Essential Tension: Selected Studies in Scientific Tradition and Change*, ed. Thomas Kuhn, 31–65. Chicago: University of Chicago Press, 1962.

———. *The Structure of Scientific Revolutions.* 1962; Chicago: University of Chicago Press, 1970.

Lake, Peter. "From Troynouvant to Heliogabulus's Rome and Back: 'Order' and Its Others in the London of John Stow." In *Imagining Early Modern London: Perceptions and Portrayals of the City from Stow to Strype, 1598–1720*, ed. J. F. Merritt, 217–49. Cambridge: Cambridge University Press, 2001.

Leong, Elaine. "Medical Recipe Collections in Seventeenth-Century England: Knowledge, Text, and Gender." Ph.D. diss., University of Oxford, 2006.

Lindberg, David C., and Robert Westman, eds. *Reappraisals of the Scientific Revolution.* Cambridge: Cambridge University Press, 1990.

Lindeboom, Johannes. *Austin Friars: History of the Dutch Church in London, 1550–1950.* The Hague: M. Nijhoff, 1950.

Lingo, Alison. "Empirics and Charlatans in Early Modern France: The Genesis of the Classification of the 'Other' in Medical Practice." *Journal of Social History* 19 (1986): 583–603.

Loewenstein, Joseph. *The Author's Due: Printing and the Prehistory of Copyright.* Chicago: University of Chicago Press, 2002.

———. *Ben Jonson and Possessive Authorship.* Cambridge: Cambridge University Press, 2002.

Long, Pamela O. "Invention, Authorship, 'Intellectual Property,' and the Origin of Patents: Notes Towards a Conceptual History." *Technology and Culture* (1991): 846–84.

———. "Objects of Art/Objects of Nature." In Smith and Findlen, *Merchants and Marvels*, 63–82.

———. "The Openness of Knowledge: An Ideal and Its Context in 16th-Century Writings on Mining and Metallurgy." *Technology and Culture* 32 (1991): 318–55.

———. *Openness, Secrecy, Authorship: Technical Arts and the Culture of Knowledge from Antiquity to the Renaissance.* Baltimore: Johns Hopkins University Press, 2001.

Loomes, Brian. *The Early Clockmakers of Great Britain.* London: N. A. G. Press, 1981.

Loomie, A. J. "Neville, Edmund (*b.* before 1555, *d.* in or after 1620)." In *Oxford Dictionary of National Biography.* Oxford: Oxford University Press, 2004 [http://www.oxforddnb .com/view/article/19927, accessed 26 October 2006].

Louis, Armand. *Mathieu de L'Obel, 1538–1616: Épisode de l'histoire de la botanique.* Ghent: Story-Scientia, 1980.

Love, Harold. *Scribal Publication in Seventeenth-Century England.* Oxford: Clarendon, 1993.

Lux, David. *Patronage and Royal Science in Seventeenth-Century France: The Academie de Physique in Caen.* Ithaca: Cornell University Press, 1989.

Lux, David, and Harold J. Cook. "Communications During the Scientific Revolution." *History of Science* 36 (1998): 179–211.

MacGregor, Arthur. "The Cabinet of Curiosities in Seventeenth-Century Britain." In Impey and MacGregor, *The Origins of Museums,* 201–15.

———, ed. *Tradescant's Rarities: Essays on the Foundation of the Ashmolean Museum 1683 with a Catalogue of the Surviving Early Collections.* Oxford: Clarendon, 1983.

Mandosio, Jean-Marc. "La place de l'alchimie dans les classifications du Moyen Age et de la Renaissance." *Chrysopoeia* 4 (1990–91): 199–282.

Mangini, Giorgi. *Il "mondo" di Abramo Ortelio: Misticismo, geografia, e collezionismo nel Rinascimento dei Paesi Basi.* Modena: Franco Cosimo Panini, 1998.

Marcus, George. *Ethnography Through Thick and Thin.* Princeton: Princeton University Press, 1998.

Marín, Francisco Rodríguez. *Felipe II y la alquimia.* Madrid, 1951.

Marlow, R. K. "The Life and Music of Giles Farnaby." Ph.D. diss., University of Cambridge, 1966.

Marotti, Arthur F. *Manuscript, Print, and the English Renaissance Lyric.* Ithaca: Cornell University Press, 1995

Martin, Henri-Jean. *The History and Power of Writing.* Trans. Lydia G. Cochrane. Chicago: University of Chicago Press, 1988.

Martin, Julian. *Francis Bacon, the State, and the Reform of Natural Philosophy.* Cambridge: Cambridge University Press, 1992.

Mathesen, R. W. "Medieval Prosopography and Computers: Theoretical and Methodological Considerations." *Medieval Prosopography* 9 (1988): 73–128.

Matthews, L. G. "Herbals and Formularies." In *The Evolution of Pharmacy in Britain,* ed. F. N. L. Poynter, 187–213. Springfield, IL: Charles C. Thomas, 1965.

McConnell, Anita. "Baker, Humphrey (*fl.* 1557–1574)." In *Oxford Dictionary of National Biography.* Oxford: Oxford University Press, 2004 [http://www.oxforddnb.com/view/ article/1123, accessed 26 October 2006].

McDermott, "The Company of Cathay: The Financing and Organization of the Frobisher Voyages." In *Meta Incognita: A Discourse of Discovery, Martin Frobisher's Arctic Expeditions, 1576–1578,* ed. Thomas H. B. Symons, 147–78. Hull, Quebec: Canadian Museum of Civilization, 1999.

———. *Martin Frobisher: Elizabethan Privateer.* New Haven: Yale University Press, 2001.

———. "Michael Lok, Mercer and Merchant Adventurer." In *Meta Incognita: A Discourse of Discovery, Martin Frobisher's Arctic Expeditions, 1576–1578,* ed. Thomas H. B. Symons, 119–46. Hull, Quebec: Canadian Museum of Civilization, 1999.

McDonnell, K. G. T. *Medieval London Suburbs*. London: Phillimore, 1978.

McElwee, William. *The Murder of Sir Thomas Overbury*. London: Faber and Faber, 1912.

McMullin, Ernan. "Conceptions of Science in the Scientific Revolution." In Lindberg and Westman, *Reappraisals of the Scientific Revolution*, 27–92.

Merrit, J. F. "Introduction: Perceptions and Portrayals of London 1598–1720." In *Imagining Early Modern London: Perceptions and Portrayals of the City from Stow to Strype, 1598–1720*, ed. J. F. Merritt, 1–26. Cambridge: Cambridge University Press, 2001.

Merton, Robert K. *Science, Technology, and Society in Seventeenth-Century England*. 1938; New York: Howard Feitig, 1970.

Meskens, Ad. "Mathematics Education in Late Sixteenth-Century Antwerp." *Annals of Science* 53 (1996): 137–55.

Middleton, W. E. Knowles. *The Experimenters: A Study of the Accademia del Cimento*. Baltimore: Johns Hopkins University Press, 1971.

Moran, Bruce T. *Distilling Knowledge: Alchemy, Chemistry, and the Scientific Revolution*. Cambridge: Harvard University Press, 2005.

———. "German Prince-Practitioners: Aspects in the Development of Courtly Science, Technology, and Procedures in the Renaissance." *Technology and Culture* 22 (1981): 235–74.

———. "Paracelsus, Religion, and Dissent: The Case of Philipp Homagius and Georg Zimmerman." *Ambix* 43 (1996): 65–79.

———. "Princes, Machines, and the Valuation of Precision in the Sixteenth Century." *Sudhoff's Archiv* 61 (1977): 209–28.

———, ed. *Patronage and Institutions: Science, Technology, and Medicine at the European Court, 1500–1750*. Rochester, NY: Boydell, 1991.

Morton, Alan Q. "Concepts of Power: Natural Philosophy and the Uses of Machines in Mid-Eighteenth-Century London." *British Journal for the History of Science* 28 (1995): 63–78.

Moss, Ann. *Printed Commonplace-Books and the Structuring of Renaissance Thought*. Oxford: Clarendon, 1996.

Muldrew, Craig. *The Economy of Obligation: The Culture of Credit and Social Relations in Early Modern England*. New York: St. Martin's, 1998.

Mullett, Charles F. "Hugh Plat: Elizabethan Virtuoso." *Studies in Honor of A. H. R. Fairchild, University of Missouri Studies* 21 (1946): 91–118.

Multhauf, Robert. "The Significance of Distillation in Renaissance Chemistry." *Bulletin of the History of Medicine* 30 (1956): 329–46.

Musson, A. E., and E. Robinson. *Science and Technology in the Industrial Revolution*. Manchester: Manchester University Press, 1969.

Neri, Janice. "Fantastic Observations: Images of Insects in Early Modern Europe." Ph.D. diss., University of California at Riverside, 2003.

Newman, William. "The Homunculus and His Forebears: Wonders of Art and Nature." In *Natural Particulars: Nature and the Disciplines in Renaissance Europe*, ed. Anthony Grafton and Nancy Siraisi, 321–45. Cambridge: MIT Press, 1999.

———. *Promethean Ambitions*. Chicago: University of Chicago Press, 2004.

Norrgrén, Hilde. "Interpretation and the Hieroglyphic Monad: John Dee's Reading of Pantheus's *Voarchadumi.*" *Ambix* 52 (2005): 217–46.

Nye, Mary Jo. *Before Big Science: The Pursuit of Modern Chemistry and Physics, 1800–1940.* Cambridge: Harvard University Press, 1999.

Oakeshott, Walter. "Sir Walter Ralegh's Library." *The Library,* 5th series, 23 (1968): 285–327.

Ogilvie, Brian W. "The Many Books of Nature: Renaissance Naturalists and Information Overload." *Journal of the History of Ideas* 64 (2003): 29–40.

———. *The Science of Describing: Natural History in Renaissance Europe.* Chicago: University of Chicago Press, 2006.

Olmi, Giuseppe. "From the Marvellous to the Commonplace: Notes on Natural History Museums (Sixteenth to Eighteenth Centuries)." In *Non-Verbal Communication in Science Prior to 1900,* ed. Renato G. Mazzolini, 235–78. Florence: Leo S. Oschki, 1993.

Orlin, Lena Cowen. *Material London, ca. 1600.* Philadelphia: University of Pennsylvania Press, 2000.

Owen, A. E. B. "Giles and Richard Farnaby in Lincolnshire." *Music and Letters* 42 (1961): 151–54.

Pagel, Walter. *Paracelsus: An Introduction to Philosophical Medicine in the Era of the Renaissance.* Basel: Karger, 1982.

Paster, Gail Kern. *The Idea of the City in the Age of Shakespeare.* Athens: University of Georgia Press, 1985.

Patterson, Annabel. *Reading Holinshed's Chronicles.* Chicago: University of Chicago Press, 1994.

Pavord, Anna. *The Tulip.* New York: Bloomsbury, 1999.

Pearl, Valerie. "Change and Stability in Seventeenth-Century London." *London Journal* 5 (1979): 3–34.

———. *London and the Outbreak of the Puritan Revolution: City Government and National Politics, 1625–43.* Oxford: Oxford University Press, 1961.

Pelling, Margaret. "Appearance and Reality: Barber-Surgeons, the Body, and Disease." In *London, 1500–1700,* ed. A. L. Beier and Roger Finlay, 82–112. London: Longman, 1986.

———. *Medical Conflicts in Early Modern London: Patronage, Physicians, and Irregular Practitioners, 1550–1640.* Oxford: Oxford University Press, 2003.

———. "Medical Practice in Early Modern England: Trade or Profession?" In *The Professions in Early Modern England,* ed. Wilfrid Prest, 90–128. Beckenham, Kent: Croom Helm, 1987.

———. "The Women of the Family? Speculations around Early Modern British Physicians." *Social History of Medicine* 8 (1995): 383–401.

Pelling, Margaret, and Charles Webster. "Medical Practitioners." In Webster, *Health, Medicine, and Mortality in the Sixteenth Century,* 165–235.

Peltonen, Markku, ed. *The Cambridge Companion to Bacon.* Cambridge: Cambridge University Press, 1996.

Pérez-Ramos, Antonio. *Francis Bacon's Idea of Science and the Maker's Knowledge Tradition.* Oxford: Oxford University Press, 1988.

Pettegree, Andrew. *Foreign Protestant Communities in Sixteenth-Century England*. Oxford: Clarendon, 1986.

Pomata, Gianna. *Contracting a Cure: Patients, Healers, and the Law in Early Modern Bologna*. Baltimore: Johns Hopkins University Press, 1998.

Pomian, Krzysztof. *Collectors and Curiosities: Paris and Venice, 1500–1800*. Trans. Elizabeth Wiles Portier. Cambridge: Polity, 1990.

Popper, Nicholas. "The English Polydaedali: How Gabriel Harvey Read Late Tudor London." *Journal of the History of Ideas* (2005): 351–81.

Porter, Roy. "The Patient's View: Doing Medical History from Below." *Theory and Society* 14 (1985): 175–98.

———. *Quacks: Fakers and Charlatans in English Medicine*. Stroud: Tempus, 2000.

———. "The Scientific Revolution: A Spoke in the Wheel?" In *Revolution in History*, ed. Roy Porter and Mikulas Teich, 290–316. Cambridge: Cambridge University Press, 1986.

Porter, Roy, and Mikulas Teich, eds. *The Scientific Revolution in National Context*. Cambridge: Cambridge University Press, 1992.

Prest, Wilfrid. *The Inns of Court under Elizabeth I and the Early Stuarts, 1590–1640*. Totowa, NJ: Rowman and Littlefield, 1972.

Preston, Claire. *Thomas Browne and the Writing of Early Modern Science*. Cambridge: Cambridge University Press, 2005.

Price, William Hyde. *The English Patents of Monopoly*. Boston: Houghton, Mifflin, 1906.

Prior, Roger. "A Second Jewish Community in Tudor London." *Jewish Historical Studies* 31 (1988–90): 137–52.

Pritchard, Allan. "Thomas Charnock's Book Dedicated to Queen Elizabeth." *Ambix* 26 (1979): 56–73.

Pumfrey, Stephen. "Who Did the Work? Experimental Philosophers and Public Demonstrators in Augustan England." *British Journal for the History of Science* 28 (1995): 131–56.

Pumfrey, Stephen, and Frances Dawbarn. "Science and Patronage in England, 1570–1625: A Preliminary Study." *History of Science* 42 (2004): 137–88.

Quinn, David B. *The Roanoke Voyages, 1584–1590*. Cambridge: Cambridge University Press, 1955.

Rabb, Theodore K. *Enterprise and Empire: Merchant and Gentry Investment in the Expansion of England, 1575–1630*. Cambridge: Harvard University Press, 1967.

Raffield, Paul. *Images and Cultures of Law in Early Modern England: Justice and Political Power, 1558–1660*. New York: Cambridge University Press, 2004.

Ramsay, G. D. *The City of London in International Politics at the Accession of Elizabeth Tudor*. Manchester: Manchester University Press, 1975.

Rappaport, Steven. *Worlds Within Worlds: Structures of Life in Sixteenth-Century London*. Cambridge: Cambridge University Press 1989.

Raven, Charles E. *English Naturalists from Neckham to Ray: A Study of the Making of the Modern World*. Cambridge: Cambridge University Press, 1947.

Read, Conyers. *Lord Burghley and Queen Elizabeth*. New York: Knopf, 1960.

———. *Mr. Secretary Cecil and Queen Elizabeth*. New York: Knopf, 1955.

Reeds, Karen Meier. *Botany in Medieval and Renaissance Universities*. New York: Garland, 1991.

Rider, Robin. "Literary Technology and Typographic Culture: The Instrument of Print in Early Modern Culture." *Perspectives on Science* 2 (1994): 1–37.

Roberts, R. S. "The Personnel and Practice of Medicine in Tudor and Stuart England: Part II, London." *Medical History* 8 (1964): 217–34.

Rose, Mark. *Authors and Owners: The Invention of Copyright.* Cambridge: Harvard University Press, 1993.

Rose, Paul. *The Italian Renaissance of Mathematics: Studies on Humanists and Mathematicians from Petrarch to Galileo.* Geneva: Droz, 1975.

Rossi, Paolo. *The Dark Abyss of Time: The History of the Earth and the History of Nations from Hooke to Vico.* Trans. Lydia G. Cochrane. 1979; Chicago: University of Chicago Press, 1984.

———. *Francis Bacon: From Magic to Science.* Trans. Sacha Rabinovitz. Chicago: University of Chicago Press, 1968.

Rudwick, Martin J. S. *The Meaning of Fossils: Episodes in the History of Paleontology.* 1972; Chicago: University of Chicago Press, 1976.

Salgado, Gamini. *The Elizabethan Underworld.* Totowa, NJ: Rowman and Littlefield, 1977.

Saunders, Ann. "Reconstructing London: Sir Thomas Gresham and Bishopsgate." In Ames-Lewis, *Sir Thomas Gresham and Gresham College*, 1–12.

Saunders, David. *Authorship and Copyright.* London: Routledge, 1992.

Saunders, F. W. "The Stigma of Print." *Essays in Criticism* 1 (1951): 139–64.

Schaffer, Simon. "Glass Works: Newton's Prisms and the Uses of Experiment." In *The Uses of Experiment: Studies in the Natural Sciences*, ed. David Gooding, Trevor Pinch, and Simon Schaffer, 67–104. Cambridge: Cambridge University Press, 1989.

———. "Self Evidence." *Critical Inquiry* 18 (1992): 327–62.

Schmitt, Charles B. *Aristotle and the Renaissance.* Cambridge: Harvard University Press, 1983.

———. "Experience and Experiment: A Comparison of Zabarella's View with Galileo's De Motu." *Studies in the Renaissance* 16 (1969): 80–137.

———. "Towards a Reassessment of Renaissance Aristotelianism." *History of Science* 11 (1973): 159–93.

Schofield, John. "The Topography and Buildings of London, ca. 1600." In Orlin, *Material London*, 296–321.

Seidel, Robert. "The Origins of the Lawrence Berkeley Laboratory." In Galison and Hevly, *Big Science*, 21–45.

Selwood, Jacob W. "'English-Born Reputed Strangers': Birth and Descent in Seventeenth-Century London." *Journal of British Studies* 44 (2005): 728–53.

Shackelford, Jole. "Early Reception of Paracelsian Theory: Severinus and Erastus." *Sixteenth Century Journal* 26 (1995): 123–35.

———. "Paracelsianism and Patronage in Early Modern Denmark." In Moran, *Patronage and Institutions*, 85–109.

Shapin, Steven. "The House of Experiment in Seventeenth-Century England." *Isis* 79 (1988): 373–404.

———. "The Philosopher and the Chicken: On the Dietetics of Disembodied Knowledge." In *Science Incarnate: Historical Embodiments of Natural Knowledge*, ed. Christopher Lawrence and Steven Shapin, 21–50. Chicago: University of Chicago Press, 1998.

———. "'A Scholar and a Gentleman': The Problematic Identity of the Scientific Practitioner in Early Modern England." *History of Science* 29 (1991): 279–327.

———. *The Scientific Revolution*. Chicago: University of Chicago Press, 1998.

———. *A Social History of Truth: Civility and Science in Seventeenth-Century England*. Chicago: University of Chicago Press, 1994.

Shapin, Steven, and Simon Schaffer. *Leviathan and the Airpump: Hobbes, Boyle, and the Experimental Life*. Princeton: Princeton University Press, 1985.

Shapin, Steven, and Arnold Thackeray. "Prosopography as a Research Tool in the History of Science: The British Scientific Community 1700–1900." *History of Science* 12 (1974): 1–28.

Shapiro, Barbara. *Probability and Certainty in Seventeenth-Century England: A Study of the Relationships Between Natural Science, Religion, History, Law, and Literature*. Princeton: Princeton University Press, 1983.

Sharpe, Kevin. *Reading Revolutions: The Politics of Reading in Early Modern England*. New Haven: Yale University Press, 2000.

Sherman, William. *John Dee: The Politics of Reading and Writing in the English Renaissance*. Amherst: University of Massachusetts Press, 1995.

Shirley, John William. "The Scientific Experiments of Sir Walter Ralegh, the Wizard Earl, and the Three Magi in the Tower 1603–1617." *Ambix* 4 (1951): 52–66.

Siena, Kevin P. *Venereal Disease, Hospitals, and the Urban Poor: London's "Foul Wards," 1600–1800*. Rochester, NY: University of Rochester Press, 2004.

Simon, Joan. *Education and Society in Tudor England*. Cambridge: Cambridge University Press, 1966.

Skinner, Quentin. "Language and Social Change." In *Quentin Skinner and His Critics*, ed. James Tully, 119–32. Princeton: Princeton University Press, 1998.

Slack, Paul. *The Impact of Plague in Tudor and Stuart England*. London: Routledge, 1985.

———. "Mirrors of Health and Treasures of Poor Men: The Uses of the Vernacular Medical Literature of Tudor England." In Webster, *Health, Medicine, and Mortality in the Sixteenth Century*, 237–73.

Smith, Pamela. *The Body of the Artisan*. Chicago: University of Chicago Press, 2004.

———. *The Business of Alchemy: Science and Culture in the Holy Roman Empire*. Princeton: Princeton University Press, 1994.

———. "Paracelsus as Emblem." *Bulletin of the History of Medicine* 68 (1994): 314–22.

Smith, Pamela, and Paula Findlen, eds. *Merchants and Marvels: Commerce, Science, and Art in Early Modern Europe*. New York: Routledge, 2002.

Solomon, Julie Robin. *Objectivity in the Making: Francis Bacon and the Politics of Inquiry*. Baltimore: Johns Hopkins University Press, 1998.

Sorsby, Arnold. "Richard Banister and the Beginnings of English Ophthalmology." In *Science, Medicine, and History: Essays on the Evolution of Scientific Thought and Medical Practice*, 2 vols., ed. E. Ashworth Underwood, 2: 42–55. Oxford: Oxford University Press, 1953.

Spiller, Elizabeth. *Science, Reading, and Renaissance Literature: The Art of Early Modern Knowledge.* Cambridge: Cambridge University Press, 2004.

Stagl, Justin. *A History of Curiosity: The Theory of Travel, 1500–1800.* Chur, Switzerland: Harwood, 1995.

Stern, Virginia. *Gabriel Harvey, His Life, Marginalia, and Library.* Oxford: Clarendon, 1979.

Stewart, Larry. *The Rise of Public Science.* Cambridge: Cambridge University Press, 1992.

Stock, Brian. *Listening for the Text: On the Uses of the Text.* Baltimore: Johns Hopkins University Press, 1990.

Stone, Lawrence. "Prosopography." *Daedalus* 100 (1971): 46–79.

Strype, John. *The Life of the Learned Sir Thomas Smith.* Oxford: Clarendon, 1820.

Symonds, H. "The Mint of Queen Elizabeth and Those Who Worked There." *Numismatic Chronicle* 76 (1916): 61–105.

Tahon, Eva. "Marcus Gheeraerts the Elder." In *Bruges and the Renaissance: Memling to Pourbus,* ed. Maxiliaan P. J. Martens, 231–33. New York: Abrams, 1998.

Taylor, E. G. R. *The Mathematical Practitioners of Tudor and Stuart England.* Cambridge: Cambridge University Press, 1954.

Tebeaux, Elizabeth. *The Emergence of a Tradition: Technical Writing in the English Renaissance, 1475–1640.* Amityville, NY: Baywood, 1997.

Thirsk, Joan. *Economic Policy and Projects: The Development of a Consumer Society in Early Modern England.* Oxford: Clarendon, 1978.

Thomas, Keith. "Numeracy in Early Modern England." *Transactions of the Royal Historical Society,* 5th series, 37 (1987): 103–32.

Trevor-Roper, Hugh. "Court Physicians and Paracelsianism." In *Medicine at the Courts of Europe, 1500–1837,* ed. Vivian Nutton, 79–94. London: Routledge, 1990.

———. "The Paracelsian Movement." In *Renaissance Essays,* 149–99. London: Secker and Warburg, 1985.

Turner, Gerard L'E. *Elizabethan Instrument Makers: The Origins of the London Trade in Precision Instrument Making.* Oxford: Oxford University Press, 2000.

Tylecote, R. F. *A History of Metallurgy.* London: Metals Society, 1976.

Urbach, Peter. *Francis Bacon's Philosophy of Science.* La Salle, IL: Open Court, 1987.

van Dorsten, Jan. "'I. O. C.': The Rediscovery of a Modest Dutchman in London." In *The Anglo-Dutch Renaissance: Seven Essays,* ed. Jan van Dorsten, 8–20. Leiden: E. J. Brill, 1988.

———. *Poets, Patrons, and Professors: Sir Philip Sidney, Daniel Rogers, and the Leiden Humanists.* Leiden: University Press, 1962.

———. *The Radical Arts: First Decade of an Elizabethan Renaissance.* London: Oxford University Press, 1970.

van Egmond, W. *The Commercial Revolution and the Beginnings of Western Mathematics in Renaissance Florence, 1300–1500.* 2 vols. Ann Arbor: University of Michigan Press, 1976.

van Leeuwen, Hendrik Gerrit. *The Problem of Certainty in English Thought, 1630–1690.* The Hague: Martinus Nijhoff, 1970.

Van Norden, Linda. "Peiresc and the English Scholars." *Huntington Library Quarterly* 12 (1948–49): 369–89.

Vigne, Randolph, and Charles Littleton, eds. *From Strangers to Citizens: The Integration of Immigrant Communities in Britain, Ireland, and Colonial America, 1550–1750.* Brighton: Huguenot Society and Sussex Academic Press, 2001.

Walton, Steven A. "The Bishopsgate Artillery Garden and the First English Ordnance School." *Journal of the Ordnance Society* 15 (2003): 41–51.

Ward, Joseph P. "Fictitious Shoemakers, Agitated Weavers, and the Limits of Popular Xenophobia in Elizabethan London." In Vigne and Littleton, *From Strangers to Citizens,* 80–87.

———. *Metropolitan Communities: Trade Guilds, Identity, and Change in Early Modern London.* Stanford: Stanford University Press, 1997.

Watson, Bruce. "The Compter Prisons of London." *London Archaeologist* 7 (1993): 115–21.

Watson, Foster. *The Beginning of the Teaching of Modern Subjects in England.* London: Isaac Pittman, 1909.

Wear, Andrew. *Knowledge and Practice in English Medicine, 1550–1680.* Cambridge University Press, 2000.

Webster, Charles. "Alchemical and Paracelsian Medicine." In Webster, *Health, Medicine, and Mortality in the Sixteenth Century,* 301–34.

———. *The Great Instauration: Science, Medicine, and Reform, 1626–1660.* London: Duckworth, 1975.

———. "Paracelsus: Medicine as Popular Protest." In *Medicine and the Reformation,* ed. Ole Peter Grell and Andrew Cunningham, 57–77. London: Routledge, 1993.

———. "Paracelsus, Paracelsianism, and the Secularization of the Worldview." *Science in Context* 15 (2002): 9–27.

———, ed. *Health, Medicine, and Mortality in the Sixteenth Century.* London: Cambridge University Press, 1979.

Weinberger, Jerry. *Science, Faith, and Politics: Francis Bacon and the Utopian Roots of the Modern Age.* Ithaca: Cornell University Press, 1985.

Werner, Alex, and Michael Berlin. "Developing an Interdisciplinary Approach? The Skilled Workforce Project." *Bulletin for the John Rylands Library* 77 (1995): 49–56.

Westfall, Richard S. "Science and Technology During the Scientific Revolution: An Empirical Approach." In Field and James, *Renaissance and Revolution,* 63–72.

Westman, Robert. "The Astronomer's Role in the Sixteenth Century: A Preliminary Study." *History of Science* 18 (1980): 105–47.

White, George. *The Clockmakers of London.* Hants: Midas, 1998.

Wilson, Catherine. *The Invisible World: Early Modern Philosophy and the Invention of the Microscope.* Princeton: Princeton University Press, 1995.

Wood, Andy. "Custom, Identity, and Resistance: English Free Miners and Their Law c. 1550–1800." In *The Experience of Authority in Early Modern England,* ed. Paul Griffiths, Adam Fox, and Steve Hindle, 249–84. New York: St. Martin's, 1996.

Woolf, Daniel. *Reading History in Early Modern England.* Cambridge: Cambridge University Press, 2000.

Wright, Louis B. *Middle-Class Culture in Elizabethan England*. Chapel Hill: University of North Carolina Press, 1935.

Young, Sidney. *The Annals of the Barber-Surgeons of London*. London: Blades, East, and Blades, 1890.

Yungblut, Laura. *Strangers Settled Here Amongst Us: Policies, Perceptions, and the Presence of Aliens in Elizabethan England*. London: Routledge, 1996.

Zagorin, Perez. *Francis Bacon*. Princeton: Princeton University Press, 1998.

Zetterberg, J. Peter. "The Mistaking of 'The Mathematics' for Magic in Tudor and Early Stuart England." *Sixteenth Century Journal* 11 (1980): 83–97.

Zilsel, Edgar. "The Sociological Roots of Science." *American Journal of Sociology* 47 (1942): 544–62.

INDEX

Note: illustrations are indicated by **boldface**